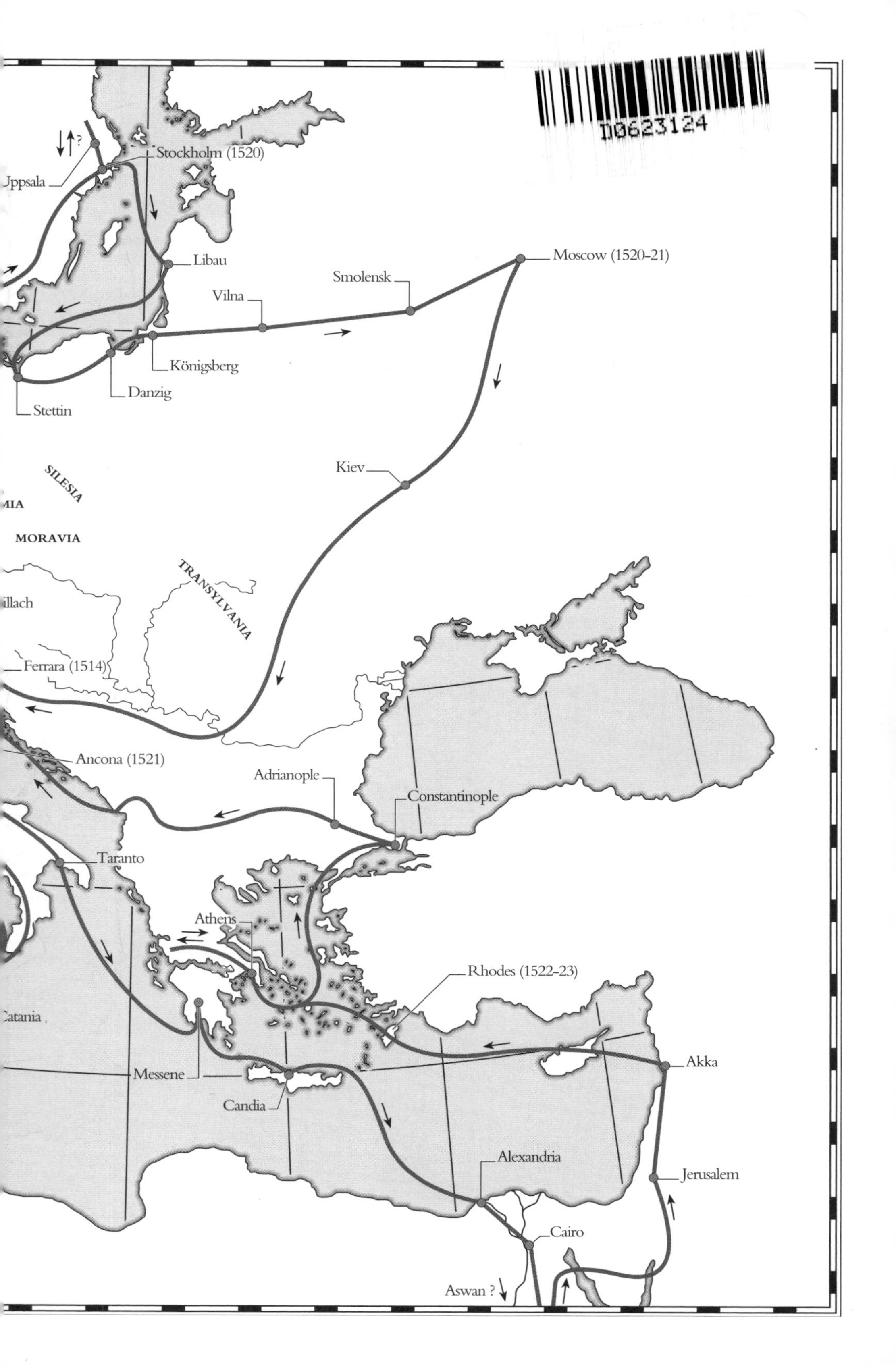

D0623124
Stockholm (1520)
Jppsala
Libau
Moscow (1520-21)
Smolensk
Vilna
Königsberg
Danzig
Stettin
Kiev
SILESIA
MIA
MORAVIA
TRANSYLVANIA
illach
Ferrara (1514)
Ancona (1521)
Adrianople
Constantinople
Taranto
Athens
Rhodes (1522-23)
Catania
Messene
Candia
Akka
Alexandria
Jerusalem
Cairo
Aswan ?

The Devil's Doctor

THE

Devil's Doctor

PARACELSUS AND THE WORLD OF RENAISSANCE MAGIC AND SCIENCE

Philip Ball

WILLIAM HEINEMANN: LONDON

Published in the United Kingdom by William Heinemann, 2006

3 5 7 9 10 8 6 4 2

Copyright © Philip Ball, 2006

Philip Ball has asserted his right under the Copyright, Designs and Patents Act 1988 to be identified as the author of this work

This book is sold subject to the condition that it shall not, by way of trade or otherwise, be lent, resold, hired out, or otherwise circulated without the publisher's prior consent in any form of binding or cover other than that in which it is published and without a similar condition including this condition being imposed on the subsequent purchaser

William Heinemann
The Random House Group Limited
20 Vauxhall Bridge Road, London, SW1V 2SA

Random House Australia (Pty) Limited
20 Alfred Street, Milsons Point, Sydney
New South Wales 2061, Australia

Random House New Zealand Limited
18 Poland Road, Glenfield
Auckland 10, New Zealand

Random House (Pty) Limited
Isle of Houghton, Corner of Boundary Road and Carse O'Gowrie,
Houghton 2198, South Africa

The Random House Group Limited Reg. No. 954009

www.randomhouse.co.uk

Maps © ML Design, 2006
The picture credits in the illustration list on pp. ix–xi constitute an extension of this copyright page

The publishers gratefully acknowledge the permission of Agencia Literaria Carmen Balcells, Spain, to reproduce an excerpt from 'Poetry' from *Isla Negra* by Pablo Neruda, translated by Alastair Reid. Translation copyright © Alastair Reid, 1981

A CIP catalogue record for this book is available from the British Library

Papers used by Random House are natural, recyclable products made from wood grown in sustainable forests. The manufacturing processes conform to the environmental regulations of the country of origin

ISBN 0 434 01134 7

Typeset by SX Composing DTP, Rayleigh, Essex
Printed and bound in Great Britain by
Clays Ltd, St Ives plc

Paracelsus, ascribed to Jan van Scorel (1495–1562) and thought to depict its subject in Alsace in the 1520s.

Contents

Illustrations

Acknowledgements

I never imagined or intended that I would write someone's life story, for I am not a biographer. But anyone with an interest in alchemy will encounter Paracelsus sooner or later, and once I did, this book became inevitable. It seemed that he left me no option but to attempt to tell his tale. And he is a difficult man and not easily denied.

Historians of science have always had to wrestle with the question of how seriously to take him. Buffoon or genius? But the question is not that simple. There was rather little in Paracelsus' life that was simple, and indeed as one begins to explore it one quickly discovers that the entire history of Renaissance science – in which Paracelsus plays a pivotal role – is not nearly as straightforward as many popular accounts would have it. Yet Paracelsus' story can help us to examine these ambiguities and complications, and perhaps to understand some of the shortcomings in our modern concept of science itself.

I am deeply grateful to Urs Leo Gantenbein for making it possible for me to visit some of the key 'Paracelsus sites' in Switzerland, and for being such a generous and companionable guide. Robin Briggs made very perceptive and valuable comments on the historical aspects of my text, and Dane Thor Daniel gave knowledgeable advice on Paracelsus. Andrea Sella nobly agreed to wade through an unwieldy first draft, and his comments helped me immensely to organize an extravagant amount of material. Much of it was gathered at the Wellcome Library for the History of Medicine, and discovering this remarkable resource has been one of the joys of this project. I am grateful for the support and advice I received from my editors Ravi Mirchandani and Caroline Knight at Heinemann, and John Glusman at Farrar, Straus and Giroux, and for continued encouragement from my agent Peter Robinson and my wife Julia, as well as that from the many friends and colleagues who share even a little of my passion for this strange and incomparable man.

Philip Ball
London, October 2005

Introduction
Fool's Quest

A Faustian Legacy

> Every thing must have a beginning, to speak in Sanchean phrase; and that beginning must be linked to something that went before.
>
> Mary Shelley, Introduction to *Frankenstein* (1831)

'I am different,' he wrote, 'let this not upset you.'

But people did get upset by Paracelsus. He upset doctors and priests, he upset town authorities, Renaissance kings and princes, Lutherans, humanists, merchants, apothecaries and theologians. He upset his friends and his assistants. He upset generations of chemists and physicians who lived under the grotesque, distorted shadow that he cast. They are still getting over it.

This self-styled doctor of medicine and theology devised an ornate and eccentric philosophy of the universe and its mechanisms that few of his contemporaries in the sixteenth century could understand and that still fewer appreciated. The enigmas of Paracelsian philosophy are reflected in the man himself. He was protean, mercurial, a mass of contradictions. A humble braggart, a puerile sage, invincible loser, courageous coward, pious heretic, honest charlatan, fuelled by profound love and by spiteful hate, dining with princes and sleeping in the ditch, both personifying and challenging the madness of his world. In his *History of Chemistry* (1843), Ferdinand Hoefer gives us an astute portrait:

> Picture to yourself a man who, in certain moments, gives evidence of a remarkable penetration, and in others raves in the most pitiable way possible; a man who, at one time, devoted to the progress of science, proclaims the absolute authority of experience, and thunders the most violent anathemas against the theories of the ancients; yet at another time, like a lunatic, seems to converse with demons and

> believe in their absolute power. Fasting in the morning, drunk in the evening, presenting exactly every idea in the order in which it came to his mind, such is Paracelsus.[1]

These contradictions are reflected in accounts of his life. As the eighteenth-century German writer Christoph Martin Wieland put it, 'What all extraordinary persons encountered, to be stupidly praised and stupidly blamed, was also the lot of Paracelsus.'[2]

The sixteenth century has often been portrayed as a period of scientific as well as religious and political reformation. Traditionally, the scientific Luthers are Nicolaus Copernicus, who transformed astronomy, and Andreas Vesalius, who did the same for anatomy: between them they reinvented the outer and the inner worlds of humankind. Scientists today are comfortable with this story, because it connects the Renaissance with the world they know, in which the earth circulates around the sun and the body's veins and organs are assigned to their proper places.

But there is another version of the story, and Paracelsus is at its centre. It is a stranger tale, for in the philosophy of Paracelsus science and rationalism do not compete with mysticism and superstition but blend with them, producing a vision of the world that now seems at the same time wonderful and bizarre. It was this vision, more than the reforms of Copernicus and Vesalius, that defied the stifling, wrong-headed certainties of the late Middle Ages, with their hide-bound and dogmatic interpretations of classical ideas about the universe. It does not diminish the genius of either man to say that Copernicus was not representative of the Renaissance astronomers, nor Vesalius a physician of revolutionary unorthodoxy. If we wish to understand what the philosophers of that age really argued over, and to take the true temperature of the intellectual ferment during the era of Luther and the Counter-Reformation, we will do better to look at the life of Paracelsus. Here is a man who genuinely acts as a prism for his time – who separates out for us, as it were, the paradoxes, the terrors, the tensions that existed between natural philosophy, religion, humanism and politics.

This is a version of his story. Versions are all we can have, since the sparse trickle of facts has become clouded by legend, diverted by hearsay, calumny and hagiography, and muddied by the many contradictory testaments of the man himself. I will do my best to clear away the fog, but as soon as you move beyond the caricature, you find yourself forced to practise some alchemy of your own, blending strange ingredients in a crucible and not quite knowing what new form will emerge.

Paracelsus lived from 1493 to 1541: a fulcrum of Western history, the dawn of the modern age. This was a world where magic was real, where demons lurked in every dark corner, where God presided over all creation, and yet it was also a time when humankind was beginning to crack nature's codes and map the geography of heaven and earth. Swiss by birth, Paracelsus roamed across all of Renaissance Europe, experiencing at first hand the wars and power struggles, the horrors and hardships, of that world. Given his argumentative nature, and since it is doubtful that he even knew how to handle the massive broadsword he grasps in many later portraits, it is a wonder that he kept his head on his shoulders throughout his travels.

He was christened Philip Theophrastus Bombast von Hohenheim; other names followed. In central Europe, an old fairy tale tells how Doktor Theophrastus communed with the devil and possessed the secret of eternal life. He was eventually killed, this doctor, poisoned by his enemies. He had a white horse, so they used to say in Transylvania, given to him by Satan, which could travel vast distances without tiring. Sometimes he is called Teofrastus, or Frastikus, or Frastus, which brings us to Faustus, the charlatan wanderer with whom Paracelsus' name has been linked, who was said to have bartered his soul for forbidden knowledge.

Or else Paracelsus takes the nickname Alpenus – a man from the Alpine slopes – which becomes corrupted to Arpenus or Arpinas, from which it is a short step to Orpinas and then Orpheus, a central character in the great tradition of natural magic, who gained victory over death.

Legend marks his every step. The broadsword holds astonishing secrets – for does he not keep his most potent and uncanny of medicines, the mysterious laudanum, hidden in its pommel? Or better still, a wily demon, according to Samuel Butler in his satirical poem *Hudibras* (part 2; 1664):

Bombastus kept a devil's bird
Shut in the pommel of his sword,
That taught him all the cunning pranks
Of past and future mountebanks.[3]

Paracelsus, aged forty-seven, in an engraving by Augustin Hirschvogel.

He was not a part of the educated elite of sixteenth-century Europe, but circled all his life on its fringes. And he was proud of his humble origins:

> I cannot boast of any rhetoric or subtleties; I speak in the language of my birth and my country, for I am from Einsiedeln, of Swiss nationality, and let no one find fault with me for my rough speech. My writings must not be judged by my language, but by my art and experience, which I offer the whole world, and which I hope will be useful to the whole world.[4]

His motto, in the archaic Swiss German that he used, reads '*Eins andern knecht soll niemand seyn, der für sich bleyben kann alleyn*', and it means: 'Let no man belong to another, who can belong to himself.' Such a blend of pride, intransigence, boastfulness, independence and wounded dignity is the essence of Paracelsus. And here at least was one boast he lived up to – for in all his wide travels, during his many appointments, throughout his fierce battles and harsh disputes, always he was no one's man but his own.

BETWEEN SCIENCE AND MAGIC

The traditional histories of science will tell you that Paracelsus helped to shape the course of chemistry and medicine at a time when both disciplines were changing from ancient to modern forms. There is, as we shall see, a case to be made for that claim, but I am not terribly interested in trying to carve out a place for him in the pantheon that takes us from Hippocrates to Crick and Watson, or from Empedocles to Linus Pauling, or Aristotle to Einstein, for there is little value in clothing him in the anachronistic garb of the proto-scientist. To find the man himself, we are forced to go out into the remote and unfamiliar Renaissance landscape to meet him. There we discover that, first and foremost, his philosophy was magical to its very core.

It is only rather recently that science has begun to make peace with its magical roots. Until a few decades ago, it was common for histories of science either to commence decorously with Copernicus' heliocentric theory or to laud the rationalism of Aristotelian antiquity and then to leap across the Middle Ages as an age of ignorance and superstition. One could, with care and diligence, find occasional things to praise in the works of Avicenna, William of Ockham, Albertus Magnus and Roger Bacon; but these sparse gems had to be thoroughly dusted down and scraped clean of unsightly accretions before being inserted into the corners of a frame fashioned in a much later period. Yet modern science did not simply spring in fledgling form into the minds of Copernicus, William Harvey and their ilk. One could easily get the impression from historical accounts that the thinking of these men was qualitatively different from that of others who went before, so that science emerged like a miraculous tree growing in a medieval wasteland. Set against this context, Paracelsus is the scientific positivist's worst nightmare. His work begins and ends in magic. Everything he writes is coloured by his religious beliefs, which create a purposeful universe full of secret signs and symbols. He claims to have made the Philosopher's

Stone, that most fantastical unicorn of early chemistry. He believes in nymphs and giants and ghosts, he tells us that men can live without food if they are planted in the soil, he interprets comets as portents, he wades in the numerological mire of the Cabbala, he claims to cure every disease. Was his not the world from which science eventually rescued us?

Well, yes, in a way. But that was never anyone's mission. If Isaac Newton was a fool and a dupe to believe in alchemy, then we are all fools and dupes and always will be, trapped in the sticky web of our own time and place. If we want to really see where science came from, it is no good starting from today's perspective. Today we have the luxury of being able to regard astrology and magic as forms of foolishness. There was very little luxury in the sixteenth century, when ideas like this were all that was on offer. Science resulted not from efforts to get rid of them, but from attempts to make sense of them.

Magic was undoubtedly bound up with medieval superstition, but it was also the precondition for science. Indeed, if the fifteenth century had a science at all, it was the 'science' of magic. John Maynard Keynes tried to imply as much with his striking description of Newton as 'the last magician', but in fact Newton was not the last in any genealogy of magic, nor the first link in its mutation into science. He shared with his contemporaries a world-view that was not a weird coexistence of magic and science, but an edifice in which all the bricks came from the same mould. Today we regard some of these as the foundations of science, and are tempted to discard the others as useless relics of another age. But if they were not there, the building would have collapsed. 'The Renaissance magus', says the historian Frances Yates, 'is the immediate ancestor of the seventeenth-century scientist.'[5]

Thus science did not emerge as a rational flight from medieval superstition; after all, the medieval scholastics were second to none in their pedantic rationality. What was needed for modern science to take shape was a renunciation of their bookish *a priorism*, with its Aristotelian notion that all things can be deduced by logical, abstract argument from (ultimately arbitrary) first principles. Before the fertile logic of a genuinely scientific rationalism could assert itself, the sterile ground of classical dogma had to give way to a form of empiricism that accepted the reality of certain unknowns and inexplicables such as the operation of occult forces. In this sense, men like Paracelsus and his fellow iconoclast Cornelius Agrippa were sceptics: they were prepared, indeed determined, to question what had gone before, to find things out for themselves rather than take someone else's word for it.

But the price they paid for that, ironically, was a greater credulity. If you renounce old certainties, you risk believing anything. Paracelsus and Agrippa had no systematic methodology to guide them towards newer, sounder knowledge, and so they had to take instruction wherever they could find it. Mostly, they succumbed to the humanist's viewpoint: that mankind had once known great things, but had been corrupted in the era that separated legendary antiquity from their own day. And so their new philosophy was derived from the jumbled ambiguities of the ancient magical literature, reforged in the fire of experience.

If the 'credulous sceptic' seems too odd, too contradictory a figure to serve as the template for the first scientists, think of Francis Bacon, of Robert Boyle, of Isaac Newton, with their visions of secret societies, mystical adepts and arcane knowledge pregnant with power. Until we make our peace with this curious and infuriating figure, we will never appreciate how the scientific world was shaped.

Paracelsus' magic was perhaps more eclectic than that of any other major thinker of the Renaissance. That is not always a virtue, nor a sign of open-mindedness, though undeniably it stems from a marked independence of mind. The fact is that he simply took what appealed to him and fitted it into his grand vision. It was sometimes a tight squeeze, and not all the parts matched, and it is scarcely surprising that 'Paracelsianism' came subsequently to mean so many different things.

Paracelsus was first of all a physician, and he regarded his own medical remedies as magical. But that was, in his view, the opposite of superstitious; the doctor systematically concentrated and manipulated the invisible, magical forces and 'virtues' of nature. And Paracelsus sought to embed this 'new' medicine within a comprehensive system of (devoutly Christian) natural philosophy, from which the doctor's art emerged naturally. In this much at least, his aim was no different from that of contemporary science: it all has to fit together. We insist that the atoms that make up genes and viruses and cells are identical to those that make up mountains and oceans, and that they are governed by the same physical forces. The laws of physics apply to everything equally, to stars as well as flowers. Botany and astronomy are separate sciences, but if they are somehow fundamentally inconsistent then there is something wrong with our theories. The need for such an all-encompassing vision was not really felt in the classical past. Aristotle wrote very widely and was happy enough to draw analogies between disparate phenomena, but he was conspicuously silent on some topics (such as what we would now call chemistry) and gives little impression of the need for congruence and continuity. For encyclopaedists such as Pliny, 'local' explanations for things were often enough: phenomena are explained largely in terms of themselves, not in terms of other things. Where do the four humours, the bodily fluids that were thought to govern health, come from? Neither Galen nor Hippocrates, the two pre-eminent physicians of antiquity, tells us that; they assume that it is just how things are.

Paracelsus' cosmology could never be truly scientific because it did not and could not exclude theology. Aristotle's world was often tautologous (objects fell to earth because earth was the natural place for them to rest), but that of Paracelsus was more explicitly teleological: his world was *designed*, and the Designer had left His mark everywhere on it. All the same, this does not mean that Paracelsus was lacking in what would later be regarded as the 'scientific spirit'. On the contrary, he firmly believed that things happen for a reason: that nature is mechanistic and follows rules, and that humankind could deduce and understand them. This is why he could never be wholly reconciled with Lutheranism, which held that God's ways were forever inscrutable and that it was impious to try to trick Him into revealing them. A mechanistic view of nature can be traced back to the great rationalist scholars of the twelfth and thirteenth centuries: men like Thierry of Chartres, William of Conches and John of Salisbury, who argued that God did not run the world with His hand constantly on the levers but instead formulated rules and then let them unfold.

The central distinction between the mystical magus like Paracelsus and the academics at the universities was that, while the latter tended to believe that these rules had already been deduced by the ancients through the power of pure thought and logic, the magical adepts thought that they could be found only by studying nature – crudely speaking (and I shall be more precise later), by experiment. Some historians have located the origins of science in the rationalism of the Greeks and their attempts to specify 'first principles'. But the spirit of empirical enquiry, without which science is impossible, began to flourish amid the natural magic of the Renaissance. The true magus, according to the Italian humanist Marsilio Ficino, is 'a contemplator of heavenly and divine science, a studious observer and expositor of divine things'.[6] The historian Wayne Shumaker makes a claim that, if extreme, nonetheless serves to remind us what many histories of science have neglected: 'what the Renaissance called magic was a more nearly direct ancestor of true science than either of the dominant philosophies, Aristotelianism and Platonism.'[7]

That magic was an *occult* art leaves many scientists uncomfortable with its role in the history of their discipline. Today 'occult' is equated with superstition, irrationality, charlatanism. But much of contemporary science is itself occult *in the Renaissance sense*, insofar as it is 'hidden' from our senses – that, of course, being the literal meaning of the word. We explain phenomena in terms of atoms or molecules too small for us to see, or electromagnetic fields that are truly (for the most part) invisible, or other fields and forces (such as gravity) that we still struggle to understand. By Renaissance standards, these things are no less occult than the astrological 'emanations' of a star or the causative agency of demons. Just as the modern engineer manipulates the forces of electricity, gravity, hydraulic pressure and so forth, so the magus manipulated occult forces by means of natural magic.

As modern science emerged, it did not banish the concept of occult forces; rather, it accommodated and formalized those that seemed useful, such as magnetism and gravity, relegating others – ghosts, telepathy, telekinesis and so on – to a ragbag of outmoded notions that, in retaining the label 'occult', gradually rendered the word disreputable. But without this belief in the occult, science would have been stymied. Before Renaissance magic stimulated a new interest in the occult, the forces of nature were dismissed as beyond man's capacity to understand. To Thomas Aquinas, magnetism is an 'occult virtue which man is not capable of explaining'.[8] This was a profoundly anti-scientific viewpoint, an admonition not to probe hubristically into the workings of the world. Isaac Newton could not have formulated his gravitational theory without a belief in occult forces – a belief surely sustained by his deep interest in magic. His rival Leibniz in fact accused him of invoking a 'scholastic occult quality',[9] whereupon his defender Samuel Clarke countered, in a suggestion that few scientists now would question, '[is] a manifest quality to be called . . . occult because the immediate efficient cause of it (perhaps) is occult?'[10] In other words, the scientific revolution was predicated on an abandonment of the Aristotelian idea that to be intelligible, a causative mechanism had to be 'sensible'. Precisely the same arguments were reprised at the turn of the nineteenth century, when some eminent scientists, including Ernst Mach and Wilhelm Ostwald, opposed the notion of atoms on the grounds that no one had ever seen one, or could adduce direct evidence of their reality. We are inclined now to regard Mach and Ostwald as pedants on this point; but it would be more appropriate to say that they were merely expressing suspicion of the occult.

MIXED FEELINGS

On the whole, the science of the Renaissance and the early Enlightenment is not so distant a cosmos that we cannot distinguish its stars. No one disputes that Copernicus and Galileo changed the shape of the universe, or that Newton revealed how it hung together. In our developing understanding of the internal world of human anatomy and physiology, the reputations of Vesalius and Harvey are unassailable. Robert Boyle's importance to chemistry is not begrudged for the sake of his fondness for a little (or more than a little) alchemy; William Gilbert's work on magnetism is obviously of profound significance.

But when it comes to Paracelsus, things are different. It is not only that some have questioned his importance in the history of science; rather, they have attempted to erase him entirely. In his own time, rather few people were convinced by him. The world, by and large, shrugged and ignored him. Some laughed and mocked; some excoriated this short, ugly man, denouncing him as Satan's disciple. But he would not go away.

We can recognize in the denunciations written during the century after his death a bile borne of battles still in progress: that Thomas Erastus in the late sixteenth century called him an 'atheist pig', and the Swiss physician Johann Georg Zimmermann a 'jackass' 200 years later, is typical of their times. But the contempt that drips from the remarks of several more recent commentators must make us suspect that Paracelsus has indeed upset and disturbed them. Take H. P. Bayon, speaking to the Royal Society of Medicine sixty years ago:

> It cannot be said that the abusive rantings of Paracelsus contributed to the general progress of science and medicine that began in the sixteenth century, principally as to the outcome of the diffusion of accurate knowledge by means of printed books. For he was a rude, circuitous obscurantist, not a harbinger of light, knowledge and progress.

He was, Bayon concludes, 'violently destructive, only rarely critically constructive and never original, if ever right'.[11] And here is John Ferguson, professor of chemistry at Glasgow, in the fourteenth edition of the *Encyclopaedia Britannica*: 'With Paracelsus' lofty views of the scope of medicine it is impossible to reconcile his ignorance, his superstition, his erroneous observations.'[12]

Anyone who has read Paracelsus' works will, if their vision has not been clouded by romanticized hagiographies, have some sympathy with these comments. His writings abound with preposterous vanity, they are often opaque if not downright incoherent, they are rambling and strange, they seem more closely allied to a world of fairy tales and superstition than science and reason. Reading them today, one cannot help but wonder from time to time: is this the work of a charlatan, perhaps even a madman? Here is a taste of the verbal thicket that the interpreter faces in a verbatim translation of his dense prose:

> One has to be greatly astonished, since God redeemed man so highly and dearly by His death and bloodshed, lets him be born an unwise man, who cannot recognize, nor understand, His name, His death, His teaching, His signs, His works, His benefactions bestowed upon man, has been robbed of all reason [and] wisdom that is required therefore, moreover, since man is a picture of God, is to be and appear thus affected with a fool, simpleton, stupid, ignorant man, whereas man is the noblest creature above all, is to stand thus like as debased before all creatures, whereas also all creatures

> are exempt and have no fools among them, only man; however, the causes are many, need not be told here in the preface.[13]

It is not just that Paracelsus is verbose, undisciplined and ungrammatical. He confounds his stylistic eccentricities with linguistic ones, inventing new words for which he provides only the vaguest definitions, if he defines them at all. The historian of chemistry James Partington is probably not far wrong when he suggests that Paracelsus sometimes invented words just to sound impressive.[14] Daniel Pickering Walker, a historian sympathetic to the significance of Renaissance magic, is yet more terse: 'I doubt whether Paracelsus' philosophical writings are in fact intelligible, that is, whether they contain any coherent patterns of thought.'[15]

Yet incoherence and linguistic incontinence don't fully explain the antipathy that he arouses, nor why the slurs of his near contemporaries were so enthusiastically repeated in later times, as in this broadside from Zimmermann:

> Further, he lived like a pig, looked like a coachman and took pleasure in the company of the loosest and lowest mob . . . During most of his notorious life, Paracelsus was drunk, and indeed, all his writings seem to have been written during intoxication.[16]

That these spittle-flecked accusations have been resuscitated and recycled throughout the ages reveals how profoundly infuriating, how *embarrassing*, Paracelsus has been to historians of science. He is, according to the historian Charles Webster, 'unique among the major thinkers of the scientific revolution [in that] while others have become painlessly absorbed into the system of modern knowledge, Paracelsus retains his status as an iconoclast and outsider'.[17] And yet the insults and defamation have consistently failed to bury him in obscurity and ignominy. Paracelsus simply cannot be dismissed as a credulous fool (at least, not all of the time). He embarrasses because he mocks the conviction, once deeply held by scientists and historians alike, that the history of ideas should be arranged into an orderly and unidirectional narrative. The personality imprinted on his writings is raw and urgent, and seemingly unconcerned about the conflicts and contradictions that they present. If we struggle with him, it is because we cannot but struggle with his times, when people asked different questions and wrestled with different dilemmas from today. The birth of the modern world, these struggles tell us, was neither easy nor painless, but rather it was turbulent, confused and violent.

ROMANTIC DELUSIONS

It is surely no coincidence that the more scientists excoriated Paracelsus, the more poets adored him. For Goethe and the Romantics, he was a noble hero. 'Any man of mechanical talents', wrote William Blake, 'may from the writing of Paracelsus or Jacob Behmen [Boehme, the seventeenth-century German mystic], produce ten thousand volumes of equal value with Swedenborg's.'[18] It was in this climate that Robert Browning undertook his epic poem about Paracelsus' life in 1834, superimposing his own agenda so that it became (somewhat absurdly) the tale of a soul's journey into love. Paracelsus' Romantic champions from the late nineteenth and early twentieth centuries made the most ridiculous claims for their hero, attributing to him insights into every aspect of modern science and medicine from 'magic-bullet' drugs to quantum physics.

It became *de rigueur* for Victorian Romantics to drop his name. In A. S. Byatt's novel *Possession* the Victorian poet Randolph Henry Ash hopes that his young correspondent and future lover Christabel LaMotte is sufficiently conversant with the works of Paracelsus to know his account of the spirits called Melusinae that 'abound in deserts, in forests, in ruins and tombs, in empty vaults, and by the shores of the sea'.[19] Indeed, says Miss LaMotte, eager to prove her erudition to Ash, the passage is known to her. He was familiar also to Ivan Turgenev's doctor Vassily Ivanovich in *Fathers and Sons* (1862): 'Old Paracelsus spoke the sacred truth,'[20] muses the elderly man as he wanders through his herbarium.

And the romantic allure of magic is surely at play in Jorge Luis Borges's story 'The Rose of Paracelsus', a parable on faith. A young man named Johannes Grisebach appears at Paracelsus' door, asking to be his disciple. But first Grisebach demands that Paracelsus prove his skill in the occult arts by conjuring back a rose that has been consumed in the fire. Grisebach casts the flower – a mystical symbol often associated with Paracelsus – into the hearth, but Paracelsus says he cannot do what Grisebach asks.

Disappointed that his host is after all no magus, Grisebach leaves forlornly. Then Paracelsus, once alone, 'whispered a single word. The rose appeared again.'[21]

The legend of Paracelsus is not always so romantically employed:

> 'I would not go that way if I were you', said Nearly Headless Nick, drifting disconcertingly through a wall just ahead of Harry as he walked down the passage. 'Peeves is planning an amusing joke on the next person to pass the bust of Paracelsus halfway down the corridor.'
>
> 'Does it involve Paracelsus falling on top of the person's head?' asked Harry.[22]

What do they make of him, this generation of children angling for the potent Paracelsus card in the Harry Potter game on their Gameboys, having learnt that he features in the roll call of Famous Witches and Wizards along with 'Hengist of Woodcroft', 'Alberic Grunnion', Circe, Merlin and Nicolas Flamel?

He has, in other words, become a fabulous creature in the true sense, a cipher for arcane and occult knowledge, for nineteenth-century romanticism, for iconoclasm and magic pure and simple. Yet this literary interest in Paracelsus has an illuminating past:

> When I was thirteen we all went on a party of pleasure to the baths near Thonon; the inclemency of the weather obliged us to remain a day confined to the inn. In this house I chanced to find a volume of the works of Cornelius Agrippa . . . My father looked carelessly at the title page of my book and said, 'Ah! Cornelius Agrippa! My dear Victor, do not waste your time upon this; it is sad trash . . .
>
> But the cursory glance my father had taken of my volume by no means assured me that he was acquainted with its contents; and I continued to read with the greatest avidity. When I returned home my first care was to procure the whole works of this author, and afterwards of Paracelsus and Albertus Magnus. I read and studied the wild fancies of these writers with delight; they appeared to me treasures known to few besides myself.[23]

Thus began the career of Victor Frankenstein, who attempted unspeakable things in the name of science and knowledge, and in the end received his Faustian retribution. To the casual reader of Mary Shelley's classic, these are but strange-sounding names from a dimly remembered era, a whispered promise of forbidden fruits. But Shelley knew better than that. Her own father, William Godwin, took a different view from Frankenstein senior: in 1834 he published *The Lives of the Necromancers*, with chapters on Paracelsus, Agrippa and Faustus, and his perception of Paracelsus was more subtle, if not exactly approving: 'the union of a quack, a boastful and impudent pretender, with a considerable degree of natural sagacity and shrewdness'.[24] Percy Bysshe Shelley wrote to Godwin in 1812, saying that he had 'pored over all the reveries of Albertus Magnus and Paracelsus'.[25] So it is not hard to guess at some of the discourse to which Mary alludes in her account of the wild revels in Switzerland that spawned her cautionary fable. 'Many and long were the conversations between Lord Byron and Shelley', she said, 'to which I was a devout but nearly silent listener. During one of these, various philosophical doctrines were discussed . . . Night waned upon this talk, and even the witching hour had gone by before we retired to rest.'[26] So when a 'supremely frightful' tale came unbidden into Mary's head as she lay on her pillow, she would have understood clearly why it should have its origin in the lives of those uncanny figures from a legendary past, Cornelius Agrippa of Nettesheim and his fellow vagabond, Theophrastus Paracelsus Bombast von Hohenheim.

I

Black Madonna

A Country Doctor

> First there is the forest and inside the forest the clearing and inside the clearing the cabin and inside the cabin the mother and inside the mother the child and inside the child the mountain.
>
> Paracelsus, physician, magician, alchemist, urge, demiurge, *deus et omnia* was born under the sign of the occult, ruled by Mars and driven by a mountain in his soul.
>
> Jeanette Winterson, *Gut Symmetries* (1997)

Every year on 14 September, it is as if the small Swiss town of Einsiedeln, in the canton of Schwyz, returns to the time when Paracelsus was born there. The town square, dominated by the vast Benedictine abbey, is dark but for thousands of candles, flickering in glass pots from every wall and window sill. As the moon rises above the wooded hill that overlooks the scene, the evening is scented with the resinous tang of incense. From out of the abbey comes a solemn procession of pilgrims, each with a flame in hand, led by priests robed in white and monks in their dark habits. The only incongruous element is a brass band in chocolate-box uniforms, lending the scene a touch of fairground baroque. It is the end of summer, and the night air grows chill. Outside, beyond the town, darkness gathers.

The candle-carrying pilgrims who parade each year in Einsiedeln mark the annual celebration of the Miraculous Consecration. The first church and monastery were erected there in 948 by the wealthy Canon Eberhard of Strasbourg, but plans for their consecration by Bishop Conrad of Constanz were pre-empted. On the eve of the ceremony, the bishop was astonished to find the chapel filled with a choir of angels, while the service was conducted by Christ Himself, assisted by St Peter, St Gregory the Great and the Four Evangelists. Not trusting his own senses, and entreated by Eberhard and the Benedictine brothers to carry out the ceremony as had been planned, the bishop began the ritual the next day only to be halted by God's voice telling him that the place was already consecrated. Hearing Conrad's account, Pope Leo VIII forbade any subsequent attempt at consecration.

Einsiedeln was already a site of pilgrimage when the abbey was built. But there had been nothing there except wild forest when, in the ninth century, the Benedictine monk Meinrad arrived from Bollingen on Lake Zurich. In a clearing by the River Sihl, Meinrad established a hermitage, where he withdrew in 829 for a life of solitary contemplation. He took with him a Madonna and Child carved from hard black wood, a gift from Abbess Hildegarde of the convent on Lake Zurich. To house this statue Meinrad made a shrine, and the 'Black Madonna' became known as Our Lady of Einsiedeln. The original figurine was destroyed by a fire, but a fifteenth-century copy still stands in a gilded case in the entrance to the abbey.

As people began to seek out this ascetic hermit and ask for his blessing, the rumour spread that Meinrad had accumulated great wealth. One day brigands turned up at the lonely hut and cut down the monk who greeted them mildly with an offer of food. But they searched in vain for treasure. According to legend, two ravens, witnessing the murder, flew to Zurich to raise the alarm, so that the thieves were caught and burnt at the stake. Meinrad, martyred in 861, was subsequently canonized. Einsiedeln does not forget its founding legend, for the two black ravens are everywhere still today: on the town's coat of arms, on the Black Madonna's shrine, even on the logo for the local beer.

By the fifteenth century the town had grown around the monastery, with a hospital where sick pilgrims were treated. Some time in the 1480s Paracelsus' father, a young medical graduate of the University of Tübingen named Wilhelm von Hohenheim, arrived in Einsiedeln and became physician for the town and hospital.

FALLEN FROM GRACE

Wilhelm may have wandered this way more or less aimlessly from southern Germany, without money and lacking even the formal qualification of a doctor, for he did not complete his post-graduate studies at Tübingen. Although his family name – von

Hohenheim – implied nobility, his father Georg had been disgraced and became impoverished, and moreover Wilhelm was illegitimate and had been raised on his uncle's farm in the village of Rieth in Württemberg.

Hohenheim Castle stood near Stuttgart in Swabia, and it was the seat (the name means 'mountain home') of Conrad Bombast, who died in 1299. Conrad was a soldier and a feudal tenant of the Count of Württemberg, with claim to the tithes of the villagers of Plieningen and Ober-Esslingen. But by the time Georg rode as a commander of the Teutonic Knights on a crusade to the Holy Land in 1468, the fortunes of the Bombasts were in decline. They had further yet to fall.

Bombast, an old Swabian name, has inevitably given rise to the idea that Paracelsus' bluster and arrogance lie at the root of the word 'bombastic'. One feels that ought to be so, but it is not. *Baum* means 'tree' in German (in the Swabian dialect it is rendered *Bom*), and *Baumbast* is the fibrous layer of a tree's bark. But in the sixteenth century 'bombast' had also come to mean cotton padding, inappropriately derived from *bombax*, the medieval Latin name for the silkworm, and it is from this origin that the connotation of 'puffed up' derives. The family line of the Bombasts von Hohenheim ended in 1574, but there were still Bombasts in Württemberg in the nineteenth century.

The story of Ritter (knight) Georg Bombast von Hohenheim has something of a Paracelsian tenor, for he was a man toppled from his position of power by his own ungovernable impulses. He fathered Wilhelm in 1457 by an unnamed mistress – no great embarrassment for its time, although disadvantageous for Wilhelm – and Georg came to hold high rank with the Order of the Knights Hospitaller of St John of Jerusalem.

In 1489 Georg became embroiled in a bitter political argument in Stuttgart, and his fierce tongue caused him to be summoned before the High Court of Justice, which decreed that Georg's estate (such as it was) was forfeit. Thereafter the disgraced von Hohenheims were a family of paltry means and little consequence. It may have been his father's downfall that set Wilhelm on his travels south. Others say that his arrival in Einsiedeln was no happenstance, but that he was summoned there from Württemberg to take up his position as town physician in 1481.

A COUNTRY DOCTOR

There is a woodcut of Einsiedeln from 1577 which shows the abbey lodged on a hill beneath forests that climb up the mountain-side. A cluster of houses is scattered further down the slopes, and the road winding up to the abbey from the valley floor is busy with pilgrims on foot and on horse. Down a ravine, steepened by artistic licence, tumbles the torrent of the Sihl, and the water is straddled at one point by the wooden Teüfelsbrücke, the Devil's Bridge. Beside this bridge is a sturdy building, somewhat larger than the humble lodgings of the town, and this, one assumes, is the inn where a footsore Wilhelm rested after his long journey.

The inn, set in green hills about three miles from the town, was run by the family of Rudi Ochsner. It stood on ground owned by the abbey, to which the Ochsners were feudally bound. Behind the inn, cattle graze in gently undulating meadows – but the hierarchy of the picture is clear enough, with the abbey dominating the vista like a lord's castle. The inn stood until the nineteenth century, when it was burnt down.

Today there is another tavern in its place, an undistinguished brick building that proudly displays a plaque claiming that it is the

Einsiedeln in the late sixteenth century.

birthplace of Theophrastus von Hohenheim, known as Paracelsus. That claim is almost certainly false, but his real homestead was indeed somewhere in these meadows, now vanished and forgotten, and the Devil's Bridge seems as good a place as any to begin his story.

At the beginning, however, there is little but legend, rumour and sheer speculation. You can tell this part of the tale any way you will, and many have done so. One version has it that the innkeeper's daughter was named Els, a matron in the pilgrim hospital. Wilhelm made the inn his home as the local doctor, and in 1491 he married Els. But other sources suggest that Els was Rudi's wife, not Wilhelm's. Paracelsus' mother has been variously identified as a member of the Grätzel family, who owned the house in which the Hohenheims lived, and as one of the Weseners of Einsiedeln. Whatever her name, Wilhelm's wife may have been no matron at all but merely a bondswoman to the abbey: this is all that might be implied by the sole remaining description of her status as a 'woman of the house of God'.

Nonetheless, the occasion of Wilhelm's marriage is purportedly recorded in a portrait painted the same year, in which he holds a carnation as a sign of betrothal. Whether this picture does indeed represent Wilhelm von Hohenheim has been disputed; but it now supplies the basis for several commemorative portraits of Paracelsus' father. Dressed soberly in black, he looks calmly out of a window at a rocky vista much like that visible from the inn at the Devil's Bridge. In the top right-hand corner of the picture are the arms of the von Hohenheims: three azure globes on a bend argent. In the opposite corner is the heraldic head of an ox – possibly a reference to his bride's family name.

If it is truly the father, this image seems that of a man with a temperament very different from his son's: by all accounts a gentle, patient man, whose years of study gave him a good working knowledge of Latin, botany, medicine, alchemy and theology. Paracelsus worshipped Wilhelm, calling him his first teacher, 'who has never forsaken me'.[1] Never one to feel bound by obligation, his visits to his father at Villach in later life were presumably made more out of genuine love than filial duty.

Wilhelm and his wife christened their son Theophrastus, although this was often prefixed by the name of his name-day saint: Philip. The Greek Christian name was an unusual choice for

A portrait painted by Geoge Stäber in 1491 and generally assumed to depict Wilhelm von Hohenheim, Paracelsus' father.

a German, and reveals Wilhelm's love of classical learning. Tyrtamos of Eresos (*c.*372–288 BC), known as Theophrastus, was a pupil of Aristotle, and succeeded him as head of the Lyceum in Athens. He inherited Aristotle's passion for natural history, but whereas his teacher had written mostly on the animal kingdom, Theophrastus compiled encyclopaedic treatises on plants and minerals. His book *De lapida* (*On Stones*), written around 300 BC, is a comprehensive study of minerals and represents perhaps the earliest known work on practical chemistry. Benefiting from knowledge that Theophrastus gathered through his interest in mining, the book is a forerunner of the great *De re metallica* (*On Metallurgy*) by the sixteenth-century German humanist Georgius Agricola. In retrospect, Wilhelm could hardly have chosen a better model for his son.

Who, really, was Wilhelm's wife? The paucity of information has driven some biographers to absurd extremes of ingenuity and invention: Josef Strebel's 1944 book attempts to suggest what she may have looked like by adding a head scarf to a portrait of Paracelsus himself. Paracelsus extols the virtues of his father, but about his mother he is silent. It has been inferred that she was a manic depressive; and certainly, Paracelsus wrote about mental illnesses with a sensitivity most uncharacteristic of his time. According to one legend, when Theophrastus was nine years old his mother walked onto the Devil's Bridge and threw herself over the parapet into the Sihl. But all we really know is that by 1502 she was dead.

In Paracelsus' time, very few children reached adulthood without experiencing death in the family. Parents expected to lose children to illness or mishap, often at birth, while the plague and other endemic fatal illnesses left countless children orphaned. Violence and murder were commonplace. But the suicide of a young boy's mother, if it really happened, would even then have had a shattering impact on the child. It is risky to speculate about the psychological consequences; but if anyone is to be permitted to do so, it might be Carl Jung, who says of Paracelsus' mother:

> She died young, and she probably left behind a great deal of unsatisfied longing in her son – so much so that, as far as we know, no other woman was able to compete with that far distant mother-image, which for that reason was all the more formidable. The more remote and unreal the personal mother is, the more deeply will the son's yearning for her clutch at his soul, awakening that primordial and eternal image of the mother for whose sake everything that

> embraces, protects, nourishes, and helps assumes maternal form . . . When Paracelsus says that the mother of the child is the planet and star, this is in the highest degree true of himself.[2]

Jung claims that Paracelsus found two substitute mothers in his life: the Church and Mother Nature. The agony of it was that these two were not always concordant, although Paracelsus laboured mightily to resolve the conflict.

As a child, Theophrastus von Hohenheim was small and frail. Despite his close relationship to his father, he recalled that 'I grew up in great misery'[3] because the poor wages of a country doctor made for a straitened existence. He suffered from rickets, a softening of the bones caused by a deficiency of vitamin D, often the result of a lack of eggs and milk in the diet. This disease produces skeletal deformities, including an enlargement of the upper head, and it has been proposed as the reason for the curious squarish profile of the balding pate evident in portraits of Paracelsus made near the end of his life. His skull was disinterred from his grave in Salzburg in the nineteenth century and inspected by an anatomist, who confirmed this diagnosis.

It was an indelicate upbringing, which Paracelsus celebrated defiantly later in his life:

> By nature I am not subtly spun, nor is it the custom of my native land to accomplish anything by spinning silk. Nor are we raised on figs, nor on mead, nor on wheaten bread, but on cheese, milk, and oatcakes, which cannot give one a subtle disposition. Moreover, a man clings all his days to what he received in his youth; and my youth was coarse as compared to that of the subtle, pampered, and over-refined. For those who are raised in soft clothes and in women's apartments and we who are brought up among the pine-cones have trouble in understanding one another well. To begin with, I thank God that I was born a German, and praise Him for having made me suffer poverty and hunger in my youth.[4]

This pride in simplicity, even coarseness, was perhaps characteristically German. (When Paracelsus thanks God for his German heritage, he obviously means this to be a label of culture rather than nationality. But when his reputation waxed in the late nineteenth century, the modern state of Germany was eager to claim him as one of its own – for after all, his was a noble family of Swabia, however low they had fallen. He is lionized there still today, where many streets and public places are named in his honour.) In 1590 the Italian painter Giovanni Paolo Lomazzo could be describing Paracelsus when he advises artists as to how the German race should be portrayed: with a 'strutting stride, extravagant gesture, wild expression, clothing all anyhow, manner hard and stern'.[5]

Although that description owes much to the Renaissance propensity for nation stereotypes, it really was a rough life in the Germanic lands, especially in a rural backwoods like Einsiedeln. One story alleges that Paracelsus was emasculated as a child in an 'operation' performed by drunken soldiers lodging at the family inn. But several such legends claim he was a eunuch, no doubt supported by his apparent abstinence from sex or romance throughout his life and his advocacy of chastity. According to another tale, he lost his manhood in a childhood encounter with a wild boar. It has even been claimed that he was castrated by his own father. Yet hermits were often thought to be eunuchs, and Paracelsus was often regarded as the former on account of both his solitary nature and the fact that Einsiedeln itself means 'hermitage'. H. P. Bayon, observing images of Paracelsus with a doctor's eye, pronounces him 'eunuchoid' (that is, having reduced sexual characteristics). Added to the evidence of his square cranium, early baldness and (Bayon asserts) premature senility, he diagnoses congenital syphilis, meaning that Paracelsus contracted the disease at birth from his mother. This is improbable, not only because he never refers to such a personal affliction despite writing extensively on syphilis (and it was to become common enough in his time as to be not especially shameful), but also because the disease does not seem to have spread through Europe until the year after his birth.

MOVING ON

Although the Hohenheims were impoverished, Paracelsus recalled his family home as 'quiet and peaceful'. What riches he experienced were in the meadows and fields. In those days in the Sihl valley and the pastures of the Etzel mountain you could find primula, gentian, daisy, ranunculus, camomile, borage, fennel, poppies, violets, belladonna, foxgloves, chicory, mint, thyme, St John's wort, mallow, azalia, saxifrage, wild plum and many other useful plants. Wilhelm introduced to his son the medicinal virtues of this natural bounty, providing the initial stock of Paracelsus' ever-expanding pharmacopoeia.

The cowbells that still clank in these meadows today may tempt us to imagine those times as idyllic. But in fact they were turbulent and uncertain, as the Swiss cantons struggled to establish their independence from the Holy Roman Emperor in Germany. In the late fifteenth century the confederacy formed by the cantons and city-states was still part of the Empire, but in 1499 it mounted an armed resistance against the Emperor Maximilian's rule. This, the so-called Swabian War, pitted the Swiss against the southern German states, and the Hohenheims were suddenly enemy aliens. The war lasted only until the following year, when the allied cantons left the Empire; but the conflict is generally cited as Wilhelm's reason for departing from Einsiedeln in 1502 and travelling east with his son (his wife, suicide or not, has vanished) to find a new home in Austria. They settled in Villach in Carinthia, where metals were made.

2

The Metal Makers

Learning the Arts

> Certainly, though it is but one of ten important and excellent methods of acquiring wealth in an honourable way, a careful and diligent man can attain this result in no easier way than by mining.
>
> Georgius Agricola, *De re metallica* (1556)

'The mountains of Carinthia are like a strong box which when opened with a key reveals great treasure,'[1] Paracelsus observed. Men drew fabulous riches from that casket: lead, iron, tin, copper, silver, gold, mercury and many other precious materials of the earth.

Villach, nestling among these mountains, thrived on their bounty. By the sixteenth century a traveller drawing nigh to this refuge in the Drau valley on the road to Venice or Vienna, Bohemia or Hungary, would have found no mere mountain village but a prosperous town, the second largest in Carinthia, making the most of its position at the nexus of trade routes. Here at the foot of the mountains, metals from the nearby mines of Bleiberg, Paternion and Hüttenberg were refined and worked by bellows, pumps and grinding apparatus driven by fast-flowing rivers from the Alpine slopes. As mining began to boom in the late fifteenth century, there were fortunes to be made at Villach. And no one took better advantage of that than the mighty Fuggers of Augsburg, the wealthiest merchants in the Empire. They effectively owned not only the silver-smelting plant (*Saigerhütten*) in Villach but also the entire silver and copper industry in the Tyrol. At Villach there stood a castle they had built with the riches amassed from their trade in metals, called the Fuggerau.

Wilhelm von Hohenheim knew enough mineralogy to find employment as a teacher in the mining school (*Bergschule*) of the Fuggers at Hüttenberg. He and his son took lodgings in Villach in a house on the main street leading to the fourteenth-century church of St Jakob. The building is still there today at Hauptplatz 18, accessed via a small passage that leads into a quiet courtyard; but now it houses a bank, plying the Fuggers' business.

Theophrastus was sent to the Benedictine monastery schools of St Paul's at Klagenfurt and St Andrae in Lavanttal, in a valley near Villach, surrounded by the smelters and laboratories of the Fuggers. Here he was instructed in the usual scholastic disciplines: Latin, grammar, logic, rhetoric. Such schoolbook stuff was never to his taste, and it shows: one can hardly claim that his writings are models of logic, and he generally found more use for vernacular invective than classical rhetoric. Although his Latin was not as execrable as some detractors have suggested, it was undeniably rather coarse.

But at Lavanttal he found a teacher with interests closer to his heart, for Bishop Erhart (Eberhart Baumgärtner) was an alchemist who conducted work for the Fuggers. Here and in the mines and Fugger workshops around and about Villach where he served as an assistant, Theophrastus gained a thorough grounding in the lore of metals and their transformations.

THE SILVER EMPIRE

Mining is an ancient art, but not until the Renaissance did the production of metals begin to slip from the grasp of kings and emperors and come under the control of merchants and private investors. With that change, men of commerce began to find themselves acquiring unprecedented political power, as well as wealth beyond the dreams even of their rulers.

This influence runs like a mineral vein through the political and social strata of the age in which Paracelsus lived. Mining forged a link between speculative science and practical technology, and it encouraged the kind of close empirical study of nature and engagement with experimentation that distinguishes the work of the later seventeenth-century scientific pioneers like Francis Bacon, Robert Hooke and Robert Boyle. It was in the crucibles of artisans and industrialists attempting to separate and purify metals that alchemy was transmuted into chemistry.

Mining has become a metaphor for (as well as being an agent of) all that is now seen as deplorable in the Scientific Revolution. The pre-eminent treatise on metallurgy in the sixteenth century,

By the early sixteenth century the machinery of mining had become awesome in its scale and complexity.

Georgius Agricola's *De re metallica* (1556), is crammed with fine woodcut illustrations of huge and complex industrial machinery designed to wrest nature's bounty from her grasp. We see nature despoiled: trees felled, rivers diverted and polluted, the earth pitted and burrowed, all in the name of trade, commerce and profit. The message is that the natural world can be subjugated and exploited by the mechanical arts for humankind's gain.

But Agricola already anticipates such criticisms, and in his book's introduction he mounts a staunch defence of mining. Doesn't the lust for gold and silver lead men into vice and

Mining laid waste to the countryside all around.

mendacity? Is the working of metals no proper employment for respectable people but rather a 'degrading and dishonourable' affair that was once the labour of slaves? Do miners not heap up riches by cunning and deception? Do their excavations not devastate the fields? Agricola rebutted these charges and argued that the immense utility of metals to man far outweighed any nuisance involved in extracting them. It is not the products of mines that cause war, he says, but the avarice of men. Precious metals make trade far easier than barter systems, because 'a small amount of gold and silver is of as great value as things cumbrous and heavy'.[2] Metals are valuable to the physician, 'for they furnish liberally the ingredients for medicines' – so that 'if there were no other reasons why we should explore the depths of the earth, we should for the sake of medicine alone dig in the mines'.[3] They yield colours for the artist, materials for the architect, and much more. 'In short, to whom are the metals not of use?'[4]

Tacitus claimed that the gods had denied Germany gold and silver, 'whether in mercy or in wrath I find it hard to say'.[5] But he admitted that no one had looked for them very hard. During the Middle Ages the Germans found major deposits of silver, and they became masters in the art of mining it. Deposits were first found in the Harz mountains of northern Germany in the tenth century. In this wild and remote region, so the legend goes, a nobleman's horse, tethered to a tree on a hill above the town of Goslar, kicked away soil to expose shining black minerals that proved to be rich in lead and silver. Soon the Harz was being plundered for its ores. In 1136 similar minerals were discovered close to Freiberg, in the hills that separate Saxony from Bohemia, triggering a 'silver rush' as frenzied as the later hunt for Klondike gold. The hardy miners of the Harz and Saxony, known as *Bergleute* (mountain people), became famed throughout Europe for their skills, and found employment as far afield as Serbia and Sardinia. The *lingua franca* of mining was German, and its imprint has left its mark in technical mining terminology even as far afield as Turkey.

The crucial innovation in silver mining was not made, however, until the mid fifteenth century. Silver deposits are rare; but miners knew that there was silver in the ores of Germany's copper mines. The problem was to separate the two metals. The process by which this was done, known as liquidation, is first described in the municipal archives of the copper and brass foundry of Nuremberg, a major centre for metal refining, in 1453. Copper ore typically contains lead as well as silver. When the ore is smelted in a furnace, the silver dissolves preferentially in the lead as the metals cool. Because lead melts at a lower temperature than copper, it can be separated and drained off while the copper is still solid. This silver-rich lead can then be resolved into its components by an ancient process called cupellation: the mixture is melted in a crucible of bone ash and the lead drains into the ash.

Liquidation, which Germans called the *Saigerprozess*, stimulated a mining bonanza that benefited Saxony and Joachimsthal, where silver-rich ores were discovered in 1516, revitalized the previously moribund mining industry in Bohemia, and spread to the mining district around Schwaz in the Tyrol. The Fugger family, Wilhelm von Hohenheim's employers, gained a hold on the Tyrolean silver mines almost by chance. Johann (or Hans) Fugger, a weaver, established a merchant business in textiles in Augsburg in 1367. At first he traded in cotton; as his business thrived, he moved into luxury fabrics such as silk, and then into spices imported through Venice. Johann died in 1408, leaving a modest fortune to his two sons, Andreas and Jakob. When Jakob made a loan to his father-in-law, a man named Basinger, he was offered in repayment a share in the silver-buying business that supplied Basinger's mint at the town of Hall in the Tyrol. The deal made Jakob rich, and in 1469 his seven sons inherited this great wealth.

The eldest three were supposed to take over the business. But when one of them died, one of the younger sons, also named Jakob, was recalled from ecclesiastical training to help run the family trade. Jakob the Younger turned out to have an excellent head for commerce, and he expanded the business into metals and money-lending.

Among the clients for the Fuggers' loans was Sigismund, Archduke of the Tyrol. In 1487 Jakob struck a deal with Sigismund much like that which his father agreed with Basinger, but on a larger scale. As security for late repayment of a loan, he was granted Sigismund's share of the silver-mining interests at Schwaz – an agreement that was subsequently expanded to include all silver production in the Tyrol.

Jakob's fortunes rose further when Sigismund passed on his title to his cousin Maximilian. In 1491 Maximilian borrowed more than 200,000 florins from the Fugger coffers to finance wars in the Habsburg territories; in return, Jakob gained control of all silver and copper mining in the Tyrol. And once Maximilian became Holy Roman Emperor two years later, his bills escalated alarmingly, for now he had to maintain a vast territory that stretched from Alsace to the Polish border, and from the North Sea to the Alps. Like an addicted gambler, he took to borrowing vast sums from the Fuggers, secured with his future share in the metals from the Habsburg mines. As Maximilian fell ever more heavily into

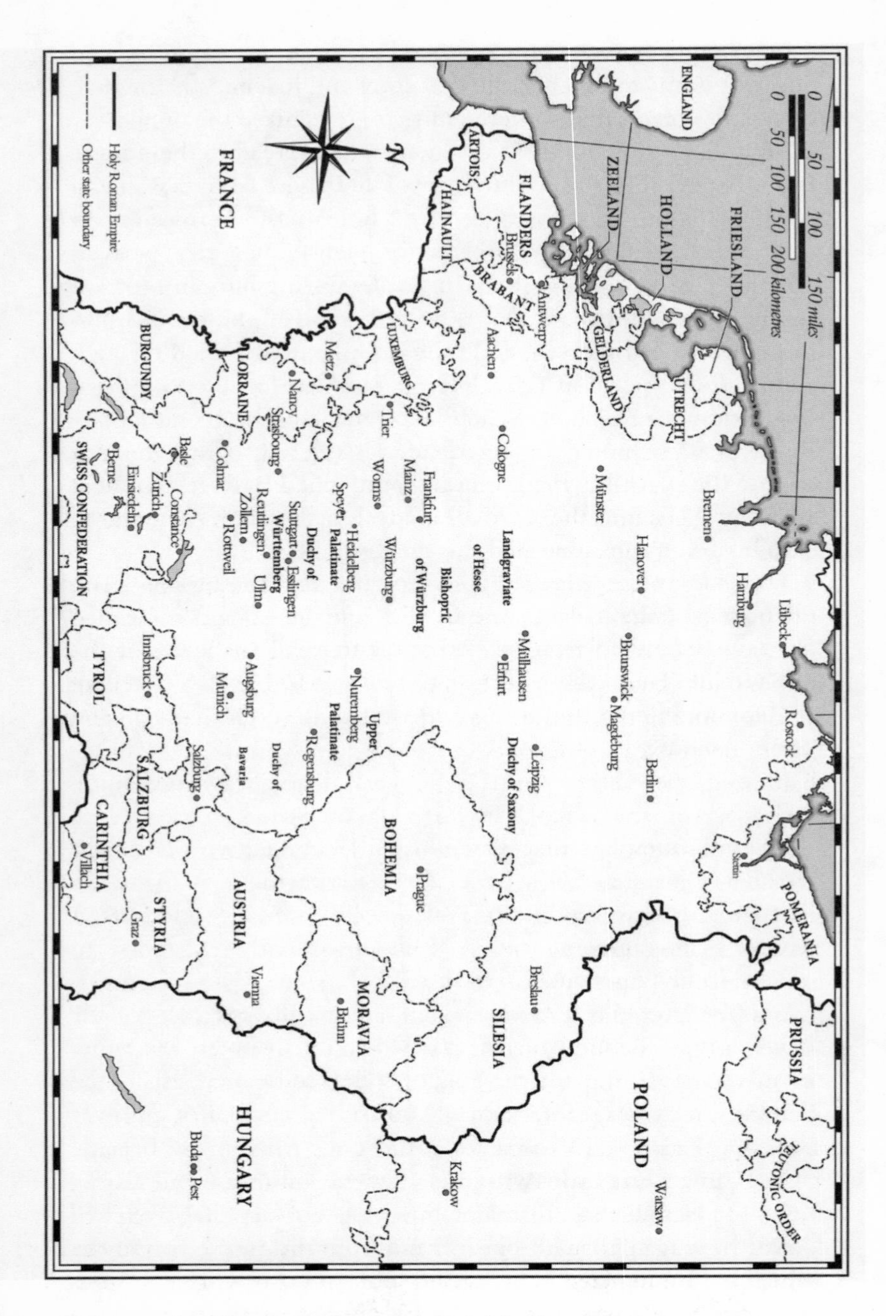

Germany and the Holy Roman Empire at the end of the fifteenth century.

debt, the Fuggers were able to expand their mining activities from Spain to Hungary, and they established an international network of banking agents for money-lending. Even the papacy farmed out the handling of its funds in Germany to the private enterprise of Jakob Fugger. His family firm was worth around two million gold florins when he died in 1525.

Maximilian himself died six years earlier, and his grandson, Charles I of Castile, realized that the fate of his candidacy as Emperor would depend on the dispensation of massive bribes to the Electors. He didn't have the money, but he knew where to get it: from his grandfather's creditor. Four years later Jakob Fugger brandished a reminder of this bribery when the new Emperor fell behind on his repayments: 'it is well known and clear as day', he warned, 'that your Imperial Majesty could not have acquired the Roman Crown without my help.'[6] There can be no clearer sign of the new power of self-made merchants.

Mining could not have undergone such rapid expansion in the fifteenth and sixteenth centuries without several innovations in technology. New methods of supporting shaft walls meant that mines could be dug deeper. Some shafts plunged to 500 feet; one at Chemnitz in the Erzgebirge (Ore Mountains) reached 660 feet. These deep excavations needed new and powerful machinery to haul up the ore and to pump water out of the tunnels. Where fast rivers were not available to turn wheels, horses were used instead.

All this is shown in extraordinary detail in Agricola's *De re metallica*. The author of this seminal treatise was born Georg Bauer in Glauchau, Saxony, in 1494 – a year after the Saxon copper smelter Hans Luther in the Harz mountains was delivered of his son Martin. 'Bauer' means 'farmer' or 'peasant', which became Latinized, after the humanist fashion, to 'Agricola'. Georg followed a typical educational trajectory: like Paracelsus, he trained in medicine at the universities of Germany (graduating from Leipzig in 1518) before studying in the great Italian cities of Bologna, Padua and Venice.

In 1527 he was appointed physician in the burgeoning mining town of Joachimsthal in Brandenburg, and in between his medical duties he began to make a study of mining lore and methods, which he summarized in his book *Bermannus* (1530). But Agricola perceived that there was a need for a more extensive reference work on mining. The classical literature on the topic was scanty

Georgius Agricola (1494–1595), the greatest Renaissance scholar of mining and mineralogy.

and of variable quality. Aristotle was still the authority, while his pupil Theophrastus provided a systematic classification of minerals in his *De lapida*, and had apparently written a treatise on metals, mines and mining that was now lost. The Greek writers Diodorus Siculus and Strabo, who flourished in the first century BC, gave little more than anecdotal travellers' accounts of mining practices. As for contemporary works, the Sienese metalsmith and military engineer Vannoccio Biringuccio wrote a manual called *De la pirotechnia* (1540) (in the Italian vernacular rather than Latin) that Agricola praised warmly; but otherwise there was nothing but a few anonymous 'skills booklets'.

From 1533 to 1550 Agricola diligently collected technical information that made his book the standard work on minerals and mining for 200 years. Because it deals with the mundane and not the theoretical, *De re metallica* has never been accorded the status in the history of science awarded to Copernicus' *De revolutionibus*, but it is surely one of the masterpieces of scientific scholarship during the Renaissance. If we were to put it crudely, we might cast Agricola as the judicious Erasmus to Paracelsus' incendiary Luther. His life and work are marked by clarity, careful judgement, moderation and integrity. The translators of *De re metallica*, Herbert and Lou Hoover (later the president and first lady of the United States), contrast Agricola's 'sober logic and real research and observation' with Paracelsus' 'egotistical ravings'.[7] Certainly there is no hint in his great book of any mysticism or occult speculation, and by all accounts he had little time for alchemists' wild flights of fantasy. He serves to remind us that the excesses of Paracelsus' magical world were not inevitable products of the times: it was possible to expound natural philosophy in a more measured way than that.

Agricola died a year before *De re metallica* was published, to the dismay of learned men of every persuasion. He was respected throughout Germany for his scholarship, and although he never formally converted to Lutheranism and had little time for Luther's theological hair-splitting, he formed lasting friendships with eminent Protestant scholars such as Melanchthon and Fabricius. He was, Fabricius wrote to Melanchthon, 'a man of eminent intellect, of culture and of judgement'.[8]

NEW METALS

Paracelsus' boyhood training in the art of metals stood him in good stead not just as an alchemist but also as a natural historian. Alchemists and miners alike recognized just seven metals at the beginning of the sixteenth century: gold, silver, copper, iron, tin, mercury and lead. But some began to suspect that the list was longer. One new metal is mentioned in print for the first time in a 1558 revision of Agricola's *Bermannus*: there it is called *zincum*, or zinc, but the text implies that *zincum* is a mineral ore (a chemical compound) rather than a metal. The earliest proper use of the word is usually attributed to Paracelsus himself, who states in his book *De mineralibus* (published in 1570) that:

> There is another metal generally unknown called *zinken*. It is of peculiar nature and origin; many other metals adulterate it. It can be melted, for it is generated from three fluid

principles; it is not malleable. Its colour is different from other metals and does not resemble others in its growth.[9]

For once Paracelsus is ahead of Agricola here, for he clearly accepts zinc as a metal in its own right. He is often accredited with coining the name himself (he was fond of naming things), an amalgam of the two metals (*zinne* and *kupfer*, tin and copper) that zinc somewhat resembles. But this is to give him too much credit, for the word *zinken* was apparently already common in Carinthia in his day. The modern Catalan and Spanish word for zinc, *cinc*, appears in some late fourteenth-century texts, where, however, it refers to brass (an alloy of copper and zinc). In fact, knowledge of zinc metal dates back way beyond the Middle Ages, even if it was not formally recognized as such. Zinc bracelets have been found in Greek ruins at Cameros dating from before 500 BC, and the metal has also been excavated at Roman sites in Europe. Strabo refers to a 'mock silver' which might be zinc. It was imported from India and the Far East in the sixteenth century before it was widely known in Europe; it was sometimes called Indian tin, and Shakespeare's reference to 'metal of India'[10] in *Twelfth Night* may allude to zinc.

Zinc minerals and ores often appear in antiquity and the Middle Ages under the vague and confusing name of *cadmia*. The Greek writer Dioscorides (*c.* AD 40–90) speaks of a *cadmium* which could be used as a balm: this was zinc oxide. It was formed in copper-smelting furnaces, since zinc is commonly present in copper ore. Later known as calamine, it is still used in medical ointments today. But there was another kind of *cadmia* called *cobaltum*, or *kobelt* in colloquial German, named after the malevolent goblins that were believed to haunt the mines. The mineral became associated with these unpleasant earth spirits because it was 'corrosive', as Agricola wrote in the *Bermannus*, eating away at the hands and feet of miners who came into contact with it. Cobalt minerals themselves are harmless enough, but Agricola's *cobaltum* appears to have been a mineral rich in arsenic, probably the cobalt ore smaltite. Cobalt itself was the key component of a blue mineral known as *zaffre*, from which comes the word 'sapphire' (which, however, is *not* a cobalt compound!). *Zaffre* was used as a blue glaze for pottery, and as a colouring agent for blue glass.

Clearly, there was a fair degree of confusion about the

components of metal ores. This is hardly surprising, as zinc, cobalt and arsenic often appear in complex mixtures with each other, with copper, and with other 'new metals' such as nickel and bismuth. Some of these substances, such as cobalt and arsenic, acquired names before being recognized and isolated as pure elements. Antimony too is often mentioned by Agricola, Paracelsus and their contemporaries, although they knew it only as a sulphide mineral (*stibium* in Latin). Paracelsus did much to advertise its supposed virtues in medicine. Some new elements, such as zinc and bismuth, could be made in pure form, although they were often mistaken for other substances. Bismuth, a soft grey metal, was probably first prepared around 1400, and by the late fifteenth century it was being used by metalsmiths who, in Germany, even formed their own specialized guild of bismuth-workers. But it was frequently confused with lead, and although Agricola argued that it was different, he had no proof.

Metals and their minerals were, as Marlowe's Faustus is told, a central component of the magic arts. They were the raw materials of the alchemist, and supplied essential ingredients for the physician's store of remedies. All of these things drew Paracelsus' interest to the earth's subterranean bounty. But his attitude is also that of a man who loved and trusted nature for its own sake, and sought to understand it as such, rather than regarding it, like the Fuggers, as merely another bank from which money can be drawn. 'All things that we use on earth,' he said, 'let us use them for good and not for evil. And never for more or for anything other than the purpose for which they exist. We should add nothing and take away nothing, spoil nothing, better nothing.'[11] While we should resist the temptation blithely and anachronistically to appropriate those words in a modern ecological context, it isn't hard to feel sympathy with the spirit in which they were written.

3
The Universal Scholar

A Renaissance Education

> All the great, daring and decisive intellectual and spiritual innovations characteristic of the culture and civilizations of the Middle Ages, the Renaissance, and even of more recent epochs, had their origin and expansion outside of organized systems of learning of a scholastic or academic type.
>
> Leonardo Olschki, 'The Scientific Personality of Galileo' (1942)

In 1507 a friend of Wilhelm von Hohenheim's came to Villach. Joachim von Watt had once taught humanities in the town; now he was rector of the University of Vienna and a well-known humanist. He called himself Vadianus, and he probably encouraged Wilhelm to send his son to be educated in medicine at the universities of Germany. And so it was that, at fourteen years of age, Theophrastus packed his bags and set out into the world.

A student did not simply enrol at a nearby university. Rather, he wandered from one institution to another, following his nose for the best teachers or for the most conducive environment. The doors of every university stood open to students, and indeed there was some competition to attract them. It testifies to the conservatism and uniformity of the academic canon that such an itinerant education could still leave students in a position to qualify at the end of it all.

Living the life of a traveller, students surely had to grow up quickly – or at least, they needed to become quickly inured to the hardships of a wandering life, which is not quite the same thing. We may assume that the young Theophrastus took to this roving, for there is soon no sign of the sickly child from Einsiedeln, and once he left Villach he was never to settle anywhere else for as long as he lived. But if travel hardened a young scholar, it was also apt to coarsen him. Students had to develop a certain brash confidence in order to survive; some earned a few coins by pulling teeth, or selling medical remedies, telling fortunes, singing at inns. At the universities, students had a reputation for unruliness – someone forced to cultivate independence will not then easily submit to the yoke of academic authorities. Students had sex in public, they threw stones or dung at their lecturers (whose own mudslinging was generally confined to the verbal), they drank and fought. In Vienna, battles between the students and the town guildsmen were serious enough to become known as the 'Latin War'. Many cities did not particularly appreciate the status a university gave them, feeling that this was hardly sufficient compensation for the trouble it caused.

Yet it was not all a life of libertine debauchery. At the university of Paris work began at six in the morning in winter, and an hour earlier in summer. Examinations could be marathons: the key test, the Cardinal Thesis, ran from five in the morning until noon. As if that were not bad enough, the student found himself confronted by such gnomic questions as 'Is the loud voice warm?' and 'Is it healthy to get drunk once a month?' Healthy or not, students rarely limited themselves so abstemiously; they were in fact obliged to attend many expensive celebratory banquets during the year, which could leave them nursing crippling debts as well as headaches.

SCHOLAR'S PROGRESS

The university was a medieval invention. The Latin word *universitas*, meaning the totality or whole, was a legal term applied in the twelfth century to guilds and corporations. When theological instructors established schools at cathedrals for the instruction of the future clergy, they began to apply the tradesman's concept of *universitas* to their own organizations. By the end of that century, the masters of a university were considered to belong to a guild-like body, the *universitas magistrorum*. Each faculty was more or less independent; indeed, each was essentially considered to comprise its own 'university'. There was not necessarily even a physical establishment as such to unify these parties; at some thirteenth-century universities such as Paris, the teaching was done in the private homes of the masters, or in rented properties, or even, it is said, in brothels.

The intimate relationship between the universities and the

Church meant that masters and students enjoyed many of the rights and privileges of the clergy. Indeed, in most cases they *were* clerics. At Paris students were not permitted to marry; even when this law was annulled for the medical faculty in 1452, the students were expected (rather optimistically) to remain celibate. But ecclesiasticism had its benefits: to assault a travelling student was tantamount to attacking a priest and carried heavy penalties. This was fortunate, for a student was expected to travel a lot, and the roads were seldom safe.

A prospective physician had to study the liberal arts for three years in order to qualify as a bachelor of science and to progress to the stage of actually learning some medicine. To this end, Theophrastus went first to the university at Tübingen, where his father had studied. He passed on to Heidelberg, Mainz, Treves and Cologne. At Freiburg in the Black Forest he found a rowdy atmosphere, 'like a house of indecency';[1] then at Ingolstadt quite the opposite, a dry, hidebound 'university of some old scholastics'.[2]

But by that stage he had already made up his mind about universities: they were the only way to gain a doctor's qualification, but they were otherwise a waste of time. 'At all the German schools', he concluded, 'you cannot learn as much as at the Frankfurt fair.'[3] Seeking open-minded debate and discussion, Paracelsus found he was expected to do little more than mouth the words – in his view, deeply misguided words – of the medieval canon, which drew from the classical medical texts of Aristotle, Galen and others and their interpretations and extensions by Islamic scholars like Avicenna and Averroës. There seemed to him to be no quest for truth; the prevailing view was that all the answers could be found in old books.

It would be unfair to suggest that all universities in the Renaissance were sterile, conformist places; they often hosted vigorous discussion and dissent. But verbal sparring matches ('disputations') could amount to point-scoring contests decided by nimble rhetoric, rather than matters of genuine enquiry and debate. Moreover, there was an unchallenged hierarchy among the sciences, according to which anything manual, be it the dissection of a corpse, the manufacture of a drug or chemical, or the construction of machines, was despised. For a young man fresh from the mining schools of Villach, this clearly seemed more than absurd; it was offensive. How was it, Paracelsus later asked, that 'the higher colleges managed to produce so many high asses'?[4]

In these early years in the academies, Paracelsus must therefore have begun to see the magnitude of his life's task looming before him. This was not where truth would be found; he would have to scour the whole world for it. 'The journeys which I have thus far made have profited me much,' he later averred, 'for the reason that no man's master is in his home and none has his teacher in the chimney-corner.'[5]

HUMANITY RECONSIDERED

In 1509 Theophrastus reached Vienna, drawn there perhaps by his father's friend Vadianus. Here he stayed for two years and gained his baccalaureate from the faculty of arts. But in 1511 Vienna was struck by plague, and in the summer he decided to move on, passing north through Bohemia to Saxony. In early 1512 he arrived at Wittenberg, where a tormented Martin Luther, now professor of theology, was struggling with his morbid fear of God. Paracelsus studied for a while at Leipzig and then at Erfurt, a well-known centre of humanism. Here at last he may have met some like-minded souls. Conrad Mutianus Rufus held court in a circle of progressive anti-dogmatists known as the Mutianiscker Bund, which included the knight and political agitator Ulrich von Hutten, the unruly poet and notorious inebriate Helius Eobanus Hessus, and the witty satirist Crotus Rubeanus.

Humanism was transforming the concept of education in the early sixteenth century. The equipment of an educated man was traditionally embodied in the seven liberal arts: the *quadrivium* of astronomy, arithmetic, geometry and music, and the *trivium* of rhetoric, logic and grammar – all anchored to the bedrock of Christian theology. But in the Middle Ages this knowledge was amassed by rote learning of a few canonical texts written many hundred of years ago: memory counted for much more than originality or critical thinking. Medieval scholarship was apt to ossify into a Glass Bead Game of hair-splitting and empty virtuosity. The scholastics, Erasmus said:

> are fortified with an army of definitions, conclusions, and corollaries . . . they interpret hidden mysteries to suit themselves: how the world was created and designed; through

> what channels the stain of sin filtered down to posterity; by what means, in what measure, and how long Christ was formed in the Virgin's womb . . . Could God have taken on the form of a woman, a devil, a donkey, a gourd, or a flintstone? If so, how could a gourd have preached sermons, performed miracles, and been nailed to the cross?[6]

This scholastic tradition allowed that one might mine the 'scientific' aspects of the Greek and Roman world – so that Ptolemy's astronomy, Galen's medicine and Aristotle's physics and logic were all standard components of the curriculum – but insisted that the poetry, moral philosophy and literature of these pagans was best left untouched. The movement known as humanism, in contrast, revealed the classical world to be a far livelier, more sensuous and more *humane* place than medieval Christian tradition would admit.

In fourteenth-century Italy, increasing civic wealth and liberty prompted some intellectuals to challenge the transcendent, ascetic attitude of scholasticism. Instead, they began to discover the worldly delights to be found in the *humaniora* (humanities) of the classical writers. At its best, humanism prompted a reconsideration of history and the relation of the contemporary to the past, of art and its objectives, and of the place of humankind in the universe. But it could also degenerate into vapid imitation of the stylistic flourishes of the classical world, a show of erudition and wit in place of wisdom and understanding.

At first the Italian humanists drew their inspiration from the Romans, but in the fifteenth century they began to search for ancient Greek texts. The works of Clement of Alexandria and his pupil Origen from the second and third centuries BC encouraged them to regard Greek philosophy as a 'handmaiden to theology' which paved the way to Christian thought. When, in the early fifteenth century, Leonardo Bruni made a Latin translation of the fourth-century work of Basil of Caesarea, *On How to Make Good Use of the Study of Greek Literature*, it provided the humanists with just the kind of instruction they needed.

Plato came to be regarded by the humanists as the pre-eminent philosopher, in contrast to the late-medieval preference for Aristotle. Florence, effectively ruled by the progressive Medici family, became a major centre for humanist thought, and it was here that the Platonic Academy was established in the late fifteenth century. But humanism found many patrons elsewhere in Italy. Even the papacy embraced this new learning: Pius II (1458–64) was a skilled poet and writer, and under Leo X (1513–21) Italian humanism is regarded as having ushered in a 'second Golden Age'.

Its acceptance in northern Europe was slower, but by the early sixteenth century Germany boasted some of the greatest humanists, including Erasmus, Johann Reuchlin and Ulrich von Hutten. Men like these could be found teaching at the universities, where they were often obliged to fight academic battles with the traditionalists. As the century progressed, these divisions commonly came to coincide with the religious fault lines that split the intellectual and the wider world.

FURTHER INSTRUCTION

By late 1512 Theophrastus was back in his father's home in Villach. At some time during this period he seems to have added 'Aureolus' to his growing list of names. No one is sure why. Nationalistic German writers keen to bring Paracelsus within the Aryan fold have implied that it alludes to his 'golden hair', although contemporary paintings offer no support to the suggestion that he was blond. (He wears a brown wig in the Scorel portrait in the Louvre, shown in the frontispiece.) Others say that Aureolus is a reference to his gold-making alchemical experiments. But Paracelsus seems to have regarded it as more than a mere nickname, for in an official letter to the Carinthian authorities in 1538 he places Aureolus before all his other titles. More likely, then, is the idea that, like Theophrastus, the name was taken (perhaps by his proud father, greeting his now-learned son) from antiquity, being derived from a writer of the fifth century BC named Caelius Aurelianus – 'Aureolus' is a plausible corruption. Very little is known of Aurelianus except that he came from Sicca in Numidia (Roman Africa) and was a translator of the Greek physician Soranus of Ephesus, who studied medicine in Alexandria and practised in Rome in the times of Trajan and Hadrian (late first and early second centuries AD). These medical works by Aurelianus were well known, and Wilhelm von Hohenheim may have possessed them in his library.

During his youthful wanderings, Paracelsus surely gained

The abbot Trithemius (1462–1516), famed and feared for his occult knowledge. He taught the occult arts to Agrippa, but his role in Paracelsus' education is less clear.

instruction from other mentors. He says that among the 'several abbots' who 'took great pains with me'[7] was the abbot of Sponheim – by whom he is generally assumed to mean one of the greatest magicians of the age, Johannes Heidenberg of Trittenheim, known as Trithemius.

In his youth Trithemius was a prodigy. He became abbot of the Benedictine monastery at Sponheim aged only twenty-two, barely a year after entering the order and before he had even been ordained. He set about transforming the humble community, turning Sponheim into a renowned centre of learning that boasted a library of about 2,000 volumes – many of which were concerned with arcane and occult matters that would have disconcerted other ecclesiastics. Trithemius fostered a progressive atmosphere, befriending humanists such as Reuchlin, an adept of the mystical Hebrew tradition called the Cabbala. But in 1506, discouraged by jealous opponents, he resigned as abbot and transferred to the monastery of St Jacob near Würzburg.

In his claims of astounding powers and revelations, Trithemius surpasses anything Paracelsus subsequently wrote. He says in a letter in 1499 that he is working on a marvellous book called *Steganographia* that is full of profound secrets, stupendous and incredible things. It will explain, for example, how 'I may express my thought to another while eating, sitting, or walking, without words, signs, or nods',[8] and how these thoughts could be sent 'by fire' over distances of a hundred miles or more, even into deep dungeons. This is done, the book reveals, through the assistance of angels. Trithemius asserts that he can learn the occult secrets of nature 'in any language of the world, though I have never heard it before'.[9]

Such knowledge, he professes, comes not from mulish study but from direct revelation. It is hardly surprising that Trithemius was accused of sorcery and trafficking with demons. Of *Steganographia*, the French mathematician and theologian Charles de Bovelles (Bovillus) confesses: 'I had the book in my hands scarcely two hours, and I threw it down on the spot because the great adjurations had begun to terrify me, as did the barbarous and unheard-of names for spirits (not to say demons).'[10] Trithemius vigorously denied charges of necromancy, denouncing black magic and witchcraft and saying that true, natural magic used only the hidden forces of nature. Nonetheless, legend has it that Trithemius was asked by Maximilian I to conjure up the ghost of his dead wife, Mary of Burgundy. He succeeded too: the Emperor recognized his wife's shade by a mole on the nape of her neck.

Trithemius was undoubtedly much sought out by students of the occult arts, although he had little time for alchemists, whom he denounced as 'fools and disciples of apes, enemies of nature and despisers of heaven'.[11] Cornelius Agrippa studied at his feet in the winter of 1509–10, and it is not hard to imagine that the young Theophrastus would have found much to fire his spirit in Trithemius' world of wonders. But we cannot be sure that the two ever really met. Paracelsus makes only a fleeting reference to the abbot of Sponheim (indeed, he acknowledges the tutelage of 'many abbots of Sponheim'), never mentioning Trithemius by name. And if he ever did seek out this mercurial magus, that could not have happened before Trithemius had moved to Würzburg. However you look at it, another part of the Paracelsus legend seems to dissolve, on closer inspection, into ambiguity and supposition. Nonetheless, the historian Noel Brann proposes – quite reasonably – that 'the two men met in their minds if not in person'.[12]

More certain, although of a quite different flavour, was his training in the metallurgical laboratories of Sigmund Füger in Schwaz, in the Inn valley of the Tyrol. Often confused with Sigismund Fugger, who owned the mines around Villach, Füger (or Fieger) was the count of Fügen, and his mines produced a rich flow of copper, zinc and silver. Paracelsus says that he gained much knowledge and experience in Füger's workshops, and praises him as 'noble and steadfast'.[13] (It is highly doubtful that his opinion of the wealthy Fugger would have been so favourable.) Here, as in Villach, Paracelsus was able to engage in practical experiments and to develop a respect for alchemy that did not depend on any vain quest for the Philosopher's Stone.

As a Bachelor of Arts, Theophrastus had completed the first phase of his studies; but to become a qualified physician he needed to return to the universities and endure more of their outmoded dogma. So he was soon on the road again.

4
The Staff and the Snake

Healing in the Early Renaissance

That was the physic! True, their patients died,
But no-one ever asked them who was cured.
So, with a nostrum of this hellish sort,
We made these hills and valleys our resort,
And ravaged there more deadly than the pest.
These hands have ministered the deadly bane
To thousands who have perished; I remain
To hear cool murderers extolled and bless'd.
Goethe, *Faust*, part I (1773–1801)

No matter who you were in sixteenth-century Europe, you could be sure of two things: you would be lucky to reach fifty years of age, and you could expect a life of discomfort and pain. Old age tires the body by thirty-five, Erasmus lamented; but half the population did not live beyond the age of twenty. There were doctors and there was medicine, but there does not seem to have been a great deal of healing. Anyone who could afford to seek a doctor's aid did so eagerly, but the doctor was as likely to maim or kill as to cure you. His potions were usually noxious and sometimes fatal – but they could not have been as terrible and traumatic as the contemporary surgical methods. The surgeon and the Inquisitor differed only in their motivation: otherwise, their batteries of knives, saws and tongs for slicing, piercing, burning and amputating were barely distinguishable. Without any anaesthetic other than strong liquor, an operation was as bad as the torments of hell. The removal of bladder stones, for example, entailed passing large instruments up the urethra or cutting into the bladder just above the pubic bone, and the French doctor Ambroise Paré tells how it typically took four strong men to hold the patient down. Many people chose to suffer from the illness rather than endure such a 'cure'.

And there was illness everywhere: plague, cholera, dysentery, tuberculosis, leprosy. The standards of sanitation, hygiene and diet were so low that between two and five children in every ten died before their first birthday. At least some of the febrile madness and apocalyptic panic of the age can be attributed to a monotonous, insipid diet that left people restless and weak: wheat, rye, barley, oats, millet. Fresh meat was rare, dairy products, salt and sugar all expensive. Drinking water was invariably dangerous, which was why ale and wine were the beverages of choice for all ages and all social strata, leading to endemic drunkenness.

In a world threatened by pain and death, stories of miracle workers are a psychological necessity, because the alternative is unmitigated horror and despair. Paracelsus may have been deemed capable of miracles, and he does not seem over-eager to deny such notions; but in his own mind his medical art was a purely rational affair. Yet whereas the conventional doctors took their methods from ancient sources and learnt them from books, Paracelsus' medicine drew on the occult forces of nature, and it was acquired by practical experience:

> The art of healing comes from nature, not from the physician. Therefore the physician must start from nature, with an open mind . . . Every physician must be rich in knowledge, and not only of that which is written in books; his patients should be his book, they will never mislead him . . . Not even a dog killer can learn his trade from books, but only from experience.[1]

But no one likes to be told they are wrong, and the doctors disparaged and jeered at this impudent dissenter. As far as Paracelsus was concerned, his opponents were motivated only by self-interest:

> There are two kinds of physician – those who work for love, and those who work for their own profit. They are both known by their works; the true and just physician is known by his love and by his unfailing love for his neighbour. The unjust physicians are known by their transgressions against the commandment; for they reap, although they have not sown, and they are like ravening wolves; they reap because

> they want to reap, in order to increase their profit, and they are heedless of the commandment of love.[2]

With such rousing and defiant words Paracelsus commands our instant sympathy: here, it seems, is a doctor who works for love and draws on gentle nature, as opposed to the dogmatic and deceitful traditionalists with their outmoded books and their fearsome scalpels and pincers. But was the difference really so stark? And who, in the end, was the more effective?

ANCIENT CURES

Plagues and epidemics repeatedly ravaged the poorly nourished people of Europe. The Black Death swept out of Asia in late 1347, and in the following two years it killed as much as a third of the population. This unspeakable blight transformed Western culture, dislocating the social structures of medieval Europe and leaving the Christian world traumatized and more than a little deranged. Death and decay were embraced and celebrated, even while they bred deep terror. The stone carvings of dead nobles on their tombs no longer showed dignified knights lying in their armour, but instead bodies crawling with worms, their flesh peeling from the bone. Artists such as Hans Holbein celebrated the Dance of Death, in which skeletal demons preyed gleefully on popes and beggars alike.

In Italy, epidemics of bubonic plague returned every ten years or so until the middle of the fifteenth century, and it is estimated that the number of households in European villages in 1470 was about half that at the beginning of the fourteenth century. 'It is safe to assume', say the historians Andrew Cunningham and Ole Peter Grell, 'that in every year between 1494 and 1649 plague was killing its thousands and its tens of thousands suddenly and horribly somewhere in Europe.'[3] Citizens of almost any town could consider themselves highly fortunate if they experienced no outbreak for two decades, and sometimes the plague would return every year. The cause was unknown, although theories were legion. The Lutheran pastor Andreas Osiander explained in a sermon after an outbreak of the plague in Nuremberg in 1533 (shortly after Paracelsus passed through the city) that 'such a scourge comes perchance from the influence of the stars, from the effects of comets, from extraordinary weather conditions and changes of the air, from southerly winds, from stinking waters, or from rotten vapours of the earth'.[4] From everything that anyone could think of, in other words – everything except the actual cause, a bacterium carried by fleas on the back of black rats.*

Most people regarded this pestilence as a manifestation of God's judgement. It was 'a punishment to mortals for our wicked deeds',[5] wrote Boccaccio in the *Decameron*, shortly after the plague struck Florence in 1348. God's wrath it might be, but the devil himself could not have devised something more horrible. Not only were the symptoms unspeakably painful, but they erupted in a fearsomely visible manner so that the victims had no doubt they were marked for death. A plague victim might first break out in red spots, but far worse were the hard, black buboes that appeared on the neck, the armpits and the groin. They caused agonizing pain and mortal fright. 'There are many', said Ambroise Paré:

> that for fear of death have with their own hands pulled away the *Bubo* with a pair of Smith's pincers: others have digged the flesh round about it, and so gotten it fully out. And to conclude, others have become so mad, that they have thrust an hot iron into it with their own hand, that the venom might have a passage forth.[6]

Doctors would treat these buboes by incision or cautery, both excruciatingly painful. Other irruptions of the skin, called pestilent carbuncles, were said by Paré to feel 'as a nail driven into the flesh'.[7]

But no medicine really worked once you had the plague, and the only protection was to quarantine the afflicted (most 'hospitals' for plague and pox sufferers were merely isolation units) or to flee the region. In plague outbreaks between 1490 and the 1520s in France, whole towns were evacuated. Without the means to earn a living, citizens would face starvation in the forests and moors.

Medicine in the early Renaissance had barely advanced since the time of the Roman Empire. To the academic physicians of the

* A recent study, however, places more emphasis on human mobility than on rats in propagating the plague from town to town.

universities it had, like every other science, already been 'solved' by the doctors of antiquity – Hippocrates (*c.*460–370 BC) and Galen (*c.* AD 129–*c.*200), whose works had been translated, augmented and honed by the great Muslim physicians such as Avicenna (Abu 'Ali al-Husain ibn Abdullah ibn Sina, 980–1037) and Rhazes (Abu Bakr Muhammad ibn Zakariya al-Razi, *c.*860–925). To become a doctor of medicine, one had simply to study these past masters and follow their recommendations. The physician's proper place was in the library, not in the surgery.

Yet it would be wrong to suggest that Paracelsus' new medicine was arrayed against a monolithic 'establishment' position in which the ideas of these ancient physicians were woven into a seamless orthodoxy. Galen did not by any means always agree with Hippocrates, and the Muslims put their own interpretations on the classical texts. One authority might enjoy favour at one time or in one institution, another elsewhere. Humanism, meanwhile, encouraged a return to the classical authors in preference to the modifications of the Islamic doctors.

Classical writers attributed the origin of medicine to Asclepius, the son of Apollo. According to Homer he was a mortal hero; later he became a god, and his emblems – the staff and the snake that were later conjoined into the physician's symbol of the caduceus – come from the culture of Egypt. Asclepius was said to have learnt medicine from the centaur Chiron, who taught him the art of *pharmaka*, the use of drugs that can heal or poison. By the time of the Greek Golden Age there existed a medico-religious cult dedicated to Asclepius, with temples throughout the Greek lands that Strabo calls Aesculapii. In these places the healing arts were practised by a brotherhood of doctor-clerics. Strabo says that Hippocrates acquired his cures and medical knowledge at the Aesculapium on the island of Cos.

Hippocrates can be justifiably regarded as the father of Western medicine, and he stands in relation to this science as Aristotle does to physics. Which is to say, he was almost entirely wrong, but he was at least systematic. As with most of the scholars of antiquity, some of the works attributed to him were surely written by later authors. In fact there is not one book in the surviving Hippocratic Corpus that is unambiguously known to have been written by Hippocrates; all were probably compiled by third-century scholars in Alexandria from a variety of older sources. But these attributions alone attest to the reverence in which Hippocrates was held; both Plato and Aristotle write of him with admiration.

Strabo's account of Hippocrates' medical training owes more to story-telling than to history – there is no real evidence that he was ever a priest of the Aesculapium, for example. It seems clear that Hippocratic medicine, which for the first time began to assign non-supernatural causes to diseases and thus to place medicine on a rational and organized footing, owed much to the traditions of the Ionian school of Greek philosophy that dates back to the sixth century BC. Just as Empedocles unified the material world with the notion of four elements (earth, air, fire, water), so there arose the idea – probably also due to Empedocles and developed by Pythagoras, although commonly attributed to Hippocrates – that illness could be ascribed to four bodily fluids called humours.

The humoral theory forms the cornerstone of Hippocratic medicine. The four humours are blood, phlegm, and black and yellow bile. For optimal health, these humours must be in balance; disease arises from an excess of one or another. Thus, for example, a person becomes choleric and bilious when his body falls under the influence of yellow bile. The goal of the physician was to restore humoral balance – by potions, diet or commonly by blood-letting, which tempered an excess of blood and gave vent to superfluous humours.

In one sense, this gave medicine a mechanical basis: health was maintained by draining or topping up the four reservoirs in the body. The doctor then needed to learn how to diagnose imbalances from a patient's symptoms. But there was also a metaphysical aspect to the system of humours. That there were four of them is no coincidence, for they mirror the four classical elements. In such correspondences we can discern an essentially magical view of nature.

The Hippocratic humours were adopted by Galen, who identified them with four basic temperamental types of person. A sanguine personality is hopeful, courageous, amorous, outgoing; the phlegmatic type is calm; the choleric angry and irascible, the melancholic prone to sadness and depression, or worse still, to guile and cowardice. (No doubt Galen would have diagnosed Paracelsus as choleric.) These character traits reflected a person's natural tendency towards an excess of one humour or another.

The humoral types became deeply ingrained in Western

The four humoral personality types, as depicted in the late fifteenth century. On the left is the choleric warrior; next to him, the sanguine courtier, then the phlegmatic merchant and the melancholic clerk. Each figure has an associated animal.

medieval culture. They are personified in the characters of Chaucer's *Canterbury Tales*: the sanguine Franklin, jovial and extrovert, the 'slendre colerik' Reeve who confesses irritably that his old heart is as mouldy as his hairs. Ben Jonson's 1598 play *Every Man in His Humour* (in the premiere of which Shakespeare is said to have acted) shows the humoral theory still thriving 200 years later. The longevity of the idea surely derived from the fact that it was, like Empedocles' classification of the elements, sufficiently broad and vague to allow any observation to be shaped to fit it. Afflictions that produced hot fever, cold shivering, inflammation and redness, pallor, tremors – all could be ascribed a cause in an imbalance of the humours. In healthy people, these imbalances were corrected by the expulsion of bodily wastes: sweat, tears, urine, faeces, pus. When that failed to do the job, it was generally the physician's task to stimulate such excretions and secretions.

Galen, born in Pergamum (now part of Turkey) in the Roman Empire during Hadrian's reign (AD 117–138), was an ambitious, irascible and immodest man. He went to Rome in AD 162, where he maintained a large practice and became physician to four successive emperors, including Marcus Aurelius and his egotistical son Commodus. He considered that he had consolidated and extended the sum of all medical knowledge of his time, and he did not hesitate to place Hippocrates a few rungs below himself. It is a nice irony that Galen, the target of Paracelsus' most vitriolic diatribes against the futile 'medicine of the ancients', shared many of his characteristics. In Rome he fell out with many of the other doctors, in part because they envied him his imperial preferment but also because Galen was never one to pass up an opportunity

Galen, whose ideas defined Western medicine until the Renaissance.

to quarrel. Yet he insisted on living simply, at least by the standards of imperial Rome: a man needs no more than two garments, he said, two sets of household utensils, and only two slaves to work with them. And one shouldn't trust what others have written before, but rather rely on one's own experience. It could easily have been Paracelsus, not Galen, who wrote:

> If anyone wishes to observe the works of Nature, he should put his trust not in books on anatomy but in his own eyes and either come to me, or consult one of my associates, or alone by himself industriously practice exercises in dissection; but so long as he only reads, he will be more likely

> to believe all the earlier anatomists because there are many of them.[8]

As with Paracelsus, Galen's success as a doctor drew accusations of witchcraft: this, he laments, is commonly the fate of the physician who diagnoses accurately. And like Paracelsus, Galen had no hesitation in throwing such charges back in the faces of his enemies, calling them 'liars or wizards or I don't know what to say'.[9] Scorning superstitions such as love potions, Galen was nonetheless prone to magical thinking, and his belief that nature was permeated by divine virtues is also one of the most marked aspects of the theories of his vociferous detractor 1,400 years later.

Galen's studies in anatomy and physiology, which relied largely on the dissection of monkeys, apes and other animals (human dissection being regarded as improper in Rome), gave his medicine an invasive aspect: the medieval enthusiasm for blood-letting can be attributed largely to him. His anatomical studies shaped most thinking until Vesalius challenged the Galenic physiology in the mid sixteenth century. Galen's reliance on animal dissections is sometimes over-emphasized, and he no doubt saw more anatomical detail than one might wish for in his job as physician to the Roman gladiators. Nonetheless, he initiated some serious misconceptions about how humans were woven together that persisted through the Middle Ages.

Galen's Hippocratic theory by no means precluded the use of drugs: he prescribed many, which became commonly known in the late Middle Ages as Galenicals. Since all ailments were traced back to just four 'causes' in the humours, however, these remedies were dished out in a rather wide-ranging and generalized manner.

If there are parallels in the lives of Galen and Paracelsus, that is not necessarily just an ironic coincidence. The unorthodox and outspoken wonder-worker is a recurrent archetype in the history of medicine. Thessalos of Tralles, Pliny tells us, was another figure of this sort: a first-century doctor who 'swept away all received doctrines, and preached against the physicians of every age with a sort of rabid frenzy'.[10] Thessalos gained such renown that a monument to him was erected on the Appian Way, proclaiming him the 'conqueror of physicians'. Surely the appeal of such stories lies with the dream, in an age when medicine was often worse than useless, that it could be transformed by an iconoclast blessed by divine revelation. Only in legend and hearsay did such prophets ever live up to their promises.

THE DOCTOR AT LARGE

In the ancient world, *physica* meant the study of nature and the cosmos – roughly speaking, what we today call physics. But in the twelfth century it became appropriated as a term for the medicine of Hippocrates and Galen: 'physic' was the work of the doctor, the physician. Moreover, that work became an increasingly secular business: once universities were given formal legal recognition around the beginning of the thirteenth century, physicians were no longer autodidact monks but men trained in an academic discipline and organized into trade guilds. Those who had not enjoyed the privilege of this training were derided as 'empirics' who practised without knowing what they were doing.

In the fifteenth and sixteenth centuries, the orthodox medicine of Hippocrates, Galen and their Islamic interpreters such as Avicenna and Rhazes was rigidly imposed by the universities. To be a better doctor, you strove to deepen your understanding of these authors, just as the theologians of the Middle Ages proved their piety by pedantically dissecting the Bible. Dissent was medical heresy. When the great anatomical textbook *Anathomia* by the thirteenth-century Bolognese physician Mondino dei Luzzi was found to contradict Galen in some point of detail in the fifteenth century, it was suppressed at some universities, including Tübingen, where Paracelsus studied. When John Geynes of the London College of Physicians criticized Galen in 1559, he was condemned by his colleagues and forced to sign a document of recantation. Even in the progressive humanist city of Padua, Vesalius voiced his disagreements with Galen only in a nervous and timid fashion.

Avicenna's *Canon*, the summation of his medical writings, established one of the most insidious notions in medieval medicine: the principles of philosophy, he said, are eternal and immutable, and cannot be overturned by mere experience. If you observe something that contradicts them, it does not prove that the theory is wrong but simply that the world is imperfect. Plato and Aristotle believed much the same thing, but its manifestation in medicine created a tremendous hurdle for 'empirics' like Paracelsus, who argued for the principle (even if he showed no rigour in exercising it) that 'practice should not be based on speculative theory; theory should be derived from practice'.[11]

So when he set out to learn the doctor's craft, the young Theophrastus von Hohenheim, accustomed to his father's eclectic approach, was in for a shock. For one thing, physicians drew a sharp distinction between themselves and surgeons. Many of them considered it beneath their dignity to get involved in the distasteful affair of actually touching a patient or performing an operation. That messy business was left to the lower ranks of surgeons, who were no better than crude, untutored barbers (the principal purveyors of blood-letting, as their traditional red-and-white sign-poles attest) and were neither expected nor supposed to know any medical theory. Some surgeons, indeed, doubled up as hangmen. This disinclination of doctors to grapple with flesh and blood was conveniently vindicated by Hippocrates, whose oath the doctor swore when he graduated. The Hippocratic oath obliged the physician never to use a knife on a patient, nor to practise cauterization (the burning of a wound to prevent bleeding or infection). At the University of Paris, medical graduates swore that they would never engage in 'manual surgery'.

This may not have been motivated by snobbery alone – surgery was a drastic and often life-threatening intervention, and a doctor who risked it could easily end up killing his patient. That might mean he forfeited his fee, or worse still, faced legal charges for negligence. Better, then, never to lift the blade at all. Moreover, until at least the thirteenth century, many who undertook the training of physician were churchmen and were therefore forbidden from practising surgery on religious grounds: the fourth Lateran Council of 1215 ruled that clergymen of the level of subdeacon or higher should not engage in any activity liable to cause bloodshed.

So if an operation was unavoidable, the wise doctor would use his knowledge of astrology to forecast the most auspicious day for the hazardous procedure, and then leave it up to the surgeon to do the slicing and sewing up. Many physicians did not even deign to see their patients in person before making a diagnosis: they did this merely from a urine sample. Uroscopy, as it was known, was a means of determining humoral imbalances based on absurdly

A hospital scene from Paracelsus' *Opus chirurgicum* (1566). The doctors stand around discussing urine samples, while the surgeons get on with the messy business all around them.

subtle and probably imperceptible gradations of colour, smell, sediment content and viscosity of urine.

Paracelsus saw it more bluntly. 'All they can do', he said of the doctors, 'is to gaze at piss.'[12] A woodcut from his *Opus chirurgicum* (*The Work of Surgery*), printed in 1566 by the Frankfurt publisher Sigmund Feyerabend, shows just what he meant: the doctors at a hospital stand around in their fine robes debating the interpretations of flasks of urine, while around them surgeons conduct the gory business of amputations and wound dressing.

Paracelsus believed that the division of medicine into the theoretical and the practical was absurd. A surgeon who knew no theory was a 'wood doctor and fool', while the resplendent, haughty physicians were 'high asses'. Medicine needed both the skills of the doctor for diagnosis and those of the surgeon for treatment: 'The patient asks for cure – 'surgery' – and not for theory – 'medicine' – it is the doctor who needs this.'[13]

Being a doctor was a fine thing. You could charge pretty much what you liked, so desperate were the rich and noble for a cure. You became an august and respected figure in society and a member of a powerful guild. You were entitled to ride on horseback, and to wear a fur hat and a splendid robe. Many doctors went about their business grandly attired – and why not, for you weren't going to risk soiling the garments with the blood of patients. Chaucer's Doctor of Physic is representative: 'in blood-red garments, slashed with bluish-grey/And lined with taffeta, he rode his way.'[14] To Paracelsus this ostentation was a sign of the corruption of the entire medical establishment, an attempt to cover over their paucity of knowledge with beautiful display. 'If disease puts us to the test,' he said, 'all our splendour, title, ring and name will be as much help as a horse's tail.'[15] Time and again he scoffs at the strutting doctors in their 'gorgeous satins, silks and velvets'.

But he was not alone in mocking the contrast between the showy appearance and grandiosity of physicians and their usually all-too-evident failure to cure the sick. 'If the world knew the villainy and knavery (beside ignorance) of the physicians and apothecaries,' the English essayist John Aubrey wrote in the seventeenth century, 'the people would throw stones at 'em as they walked in the streets.'[16] Many people felt that doctors took advantage of their patients' vulnerability to prescribe useless and protracted cures at inflated prices. In 1408 one John Clotes of London won a court case against the Flemish physician John Luter, who was convicted of fraud for demanding fifteen semi-precious jewels, a gold tablet and a sword from Clotes after diagnosing him inaccurately as suffering from leprosy.

It would be wrong to imagine that the surgeon's lot was, in comparison to the physician, necessarily humble. Some obtained lucrative tenure with wealthy clients. And some even had a formal university education: at the College of Saint-Cosme in Paris they were accepted as students of the faculty of medicine from 1436. The 'craft' aspect of the surgeon's task was always explicit, however, and inevitably gave the profession a lower status than that of the 'philosophical' doctors. In 1519 Thomas Ross, warden of the fellowship of surgeons of London, described what it entailed; it was clearly an occupation for the stout-hearted:

> It rests most principally in manual application of medicines; in staunching blood, searching wounds with irons and with other instruments, in cutting the skull in due proportion to the pellicules of the brain with instruments of iron, couching cataracts, taking out bones, sewing the flesh, lancing boils,

> cutting apostumes, burning cancers and other like, setting in joints and binding them with ligatures, letting blood, drawing teeth, with other such like, which rests only in manual operation, principally with the hands of the workman. And surgery is in comparison to physic as the craft of carpenter is compared to geometry: for just as the geometer considers causes of compasses, quadrangles, triangles and counterweights, and as his knowledge serves for building . . . the carpenter occupies it manually to his own profit and of necessity profitable to man, wherefore it is called *ars mechanica*.[17]

ITALY IN ECLIPSE

The best place to learn medicine in the sixteenth century was in Italy, and in late 1512 Paracelsus said his farewells to his father in Villach and set out across the Alps towards the cradle of the Renaissance.

There was a host of others making the same journey from the Swiss cantons to Italy that winter: the Swiss mercenaries, the fierce *Landsknecht*s whose pikes were feared throughout Europe. Italy was in turmoil, and that was good news for the soldiers who sold their services to the highest bidder.

For all the glories of its Renaissance culture, sixteenth-century Italy was a battleground for foreign sovereigns playing out power games, aided and abetted by the intrigues and disputes between the Italian states themselves. Italy had become a bewildering maelstrom of conflicting interests, and more blood was shed there than anywhere else in Europe. The end of the *quattrocento*, said the Italian chronicler Francesco Guicciardini, was 'truly the beginning of years of wretchedness' for his country, opening the way for 'innumerable horrible calamities'.[18]

There were five major Italian powers in the late fifteenth century. Venice, the trading gateway to the East, was a wealthy and relatively stable oligarchy ruled by a Doge elected from among several of the city-state's leading families. The lagoon kingdom laid claim to territories stretching as far as Asia Minor on the eastern coast of the Adriatic. The kingdom of Naples stretched from Sicily and Calabria across all of southern Italy up to the Abruzzi. The Papal States had a reputation for unruliness, and

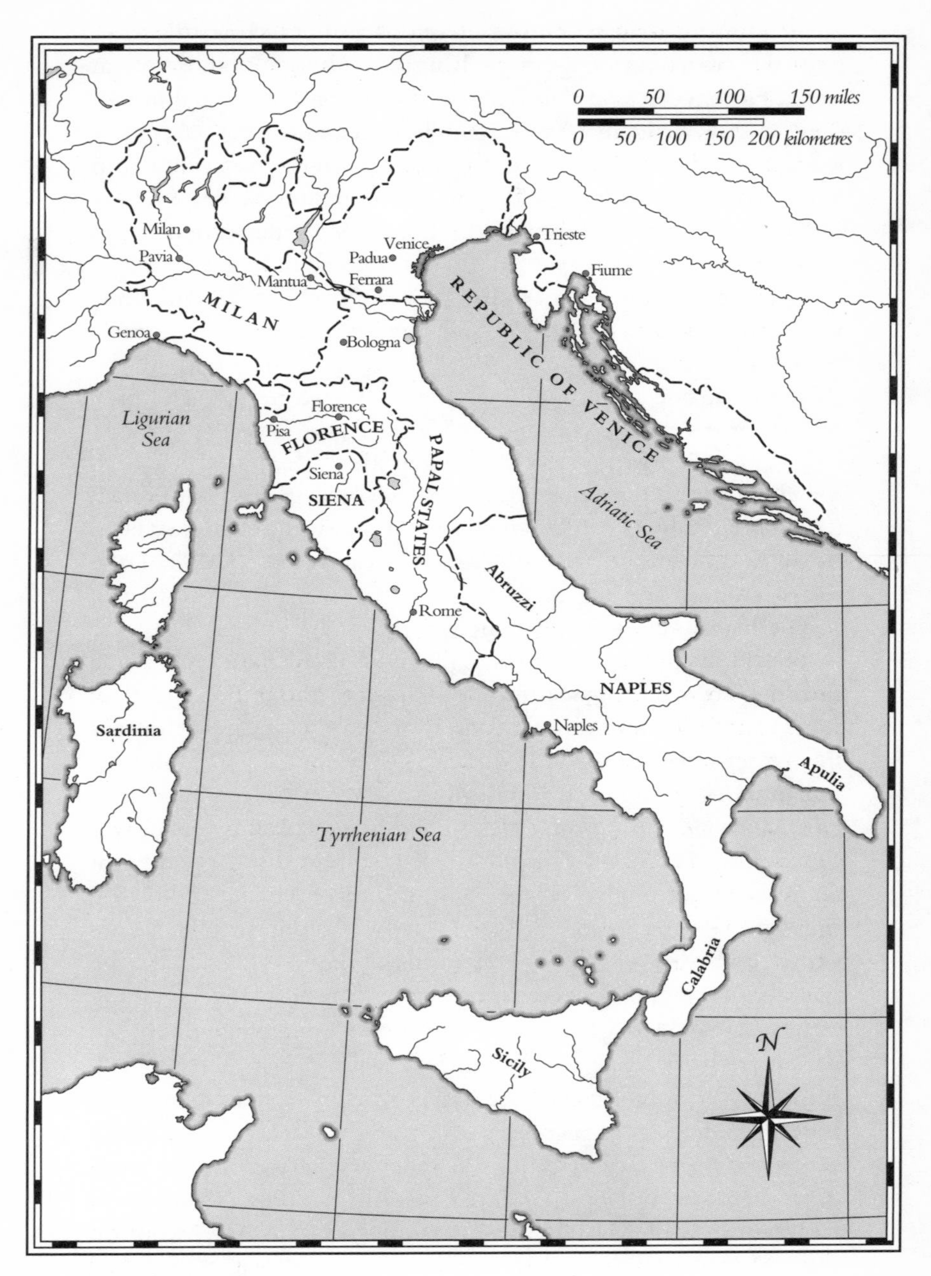

Italy in the sixteenth century.

never more so than when the infamous Borgias came to power with the crowning of Rodrigo Borgia as Pope Alexander VI in 1492. Florence was dominated by the Medici family, although they were supplanted by a republic between 1494 and 1513 following a revolt instigated by the puritanical friar Girolamo Savonarola. The fifth big player in Italian politics was the duchy of Milan, which encompassed the flourishing cities of Pavia and Mantua.

The Italian states agreed in the Peace of Lodi in 1454 that they would respect one another's boundaries. For nearly half a century, this peace persisted. But when Lodovico Sforza usurped the Milanese throne of his nephew Gian Galeazzo in 1480, he provoked a war with the young duke's in-laws, the house of Ferrante I of Naples. Knowing that the French king Charles VIII had an old claim to the Neapolitan throne, Lodovico invited the French troops into Italy. But Charles had more ambitious plans. Swelled by Swiss mercenaries, his army advanced across the Alps in 1494, seized Genoa, and began to overrun Florence. By February of 1495 Charles found himself king of Naples.

The Holy Roman Emperor and the Spanish monarchs were unhappy to see France spreading its power so far afield, and so Lodovico Sforza's family feud drew the rest of Europe into the fray. Allegiances between France, the German Emperor, and the Italian states shifted back and forth as each of them sought to gain an advantageous position. Charles died in 1498, but his successor, Louis XII, Duke of Orleans, had a still greater appetite for conquest. Claiming to be the rightful lord of Milan, he invaded through Lombardy and defeated Lodovico, whose treacherous nature had lost him any hope of gathering Italian allies. Lodovico died in a French dungeon in 1508.

In 1502 Spain and France carved up Naples between them, but they were soon at each other's throats again. It was then Venice's turn to face the vultures. Maximilian I believed that Trieste and Fiume, under Venetian rule, were by rights a part of Carinthia. Louis laid claim to part of the Po valley in the west of the kingdom; Ferdinand of Spain wanted the Apulian ports that Venice had seized while Naples was occupied by the French. Pope Julius II (1503–13) joined an alliance between Louis, Maximilian and Ferdinand to attack Venice. The Venetian Wars began with a victory for Louis's troops in 1509, but Venice was saved from total defeat by squabbles that inevitably ruptured the alliance they faced. By 1510 Julius was at war with Louis, and in 1511 it was France who stood alone against the Holy League of the Papal States, Maximilian, the Swiss cantons and later England too.

From the distance of several centuries the futility of it all is painfully apparent, and the plans of conquest formulated by Maximilian, Louis, Ferdinand, the popes and the Medici are plainly vain dreams. No one owed any allegiance or natural loyalty to anyone else: each player followed a strategy of expedience. Mortal enemies one year might be allies the next. The game was potentially interminable, since no single nation had the power to dominate the others. In his *Utopia* (1516), the English humanist Thomas More provides a razor-sharp satire of this mindless warmongering when he has himself ask the returning explorer Hythlodaeus why he does not put to good use the knowledge he has gained in Utopia by making it known to kings. Suppose, says Hythlodaeus, that while the French king goes about his manoeuvres and strategies, he were to remind the monarch that 'even if he does start all these wars and create chaos in all these different countries, he's still quite liable to find in the end that he has ruined himself and destroyed his people for nothing. I therefore advise him', says the adventurer, 'to . . . give up all ideas of territorial expansion, because he has got more than enough to deal with already. Now tell me, my dear More,' Hythlodaeus asks, 'how do you think he'll react to my advice?' [19] 'Not terribly well, I must admit,' the fictional More acknowledges meekly.

Thus not everyone in the sixteenth century accepted war as an inevitable and even necessary state of human affairs. For rulers of nations the battlefield was the natural arena for settling their differences, and this was much to the liking of belligerent warrior-knights and mercenaries. War was 'the royal sport',[20] according to Galileo. But thinkers such as More and Erasmus deplored this sport, and advocated pacifism on humanitarian grounds. 'The common folk', said More, 'do not go to war of their own accord but are driven to it by the madness of kings.'[21] Erasmus concurred: 'War is a fine thing', he wrote in the pacifistic *The Complaint of Pity*, 'to those who know it not.'[22] Paracelsus loathed warfare too (even though, or perhaps because, he was drawn into it on several occasions): it is caused, he said, 'by wilfulness – since all might hails from evil and is illegitimately born'.[23] Like Erasmus, he retained the vision of a just, equitable and peaceful Christian community.

Soldiers were thieves, bandits and murderers legitimized by the state. They commandeered all food supplies, since their own army provisions were scant and unreliable, and they occupied any house they chose. Brutalized by their occupation and lifestyle, they might torture and kill peasants just for the fun of it. Some nations swelled their ranks with troops pulled from the prisons; Erasmus considered mercenary soldiers 'the dregs of all men living'.[24] One French nobleman joked that soldiers wore their hair long to disguise the mark of the convict: ears lopped off by the hangman. Even Luther, who was content to regard war as 'the punishment of wrong and evil' and 'as necessary as eating, drinking, or any other business', admitted that 'a great many soldiers belong to the devil'.[25] If they did not start out that way, the brutalities of military life soon corrupted them. 'Out of this fountain [of war]', Erasmus proclaimed, 'springs so great a company of thieves, robbers, sacrilegious men and murderers.'[26] When Rome was sacked by German forces in 1527, their leader Charles V was dismayed by the atrocities they committed; it shocked even those for whom the pope had become an Antichrist. 'This whole age', wrote Melanchthon, 'by what fate I do not know, is more inimical to culture than any other ever was.'[27]

OPENING UP THE BODY

When Paracelsus set forth to Italy in late 1512, France was floundering against the Holy League. That Christmas, Swiss mercenaries expelled the French troops from Milan and returned Lodovico's son Massimiliano Sforza to power. The Swiss domination of Lombardy meant that there was a safe route for a Swiss student across the Brenner Pass and through Bolzano, Trento and Verona all the way to Milan. From there Paracelsus made his way to Pavia, Mantua and into Swiss-controlled Venice. He went to Padua before reaching Ferrara some time in 1513.

Ferrara was ruled by Duke Alfonso I d'Este, whose wife was the infamous Lucrezia Borgia, sister of Cesare. Intellectual life in the town was humanist and progressive. On the medical faculty of the university was the renowned Niccolò Leoniceno, a specialist on syphilis. He was familiar with the Cabbala and the mystical doctrines of Neoplatonism, having been taught by the great magus Michael Savonarola, grandfather of the Florentine revolutionary Girolamo. As an ardent humanist, Leoniceno called for scholars to return to the original texts of the ancients, and he was critical of the commentaries of Avicenna and the other Islamic writers. But he was not an indiscriminate disciple of antiquity, famously attacking Pliny in 1492 for all manner of errors in his *Natural History*. He deplored the tendency of doctors to learn their art by soaking up old books, denouncing those who 'sit in school discussing things of no relevance to life and men's fates, placing faith in others rather than themselves'.[28]

By the time Paracelsus reached Ferrara, Leoniceno was a very old man: he was to die in 1524 at the age of ninety-six. So Paracelsus may not have actually heard the octogenarian lecture, but his broad-minded, inquiring attitude had impressed itself strongly on the medical faculty. The notion that one might challenge so great an authority as Pliny was daring in the 1490s, and Leoniceno himself recorded that his book incited protest and personal vilification. Yet it prompted scrutiny and reappraisal of other classical sources. Cornelius Agrippa mentions in 1527 how Leoniceno exposed many cases of drugs misnamed by apothecaries because of spurious identifications in ancient texts, and in 1534 Antonius Musa Brasavola, a pupil of Leoniceno, wrote a sustained critique of the errors and credulity in the classical works of medicine and their Islamic translations. There is a Paracelsian ring to Brasavola's claim that experience is the 'mistress of all things'[29] and to be trusted above the word of the Greeks, the Romans and the Arabs.

Another of Brasavola's teachers at Ferrara, and a man whom Paracelsus probably encountered there, was Johannes Manardus (Jean Manard), who made a careful study of the preparation of drugs and questioned the use of astrology in medicine. Both Leoniceno and Manardus were Galenists at root, but not, it seems, of the complacent and inflexible sort.

However much medical students in the Renaissance were encouraged to venerate Galen, the sharp-eyed among them found ample reason to doubt him. For they had access to a resource that Galen did not: the human body. The taboo against human dissection loosened in the early Renaissance: in Italy the practice of cutting up corpses had been approved – one might better say tolerated – from the beginning of the fourteenth century, and

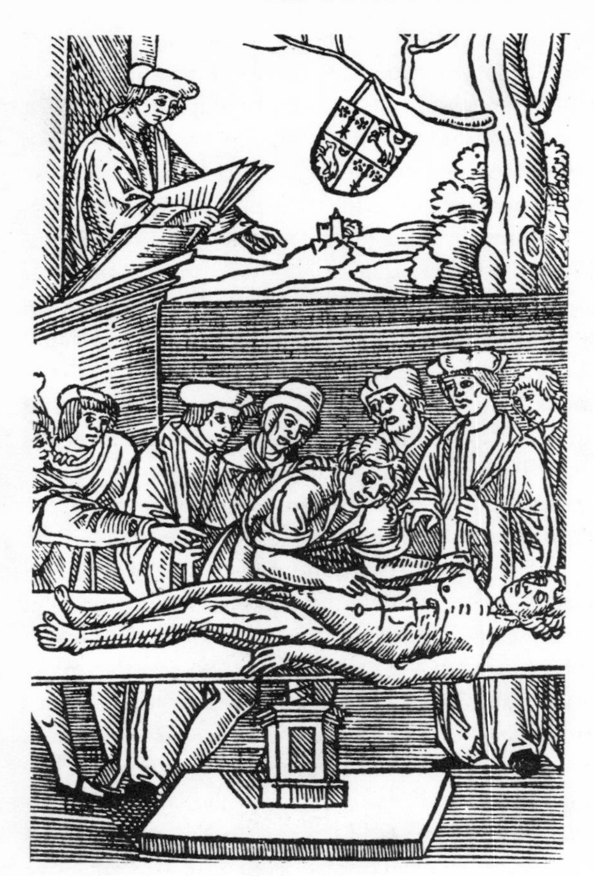

A dissection depicted in Mondino dei Luzzi's *Anathomia.*

since 1340 an annual dissection took place at the University of Montpellier in France. The standard dissectors' manual *Anathomia* (literally, 'cutting up'), published in 1316 by Mondino dei Luzzi, shows the kind of scene that Paracelsus would have encountered regularly during his university days. It is an anatomy class, and a human corpse is laid out on the table. The students gather round to watch the dissection. But the doctor is not there with them; he stands at a lectern high above, a book by Galen or Avicenna open before him, and he reads out the ancient knowledge while some menial surgeon provides illustration by opening up the body.

Many physicians lamented the lack of hands-on experience in their training. But quite aside from the prejudice against surgery, bodies were in short supply. The lack of cadavers prompted all kinds of subterfuge. Andreas Vesalius confessed that as a student, desperate for a real human skeleton to study, he sneaked out night after night from the city of Louvain in 1536 to pull the bones one by one from the charred corpse of a criminal burned at the stake outside the walls. He later pretended to have obtained the skeleton by legitimate means in Paris. The sixteenth-century German doctor Felix Platter would plunder the cemetery of Saint-Denis beyond the walls of Montpellier to obtain fresh bodies for dissecting.

Grave-robbing was, needless to say, a risky business, especially once friends and relatives of the deceased became aware of the danger. Vesalius tells of one close shave (quite literally, you might say):

> The handsome mistress of a certain monk of San Antonio here [in Padua] died suddenly as though from strangulation of the uterus or some quickly devastating ailment and was snatched from her tomb by the Paduan students and carried off for public dissection. By their remarkable industry they flayed the whole skin from the cadaver lest it be recognized by the monk who, with the relatives of his mistress, had complained to the municipal judge that the body had been stolen from its tomb.[30]

And that was in Padua, where Vesalius found things much easier than in France: the judge Marcantonio Contarini was even prepared, on occasion, to move the date of execution of a criminal so as to comply with Vesalius' need for a fresh corpse.

All this effort could be of limited value in the end, however, since dissections were such chaotic affairs that one wonders if anything was ever learnt from them at all. The prospect of a grisly spectacle could be guaranteed to draw leering crowds of idle, curious and ghoulish gawpers: Platter records that a dissection in Montpellier was attended not only by the students but also by 'many people of the nobility and the bourgeoisie, and even young girls, notwithstanding that the subject was a male. There were even', he adds indignantly, 'some monks present.'[31]

To Vesalius, this made public dissections pointless and unpleasant affairs, as well as emphasizing the snobbish distinction between the physician and the surgeon. Dissections, he said, were:

> that detestable procedure by which usually some conduct the dissection of the human body and others present the account of its parts, the latter like jackdaws aloft in their high chair, with egregious arrogance croaking things they have never investigated but merely committed to memory from the books of others, or read what has already been described. The former are so ignorant of languages that they are unable to explain their dissections to the spectators and muddle what ought to be displayed according to the instructions of the physician who haughtily governs the ship from a manual since he has never applied his hand to the dissection of the body. Thus everything is wrongly taught in the schools, and days are wasted in ridiculous questions so that in such confusion less is presented to the spectators than a butcher in his stall could teach a physician.[32]

Vesalius was a practical man with little interest in philosophy; he wanted simply to discover how things in the world were arranged. He appears in person in an illustration for his seminal *De humani corporis fabrica* (*On the Fabric of the Human Body*) (1543), a good-looking bearded young man stripping the veins from a flayed arm with his own fingers. This book – published, two years after Paracelsus' death, by his former pupil Oporinus in Basle – is often celebrated as the first to mount a serious challenge to Galen's errors of anatomy. In fact, Vesalius was by and large a Galenist and was hesitant to contradict the ancient authorities. 'Not long ago', he confessed, 'I would not have dared to diverge a hair's breadth from Galen's opinion'; and when he now did so, it was with the caveat that 'I still distrust myself.'[33] The novelty and impact of *De fabrica* were due mainly to Vesalius' recognition that proper instruction in anatomy demanded good diagrams.

And those in *De fabrica* were exceptional. The woodcuts are thought to have been prepared mostly by Jan van Calcar, a pupil of Titian. Whereas drawings of the body and organs in earlier texts had been schematic at best, those in Vesalius' book reflected the spirit of the Renaissance artist with their dramatic realism. The handiwork of a master is evident in the classical postures of many of the figures, based on the sculptural tropes of the ancient world. *De fabrica* spoke not only to physicians but also to artists, whose new-found concern for anatomical rectitude led Leonardo da

Vesalius (1514–64) takes centre stage in the frontispiece of his *De Fabrica*.

Vinci and Michelangelo to carry out dissections of their own (Leonardo claimed to have dissected 'more than ten human bodies'[34]). It was one thing to *say* Galen was (sometimes) wrong; but illustrations of that fact as finely observed as those in *De fabrica* seemed to present an unanswerable case. The transformation of medicine, and in particular the undermining of its classical basis that was so central to the spread of Paracelsian ideas towards the end of the sixteenth century, must be attributed in large degree to Vesalius' careful eye and to the skill of his illustrators.

PARACELSUS GETS HIS NAME

If Paracelsus had lived to see *De fabrica* published, it is not obvious that he would have been impressed by it, despite its occasional anti-Galenism – for he was no enthusiast of dissection. Indeed, he disliked anatomy intensely (and was later to be embarrassed by his ignorance of it), insisting that to understand medicine one had to study man himself, not his empty husk. 'It is the living body that teaches the anatomist health and disease, not the dead one: he requires therefore a living anatomy.'[35] This was not simply the pragmatism of one who wishes to see the machine in motion before diagnosing how, and how well, it functions; Paracelsus' vitalistic philosophy held that, without the animating spark of life, the flesh was just barren matter fit for the worms. Dissecting a cadaver was like studying a natural spring after the sun had baked it dry.

Paracelsus studied at Ferrara until 1515, and it was here that he is generally believed to have adopted his pseudo-humanist name. It was customary for humanists to display their kinship with the ancients by taking a Latin or Greek version of their name: Vadianus, Melanchthon (Philip Schwarzerd, whose family name means 'black earth'), Agricola. But the derivation of 'Paracelsus' has been the subject of much debate. Some claim that, in the typical fashion of adapting a family name, it is simply a (very loose) translation of Hohenheim ('high home'), for *celsus* in Latin means a high place, and *para* can indicate 'beyond'. After all, by the time Theophrastus became Paracelsus, his family were indeed 'beyond Hohenheim'. But the usual explanation sounds more authentic: interpreting the name as 'beyond Celsus' suggests that its owner is a better physician than the Roman writer Aurelius Cornelius Celsus of the first century AD. Celsus, like Pliny, was an encyclopaedist who wrote widely, encompassing agriculture, military arts, oratory, philosophy and jurisprudence. He seems not to have been a practising physician at all, but in his sole surviving work of any kind, *De re medicina*, Celsus collated the sum of medical knowledge of his time. This text was one of the earliest 'discoveries' of the humanists, unearthed by Guarino de Verona in 1428, and its reputation within the Renaissance medical community is attested by the fact that the first printed edition, published in 1478, predates printed volumes of any of the works of Hippocrates or Galen. *De re medicina* was, naturally, widely read at Ferrara. What better way to announce the beginning of a great doctor's career than by his telling the world that he was destined to advance medicine far beyond this antique tome?

Well, perhaps. But 'Paracelsus' could also mean 'alongside Celsus': an expression of fellowship rather than supremacy. Moreover, it isn't clear that Paracelsus was much inclined to the humanists' affectation of renaming themselves, and he rarely used this name himself. It first appears on an almanac of political forecasts printed in Nuremberg in 1529, but more prominent is the attribution of his *Greater Surgery* (1536–7) to 'Doctor Paracelsus' (itself rather more reminiscent of the way Roger Bacon and Albertus Magnus had become known as Doctor Mirabilis and Doctor Universalis three centuries earlier).

Equally vexing is the issue of whether Paracelsus actually graduated as a doctor at all. On the one hand, Paracelsus' eulogists have tended to gloss over this uncertainty as though it were beneath consideration; his detractors, meanwhile, have suggested that, as there is no definitive evidence of his qualification, Paracelsus styled himself a 'doctor of medicine' entirely without official sanction. The truth is that we do not know either way. It is only marginally relevant in any event, both because unlicensed medical practitioners were common enough, and because Paracelsus had so low an opinion of the medical professionals and so little regard for their conventional theories that he was to all intents and purposes not one of them whatever the official records might once have said. If Paracelsus lacked the doctor's formal qualification, that did not prevent him from gaining numerous medical appointments and offices throughout his life. If he possessed such a qualification, it did not protect him in disputes and arguments against conventional physicians. All the same, the prevalence of unqualified quacks could foster considerable antipathy towards those who professed medical skills without appropriate qualification, as the Swiss traveller Thomas Platter (father of Felix) indicated:

> If an unlicensed doctor is found, or a hawker of ointments, the doctors and the students have the right without further ado to set him backwards on an ass with the tail in his hands for a bridle, and to drive him around the town; which delights the populace, who pelt the fellow with mud and rubbish until he is filthy from head to foot.[36]

Paracelsus seems never to have suffered quite this level of indignity, although there was to be abuse enough in store for him.

PHYSIC IN THE FIELD

Paracelsus left Ferrara in 1515, convinced at least in his own mind that he was now a doctor of medicine. Perhaps he simply considered his work there to be finished; but it is also possible that political events hastened his departure. In 1513 Venice went to war against Milan and the papacy, its former allies in the Holy League. Venice joined forces with France, and Louis XII invaded Lombardy, only to be repulsed by Massimiliano Sforza's Swiss mercenaries at Novara. Louis's death in 1515 did nothing to end the struggles, for his successor Francis I at once crossed the Alps with 30,000 troops and in September, with Venetian aid, he defeated Massimiliano's army at Marignano (where the Swiss Reformer Ulrich Zwingli served as a chaplain). Lodovico's son was captured and sent to France with a royal pension, while Francis ruled Milan.

The defeat of Massimiliano's Swiss troops by the French and their Venetian allies may have made life somewhat uncomfortable for a young Swiss man in Venice in 1515. At the same time, there were ample opportunities further afield in Italy for a Swiss, and thus politically neutral, doctor. After passing through Bologna, Florence, Siena, Rome (now ruled by the Medici pope Leo X) and Capua, Paracelsus found employment as a military surgeon in the army of Charles I of Spain, who had invaded Naples to restore a balance of power dominated by the French. Paracelsus went first to the city of Naples itself, full of Spanish soldiers stricken with syphilis (the 'French disease'), and then to the famous medical school at Salerno in Sicily. Salerno had been the first medical school in the West, famed already by the ninth century; in the eleventh century its teachers included Constantine of Africa, a

North African Muslim who became a Christian monk and a great Latin translator of the Arab doctors.

Life as a military surgeon was bloodier than any dissection the doctor could have encountered at the universities, as well as being a perilous business in itself. Army camps, swelled by great crowds of camp followers and lacking even the most basic sanitation, frequently fell prey to diseases such as cholera and typhus. When an army stagnated as it laid siege to a town, 'camp fever' or 'siege disease', as typhus was known, was apt to be more lethal than the weapons of the enemy. And a victorious army did not trouble too much to distinguish, in the heat of battle, an enemy soldier from a non-combatant.

The Venetian Alessandro Benedetti described what the military doctors had to cope with after the French were defeated by the League of Venice at the battle of Fornovo in 1495:

> Very many wounded were found naked among the corpses, some begging aid, and some half-dead. They were weakened by hunger and loss of blood and wearied by the heat of the sun and thirst . . . [They were] attended by the [Venetian] surgeons at public expense. Some still breathing after hands and feet had been amputated, intestines collapsed, brains laid bare, so unyielding of life is nature.[37]

The slaughter became even worse in the sixteenth century. For one thing, the scale of warfare ballooned: armies grew markedly in size, while rulers took advantage of growing economies and the availability of credit facilities from bankers like the Fuggers to hire troops and finance wars. And the nature of combat changed as the archers and knights of the Middle Ages were replaced by ranks of pikemen and gunners. The arquebus of the early sixteenth century fired heavy bullets that could penetrate armour. Detonated by a match which ignited a bowl of powder, this 'alchemist's weapon' was a cumbersome and rather haphazard affair that jammed easily and could be as dangerous to the arquebusier as to his opponent. Nevertheless, the arquebus and the light, portable cannon of iron or bronze represent the beginnings of a new kind of warfare, one dominated by machines that surpass the capability of mere humans to maim and kill.

Ambroise Paré, one of the greatest surgeons of the late

Renaissance, served in more than forty campaigns with the French army, and he came to loathe the new gunpowder-fired weaponry. One of his innovations was an ointment of crushed onions and salt to treat burns, a remedy that was still in use during the Second World War. Paré's most important book was concerned with the damage caused by firearms: *Method of Treating Gunshot Wounds*, published in 1545. Surgeons often made such wounds worse than they already were, since it was believed that gunpowder was a poison borne into the wound by the bullet. This necessitated that the bullet be dug from the flesh, however deeply lodged. Moreover, wounded soldiers generally faced the appalling prospect of having their wounds cauterized with boiling oil. Paré realized that this could do more harm than good. After one battle in 1536 he ran out of oil, and so concocted instead an ointment from egg yolk, oil of roses and turpentine, which he administered cold. The following day, the patients who had received this soothing balm were in far better shape than those cauterized with hot oil. 'I resolved never so cruelly to burn poor wounded men,'[38] he said. Paré used ligatures to improve a patient's chances of surviving amputation (a common procedure on the battlefield), and he found ways to reduce the risk of a wound becoming infected. In many ways, Paré's innovations mirrored those introduced by Paracelsus in surgery, insofar as they advocated minimal intervention, keeping the wound clean and giving the body the best chance to heal itself.

But Paré would have seen many sober reminders of how little a military doctor could do in the face of the stark and bloody realities of warfare. Once, when in the field, he was treating two badly maimed soldiers when a grizzled veteran approached and asked if the men had any hope of recovery. Sadly not, Paré replied. Then he turned to treat other casualties in need of his attentions. Glancing back at the wounded men, he was horrified when he saw that 'the old soldier had without malice and quite gently cut both their throats.' What are you doing? Paré cried. The old dog looked at him and said, 'I pray God, if I am ever in similar circumstances, someone will be good enough to do for me what I have just done for these men.'[39]

Few soldiers, however, were lucky enough to benefit from the treatments of a doctor as skilled as Paré. Those who returned home lacking the use of a limb (like Cervantes, who wrote *Don*

Quixote one-handed) were as likely to have the surgeon to blame for their disfigurement as the weapons of the enemy. Indeed, the English doctor William Clowes asserted in 1591 that military surgeons were more lethal than the combat of the battlefield.

Notwithstanding the hazards of being attached to an army, Paracelsus' days as a military surgeon in Italy launched him on his travels across Europe and beyond. By sea he passed from Sicily to Liguria in northern Italy, alighting at Genoa, from whence he made his way to Marseilles and Montpellier, where the medical faculty of the university was considered one of the best in France. Yet from the insults that he later hurled at the doctors of Montpellier, we may assume that Paracelsus found it as steeped in benighted tradition as elsewhere. He made his way along the Mediterranean coast into Catalonia, passing over the eastern Pyrenees, and came to Barcelona in late 1517 or early 1518. At the port of Cartagena he joined the Spanish troops on a galley headed for North Africa, travelling first to Oran and then to Algeria.

When the Moors were forced out of Granada in 1492, they settled in Algeria, from where the Sultan of Algiers continued to wage war against the Spaniards. The Marquis de Gomarez launched an assault on Algiers in 1517, and Paracelsus resumed his career as army physician during this campaign, which ended in 1518 with the capture and execution of the Sultan.

Paracelsus returned from the shores of 'Barbary' with the Spanish forces, going first to Granada and then to Cordoba and Seville, which their remnants of Arabic culture made an attraction for anyone interested in alchemy. It seems that he had now decided to scour the world for arcane knowledge, which he began to assemble piece by piece into a theory of all creation.

5
Intellectual Vagabonds

Walking the Pages of the Book of Nature

> If someone complained to [Monsieur de Montaigne] that he often led his party, by various roads and regions, back very close to where he had started (which he was likely to do, either because he had been told about something worth seeing, or because he had changed his mind according to the occasions), he would answer that as for him, he was not going anywhere except where he happened to be, and that he could not miss or go off his path, since he had no plan but to travel in unknown places; and that provided he did not fall back upon the same route or see the same place twice, he was not failing to carry out his plan.
>
> Secretary of Michel de Montaigne, *Travel Journal* (1580–81)

Between 1517 and 1523 Paracelsus embarked on a series of extraordinary journeys. We don't know for sure exactly where they took him, but he seems to have covered thousands of miles (a tentative route is shown on the endpapers). There are stories of his appearance in Ireland, in northern wastes where the sun shines at midnight, in crocodile-infested African swamps, in all kinds of remote and lawless places from where it is a wonder he ever returned. If these stories are true, his explorations of the Renaissance world were the equal of any voyages made by Columbus, Walter Raleigh or Vasco da Gama; and if he discovered no new lands, that is because he was not looking for them. He was looking for knowledge, and he believed it was to be found in every nook and cranny of the wide world. 'Diseases wander hither and thither throughout the breadth of the world,' he said. 'If a man wishes to recognize many diseases, let him travel.'[1]

'I am so much harangued to, to vex me and ridicule me too,' he lamented later in his life, 'because I am a wayfarer and as though I were therefore the less worthy.'[2] But travelling, he argued, is an obligation for all genuine physicians. Because he believed that nature, not Galen or Aristotle, was the surest teacher, anyone who would learn the art of healing must scrutinize nature in all its aspects. So there was no alternative but to shoulder one's pack or saddle one's horse and take to the roads, forever searching and forever learning. As he said:

> The arts are not all confined within one's fatherland, but they are distributed over the whole world. Not that they are in one man alone, or in one place: on the contrary, they must be gathered together, sought out and captured, where they are . . . Is it not true that art pursues no man but must be pursued? Therefore do I have authority and reason to seek her, and not she me. Take an example: If we would go to God, we must go to Him, for He says: Come to me. Now since this is so, we must go after what we want. Thus it follows: if a man desire to see a person, to see a country, to see a city, to know these same places and customs, the nature of heaven and the elements, he must go after them . . . Thus the way for anyone who would see and experience something is that he go after the same and competently enquire; and when things go best, move on to further experiences.[3]

Paracelsus insists that such wandering is essential not merely because there is so much in the world to see and learn, but because the specific attributes of each region shape its own medicine. That is to say, cures may be specific to the location in which they are found, and useless elsewhere. This is because each land is subject to different astral influences (which accounts for its particular climate), and so the occult forces on which the physician draws vary from place to place. Every land is a leaf of the Codex of Nature, and 'he who would explore her must tread her books with his feet'.[4] He explains that 'English *Humores* are not Hungarian, nor the Neapolitan Prussian'[5] – and so medicine that works in one place may be ineffective in another:

> The physician must give heed to the region in which the patient lives, that is to say, to its type and peculiarities. For one country is different from another; its earth is different, as are its stones, wines, bread, meat, and everything that grows and thrives in a specific region . . . The physician should take

> this into account and know it, and accordingly he should also be cosmographer and geographer, well versed in these disciplines.[6]

This stands in stark contrast to the growing nationalism of the sixteenth century, which tended to insist that one's own ways and customs and religion were the best. Paracelsus' metaphysical view, in which the virtues and *inclinationes* of nature are distributed evenly throughout the cosmos, forces a kind of geographical parity that accords equal value to every place and race.

The learned doctors, he says, know nothing of the wider world but sit at home with their fatuous books and never broaden their minds. These 'cushion-sitters':

> without a sledge, carriage or wagon cannot go outside the gates and know not with their art how to get to a shoemaker's for a pair of shoes except on an ass and for a ducat . . . fashioned as they are in the world in my times, they like neither to travel nor to learn. To this the people bring them by continually giving them more money, although they know nothing. When they notice that the peasants know not how a physician should be, they stay in the chimney-corner, seat themselves in the midst of books, and ride thus in the Ship of Fools.[7]

It is because of his constant travelling, Paracelsus explains, that he is so untidy and ragged of appearance, so frequently poor, so rough and unrefined:

> Those who sit in the chimney-corner eat partridges and those who pursue the arts eat milk-soup. The corner-trumpeters wear chains and silk; the wanderers can scarcely pay for ticking. Those within the walls have it cold or warm according to their wishes; those in the arts, were it not for a tree, would have no shade.[8]

But Paracelsus self-righteously claims that simplicity and poverty are good – 'for Juvenal has written that only he roams happily who has nothing . . . I think it praiseworthy and no shame to have thus far journeyed cheaply.'[9]

One should be humble enough, Paracelsus says, to learn from anyone and anything, and not to turn up one's nose at folk remedies or the 'ignorant' ways of peasants and journeymen:

> Wherever I went I eagerly and diligently investigated and sought after the tested and reliable arts of medicine. I went not only to the doctors but also to barbers, bathkeepers, learned physicians, women, and magicians who pursue the art of healing; I went to alchemists, to monasteries, to noble and common folk, to the experts and the simple.[10]

Today it is acknowledged that folk medical tradition, tested in the crucible of experience, has identified numerous cures of genuine efficacy, such as the use of willow bark (containing salicylic acid, the precursor compound for aspirin) as an analgesic. By refusing to judge such remedies according to their conformity with some ancient and irrelevant theory of medicine, Paracelsus is likely to have acquired valuable knowledge on his journeys.

THE PEOPLE'S MAGIC

There is no question, however, that he also picked up a lot of superstition, for he was not above such delightful nonsense as this:

> It happens that in the Alps cattle are able to remain the whole summer without drinking; the air is drink for them, or supplies its place: and the same should be judged with regard to man. The nature of man, too, may be sustained in the absence of food, if the feet are planted in the earth. Thus we have seen a man who lived six months without food and was sustained only by this method: he wore a clod of earth on his stomach, and, when it got dry, took a new and fresh one. He declared that during the whole of that time he never felt hungry.[11]

And then there was the weapon salve, Paracelsus' cure for wounds. It was arguably of little use to the soldier injured in battle, because it demanded that you had access to the weapon that cut you. But what a boon it was for the clumsy butcher, carpenter or blacksmith. To heal the wound, one should apply a specially prepared ointment – not to the wound itself, but to the blade that caused it.

At face value this is ridiculous. The only conceivable benefit of this treatment was to spare the wound itself from a doctor's noxious poison or from further tampering that might have led to infection. But to Paracelsus there was a perfectly rational, mechanical explanation: the salve helped the vital spirits in the blood on the blade to reunite with the spirit in the victim's body. If you allowed that invisible influences can act at a distance (and try explaining magnetism otherwise), it all made perfect sense. That is why this apparently absurd remedy was still being debated furiously by physicians in the 1630s, when the Englishman William Foster published *Hoplocrisma-spongus; or a Sponge to Wipe Away the Weapon-salve* – only to be energetically refuted by the Paracelsian doctor Robert Fludd. Even then, Foster did not denounce the weapon salve as inefficacious, but rather as the product of witchcraft.

The weapon salve was not Paracelsus' invention. It was one of his many appropriations from folk magic, the quotidian superstition embraced by common people the length and breadth of Europe. Paracelsus would readily adopt such remedies as he encountered them on his travels, sometimes giving them a gloss of 'philosophical' interpretation. There is some justification in the charge that he and his fellow-minded wandering doctors had a tendency to believe everything they heard and to then find a way of rationalizing it.

There was a magician in nearly every town or village, and he – or just as likely, she – was a busy person. If you were ill, if something of yours was lost or stolen, if you feared that you had been placed under a curse, if you sought protection from evil spirits, bad luck, or pestilence, where else could you go? Only a rich man could pay the doctor's fees. The poor were beset with countless dangers, misfortunes and worries, and they were essentially powerless against them. What slender hope they had depended on the local sorcerer or wise woman.

The Church did not approve. Man suffers, according to the priests, because of sin, and for that there is only one remedy: God's grace, which is what they were here to dispense. The magicians undercut their trade, which not only discouraged piety and penitence but also robbed the church of the fees that the clergy could charge for their services.

But for the common person there was a big difference between a prayer and a spell. The first was a plea that God might grant or refuse – as the priests reminded their congregation, you could not make demands on the Lord. A magic spell, on the other hand, was a formula that was guaranteed to work. It didn't matter whether you were devout or wicked, so long as you followed the formula. Magic was regarded as a mechanical process, a matter of pushing the occult buttons of nature.

Needless to say, those manipulations often failed to produce the desired result. But magic was immune to its failures. If it didn't work, that was because the incantation had not been conducted properly; it merely went to show how *difficult* it was to get right. The fault lay with the magician, not the magic – so the solution was to seek out a better magician. And a single apparent success would erase from memory a hundred failures. 'Magic', the historian Stephen Sharot points out, 'is a cosmology or worldview that has an explanation for everything, including failed magic.'[12]

The persistence of popular magic can thus be seen as a necessary self-delusion. In the face of so much uncertainty, hardship and, in much of Europe, the threat or reality of war and brigandry, magic offered some hope of gaining control over the chaos. It provides an antidote, however illusory, to impotence. In the words of the anthropologist Bronislaw Malinowski, 'it ritualizes man's optimism'.[13]

Unable to suppress the popular demand for the services of magicians, the Church was pragmatic. On the one hand condemning wizardry as demonic witchcraft, with all of the terrible consequences that entailed, on the other hand it was ready enough to let the flock believe in holy miracles. The blessing of a priest was akin to a ritual charm, and was used to sanctify ships, armour and buildings, and even to protect cattle and to ensure a good harvest. Holy water was a potent charm, invested with the power to drive away pestilential vapours and to ward off disease. When a storm approached a village, people considered that it signified a gathering of demons in the air, and the church bells were rung to ward them off.

Peasants brought up on a tradition of superstitious magic could hardly be expected to distinguish between such ostensibly Christian rituals and the mumbled incantations of the local wizard. And so, to the discomfort of the priests, many came to regard elements of Christian devotion as simple magical spells. The Latin mass was, after all, incomprehensible to the common people, so it already had the aspect of an occult formula. It came to be seen, like magic, as an essentially mechanical rite through which absolution was achieved by observing the correct procedures. In that case, there was no real need for faith. People took to retaining the host under their tongue so that they might remove it after the service and use it as a charm. There were even instances of theft of the host from churches for use in magical rites. This was not wilfully blasphemous or diabolical; the accoutrements of the church were simply desired as objects of occult power.

THE NECROMANCERS

The figure of the itinerant magician-healer was familiar in the Renaissance. Lynn Thorndike, a historian of the magic tradition, speaks of the 'wayward geniuses and intellectual vagabonds so common in the later fifteenth and early sixteenth centuries'.[14] He places Paracelsus in this category, alongside another iconoclastic wanderer whose life and career offer a remarkably close parallel: Heinrich Cornelius Agrippa of Nettesheim.

The two men clearly knew of one another, although there is no record of their ever having met. We can picture each of them weaving his tangled web through sixteenth-century Europe like wary yet respectful spiders, occasionally detecting one another's traces as they pass through this town or that one. Like Paracelsus, Agrippa was descended from a family of German nobles. Born in Cologne on the Rhine, he studied medicine and law, and throughout most of his life he earned his keep as an unorthodox physician – although, like Paracelsus, it is not clear whether he ever gained a medical degree. Between 1511 and 1517 he lived in Italy, where he became engaged as a diplomat in the service of Maximilian I. In 1518 he returned north to take up a restless peripatetic life, drifting from one position to another in the Low Countries, Germany and France.

Agrippa, perhaps even more than Paracelsus, was the archetypal Renaissance magus. He was an adept of the occult, and sought out Trithemius of Sponheim for instruction in 'divers things concerning chemistry, magic and Cabbalism, and other things which as yet lie hidden in secret sciences and arts'.[15] He shared

Heinrich Cornelius Agrippa von Nettesheim (1486–1535).

Paracelsus' belief in a rational natural magic, free from superstition and religious dogma, yet rooted in the Christian tradition. As Agrippa wrote:

> Magic comprises the most profound contemplation of the most secret things, their nature, power, quality, substance, and virtues, as well as the knowledge of their whole nature. It instructs us concerning the differences and similarities among things, from whence it generates its marvellous effects by uniting the virtues of things by the application of one to another, joining and knitting together appropriate inferior subjects to the powers and virtues of superior bodies.[16]

The magus, Agrippa explains, accomplishes nothing more than a hastening of processes already immanent in nature. The magician's art is no more than a servant of nature, he says, and it merely brings things into being before the time that nature had appointed for them – like a man producing roses or ripe grapes in March, or growing parsley from seed within just a few hours.

Yet this was not how many men of the church saw things; they considered that magic involved satanic forces. Magic, according to the twelfth-century Saxon theologian Hugh of St Victor, is:

> the mistress of every form of iniquity and malice, lying about the truth and truly infecting men's minds, it seduces them from divine religion, prompts them to the cult of demons, fosters corruption of morals, and impels the minds of its devotees to every wicked and criminal indulgence.[17]

The Italian Neoplatonist scholar Marsilio Ficino went to great pains to refute such accusations and to remove the demons from magic. He did not deny that they existed and could be summoned, but he rejected the suggestion that all magic is necessarily demonic. On the contrary, he said, there is a 'good', *natural* magic that is consistent with Christian theology and entirely separate from the powers of the devil. Paracelsus' defence of magic was much the same: 'Natural magic', he said, 'is the use of true, natural causes to produce rare and unusual effects by methods neither superstitious nor diabolical.'[18]

The belief that magic bore an evil taint was reinforced by the reputation of perhaps the most important and notorious book of medieval occultism, called *Picatrix*. Written in tenth-century Arabic in the city of Harran in south-east Turkey, it was brought west along the caravan trail of the Mediterranean. It was known to Agrippa and Trithemius, and surely to Ficino and Paracelsus too. The reader of the *Picatrix* could learn how to capture and channel the astrological influences of the planets. Although some of the uses of this power were mundane to the point of bathos – to cure toothache, to attract a lover, to escape from prison – the book became associated with rumours of necromancy, and was thought in the fifteenth century to be too incendiary to be committed to print. As an example of the kind of dark arts that books like the *Picatrix* were supposed to encourage, Ficino cites the case of Persian magicians who used magic to create a creature like a blackbird with a serpent's tail, which they burn to ashes. When these ashes are poured over a lamp, the room is suddenly filled with snakes.

Thus the suspicion always hounded both Paracelsus and Agrippa that they worked their wonders by demonic means. Agrippa's magic in particular had an air of propitiation and coercion about it: a sense that one needed to cajole and persuade the reluctant forces of nature to do one's will, as though dealing with intelligent spirits.

Like Paracelsus, Agrippa could not restrain his tongue before the great and the powerful: in Metz he acquired the enmity of the Dominican Inquisitor Nicolaus Savini when he defended an old peasant woman against charges of witchcraft. And like Paracelsus, Agrippa had instinctive sympathies towards the humanists and reformers – but he too never formally converted to Luther's theology. He sympathized with calls for religious change, but had no taste for revolution.

No one personifies the great paradox of Renaissance science better than Agrippa. One moment he exhibits robust scepticism and contempt for superstition; the next, he comes across as the most credulous dupe. These two aspects of his personality are represented in Agrippa's two great works, published within a year of one another, and at face value in total contradiction. They have stimulated a debate about the 'true Agrippa' that has never yet been resolved, and probably never will be.

Agrippa wrote *On Occult Philosophy* as a young man in 1510, but he heeded the advice of his mentor Trithemius and refrained from publishing it. The work circulated in manuscript form for the next twenty-three years, helping to build Agrippa's formidable reputation: according to the historian Frances Yates, it was 'the best known manual of Renaissance magic'.[19] Paracelsus is said to have read it and learnt much from it. It was not until 1532 that Agrippa attempted to put the book through the press in Cologne; but such was its notoriety (and that of its author) that the Dominicans in the city managed temporarily to halt publication. The book finally appeared in the summer of the following year, while Agrippa was working as an archivist and historiographer for Margaret of Austria, Habsburg governor of the Low Countries. She was outraged at its impiety, and Agrippa was dismissed from his post. The University of Louvain was asked to assess the work and pronounced it scandalous and heretical.

These charges were in no way mitigated by the bizarre fact that Agrippa had only just published a book refuting the very ideas espoused in *Occult Philosophy*. In *On the Vanity and Uncertainty of the Arts and Sciences*, written around 1526 and published in 1530, he appears to pour scorn on the entire programme of magic – indeed, there is barely any aspect of Renaissance culture and society that escapes censure in this bitter, despairing diatribe. *Vanity and Uncertainty* makes sure to offend everyone: clerics and aristocrats, doctors and apothecaries, astrologers and kings, all are accused of abusing their positions of power. All of them claim the privilege of truth, said Agrippa, and all are deceived – for 'the way of truth is shut up from the senses . . . Wherefore all these derivations and sciences, which are fast rooted in the senses, shall be uncertain, erroneous and deceitful.'[20]

It is a devastating litany of despondency and cynicism; Goethe admits that in his youth Agrippa's book threw him into a despairing intellectual crisis. According to Agrippa:

> The knowledge of all sciences is so difficult (I will not say impossible) that all man's life will fail before one small iota of learning may be found out . . . Moreover, all Sciences are nothing else but the ordinances and opinions of men, as noisome as profitable, as pestilent as wholesome, as ill as good, in no part perfect, but doubtful and full of error and contention.[21]

For the conventional occultist, the way through this fog of deception was illuminated by the 'light of nature', a kind of direct revelation that unlocked all the hidden codes and symbols. But that was not the answer for Agrippa, or at least not here: original sin, he said, has obscured that divine light, and not without immense effort and contemplation can man hope to see its dim glimmer again. So all occult practices are to no avail, and it would be better if they were not pursued at all. This once-avid alchemist calls for alchemy to be banned. Of the Jewish Cabbala, that 'divine science sublimer than all human striving'[22] according to *Occult Philosophy*, he says in *Vanity and Uncertainty* that it 'is nothing but a certain most pernicious superstition'.[23] Just in case the magicians should not get the point, Agrippa added this comment on his own contribution to magic:

> I while yet a youth wrote in a quite large volume three books of magical things, which I called *De occulta philosophia*, in which whatever was then erroneous because of my curious youth, now, more cautious, I wish to retract by this recantation, for formerly I spent much time and goods on these vanities.[24]

Were the *Occult Philosophy* to have appeared first, we might be inclined to see *Vanity and Uncertainty* as a genuine retraction. But why would Agrippa consent to – no, even enthusiastically campaign for – the publication of his book of youthful folly a few years *after Vanity and Uncertainty*? Conceivably, the latter may have been a bluff, designed to ward off attacks from the Church over his book on occult arts. Yet Agrippa was not averse to controversy, and was quite capable of standing on principle even when it made life uncomfortable for him. In his introduction to the *Occult Philosophy* he acknowledges the apparent contradiction, but maintains that he has simply 'corrected' his youthful work and that it is better that it comes amended from his own hand than 'torn and in fragments' from the hands of others. 'Wherefore now I pray thee, Curteous Reader,' he appeals, 'weigh not these things according to the present time of setting them forth, but pardon my curious youth, if thou shalt findd any thing in them that many displease thee.'[25]

So perhaps both *Occult Philosophy* and *Vanity and Uncertainty* are honest expressions of what Agrippa believed – or at least, perhaps both were published in good faith. As we have seen, this apparent coexistence of contradictory extremes in a single individual is a recurring feature of Renaissance thought – we see it very clearly in Paracelsus – and the sceptical mystic is one of the difficult personality types of that age, with which anyone seeking for the origins of scientific thought must at some point wrestle.

In any event, it is clearly unrealistic to demand too much coherence from a character as contrary as Agrippa of Nettesheim. He would be apt to say whatever came into his head, and if it was not the same thing as he said the day before, it was nevertheless equally likely to offend someone. He consistently raised the hackles of the conventional physicians, calling them 'envious pigs' when he defended the unconventional doctor Jean Thibault against their attacks in Antwerp in 1530. His life was thus plagued by acrimony and condemnation: it is said that Charles V banished Agrippa from all his territories. As he began to lose favour in the French royal court at Lyons in the 1520s, he quickly ran out of tact, so enraging Louise of Savoy and her son that he was forced to flee from French soil without his family.

Agrippa's name attracted legends to rival those associated with Paracelsus. He had a magic glass, people said, that enabled him to see things distant in space and time. Before a distinguished audience including the Elector of Saxony, the Earl of Surrey and Erasmus, he conjured up the shade of Cicero. At Louvain, a young pupil of Agrippa's entered his chamber to study a book of incantations that summoned demons. Opening the book at such a spell, the young man inadvertently caused a demon to appear there and then, whereupon he died of fright. Returning to his room, Agrippa found the body and realized he would be suspected of murder. So he called the demon back and compelled it to reanimate the student's body, enabling the corpse to walk several times around the courtyard so that others might imagine it was still alive. Eventually the demon fled and the body collapsed, but Agrippa was absolved of suspicion for the boy's death.

And then there was Agrippa's dog: a black beast called Monsieur, which was rumoured to be a demon and was accompanied by a bitch called Mamselle. The relationship between animal and master was the subject of some speculation; Agrippa let Monsieur eat by his side at the table, and they even slept side by side (at least at such times when the twice-married Agrippa had no wife). 'When Agrippa and I were eating or studying together,' his pupil Johann Weyer testified, 'this dog always lay between us.'[26] The demon dog was said to inform Agrippa of what was happening in distant lands. When Agrippa died, in the region of Grenoble and apparently in poverty, Monsieur is said to have hurled himself into a river, never to be seen again. His master was rumoured to have expired with his face turned to the earth: the posture expected of a disciple of the devil. Here, clearly, was another contribution to the Faust myth. Faust too was said to have been accompanied by a dog, which (since it was in reality the demon Mephistopheles) could take the form of a servant and wait on his master at meals. Marlowe's Doctor Faustus aspires to be 'as cunning as Agrippa was, whose shadows made all of Europe honour him'.[27]

Yet in the long term Agrippa's influence was never the equal of

Paracelsus' – although perhaps it is more surprising, given his modest literary output and achievements, that he was influential at all. To the great astronomer Tycho Brahe, a supporter of Paracelsian science, Agrippa was nothing but a 'most worthless fabricator of vanities'.[28] But the fact that Victor Frankenstein sees fit to rate Agrippa before Paracelsus tells us something about their relative standing in the early nineteenth century. And Agrippa was the inspiration for Frankenstein's notorious predecessor, the tragic incarnation of Faust depicted by Goethe.

TALL TALES

But Goethe's hero was far nobler and more scholarly than the 'original' Faust seems to have been. The image of the intellectual vagabond of the Renaissance had no more raffish a personification than this legendary wanderer who, alongside Paracelsus and Agrippa, is said to have plied the roads of sixteenth-century Germany.

The Swiss encyclopaedist Conrad Gesner considered Faust and Paracelsus two of a kind:

> [Paracelsus] certainly was an impious man and a sorcerer. He had intercourse with demons . . . His disciples practice wicked astrology, divination, and other forbidden arts. I suspect they are the survivors of the Celtic Druids who received instructions from their demons . . . This school also is responsible for the so-called vagrant scholars, one of whom, famous Dr Faustus, died only recently.[29]

With justification does the Canadian playwright George Ryga have the eponymous hero of his *Paracelsus* (1982) complain of 'the legend of a Doctor Faustus branded on my face'.[30]

So many stories and rumours have interposed themselves since Christopher Marlowe popularized the figure of Faust in the English-speaking world that it is no longer possible to distinguish truth from myth. Even the full name of Faust is blurred: sometimes he is christened Johannes, sometimes Georg. Were there perhaps two Fausts, the historian Eliza Butler wonders – brothers, or even twins?

Johannes Faust allegedly gained a degree in theology at the University of Heidelberg, but his most famous exploit is recounted by Adrien de Jonghe, rector and teacher in Haarlem, who relates in his *Batavia* (1588) how this Johannes stole the secret of printing from its inventor, Laurens Janssen of Haarlem, known as Coster. Coster's workmen all swore never to reveal their master's secrets; but at midnight one Christmas Eve, one of them slipped away to Mainz, where he told all to Johann Gutenberg, a local goldsmith. This traitor's name, according to one of Coster's servants, who told the tale as an old man to Jonghe's teacher, was Johann Faust.

Does this not after all fit with the story that Gutenberg was bankrolled in setting up his press by a financier named 'Johann Fust'? Indeed, isn't there something Faustian about the entire enterprise of printing, which traps ancient secrets and divine wisdom on the page and scatters it across the world for all and sundry to see? In an age when machinery was allied to magic, there was something disreputable, even heretical, about printing. According to one story, Johannes Faust went to Paris to sell his printed bibles, which the rich and powerful mistook for manuscript copies and for which they paid the high prices that handwritten books commanded. Subsequently Faust was lucky to escape alive:

> The buyers finding a greater number upon him, than it was possible for several men to transcribe in their whole life, and the pages of each copy so exactly alike, . . . he was seiz'd, try'd and condemn'd for *Magick* and *Sorcery*, and was accordingly dragg'd to the stake to be burnt; but upon discovering his Art, the parliament at *Paris* made an act to discharge him from all prosecution, in consideration of his admirable invention.[31]

In any event, it was Gutenberg who reaped the greatest benefit. He was reputedly rich by 1455 (his best-selling Bible appeared two years later), while Coster saw his business founder and died of grief.

It is surely no coincidence that Jonghe's *Batavia* appeared just a year after one of the first major accounts of the Faust legend – the first so-called *Faustbuch*, of which many followed. Entitled *Historia von D. Johann Fausten*, it was published anonymously by the Frankfurt publisher Johann Spiess in 1587. It was a sound commercial decision to put a popular folk legend into print: Spiess's book was a runaway success and was translated into just about every tongue in Europe within five years. Marlowe wrote his *Tragedie of Dr Faustus* two years after the English translation of the Spiess book appeared in 1592.

But Spiess, Marlowe and the *Faustbuchs* tell a rather different story from Jonghe. Spiess's Faust was, according to a contemporary source, 'such a celebrated name among the common people that there can hardly be found anyone who is not able to recount some instance of his art'.[32] He was supposedly born around 1480, in Knittlingen in Germany, and he styled himself Magister Georg Sabellicus, Faustus Junior. For years he roved in the lands of the lower Rhine, cheating and swindling and duping lords and princes to whom he promised 'mountains of gold'. The archetype of the mountebank alchemist goes back further, as we shall see; but Georg Faustus became its most notorious representative.

Faust, as depicted in an etching by Rembrandt.

For a time he enchanted the humanist knight Franz von Sickingen, who helped Faust obtain the post of schoolmaster at Kreuznach. According to Trithemius of Sponheim, Faust indulged in 'the most dastardly kind of lewdness with the boys',[33] and he was forced to flee one night in disgrace. Thereafter he went to Heidelberg, where in 1513 he called himself Helmitheus Hedebergensis, the 'demi-god of Heidelberg'. His outrageous bragging knew no bounds of either proportion or decency: he claimed he could work better miracles than Christ, and could recount all the works of Plato and Aristotle from memory, while improving on them in the process. In Cracow he was said to have turned lead into gold and water into wine. At Erfurt he lectured at the university, and as he spoke about Homer he conjured forth each of the heroes of the epic poems. The cyclops Polyphemus was particularly terrible to behold, summoned in the act of eating a man whole, and the monster almost devoured a few students before Faust, with great effort, dismissed him.

The *Erfurt Chronicle*, written in the mid sixteenth century, contains several such tales. Faust was said to have met the humanist and reformer Philip Melanchthon, who found the trickster despicable and scolded him for his evil ways – whereupon Faust flew into a rage and threatened Melanchthon that he would make 'all the pots in your kitchen fly out of the chimney, so that you and your guests will have nothing to eat'. The stolid Reformer was unmoved. 'You had better not', he retorted. 'Hang you and your tricks.'[34] Yet Melanchthon went on to write a biography of Faust in 1562 that went through nine editions by the end of the century and no doubt alerted the likes of Spiess to the lucrative market for *Faustbuch*s.

Paracelsus was on those same German roads in the 1520s, and he would presumably have heard tales of this malicious prankster. Agrippa met Faust, probably in Paris in 1528, when Francis I sent for him and saw him conjure the royal princes out of thin air even though they were many miles away. Before the king's eyes, Faust filled all the hills with spectral armies, horses and chariots. In the same year, he turned up in Hallestein in Germany, claiming to be commander of the Knights of St John. Soon after, he was banished from Ingolstadt, and four years later the city of Nuremberg barred him too. His wandering, boasting and prophesying continued until his death around 1539–40.

All of this vagabondry is echoed in the story of Paracelsus. Magic powers, trickery, bravado and bluster: all help to explain why, in the words of Carl Jung's pupil Jolande Jacobi, 'Paracelsus and this "original Faust" have long since been inseparably linked.'[35] But there was more to the Spiess *Faustbuch* than a tale of a scoundrel. His Faust meets a horrible end, for he has made a pact with the devil. One night awful screams are heard, and in the morning Faust's body is found dreadfully mutilated, his brains spattered over the wall. Johann Weyer added to this legend, saying that on the night of Faust's death the house in which he lodged, in a village in Württemberg, was shaken violently, and in the morning he 'was found dead beside his bed with his face twisted back'.[36] Others say that his corpse kept turning face down – a sign of a soul headed for hell – though placed five times on its back.

The trickster archetype has here turned into something darker. For Spiess, a deeply religious Lutheran, the *Faustbuch* had a serious purpose: it was not just a rumbustious yarn after the manner of Rabelais, but a moral tale. The sixteenth-century reader hardly needed telling, of course, that a pact with the devil left hell to pay, but the implication was that dabbling in magic was tantamount to consorting with demons.

The moral overtones of the Faust legends point us back to their true source: not in sixteenth-century Europe but in the Holy Land, where St Peter on his missionary journeys met a sorcerer named Simon Magus who had 'amazed all the people of Samaria'.[37] He could fly, they said, he could melt iron and make dishes move. He possessed the secret of the alchemists, being able to produce gold in abundance and even to turn himself into the yellow metal. But this Simon was also a boaster who claimed to be 'the Great Power' – a Messiah, if not a god himself. Seeing how Peter and John could, with a touch of their hands, convey the power of the Holy Spirit, Simon offers the apostles money if they will share with him this miraculous ability, thus originating the sin of simony: attempting to purchase God's power for selfish ends. Peter berates him for this crass demand, whereupon Simon Magus challenges him to a magic duel. Each of them has to outdo the miracles of the other. But Simon's marvels are exposed as fraudulent, and he falls flat on his face – literally, for his attempt to fly fails.

Simon Magus called himself *faustus*, as it is rendered in Latin, meaning 'the favoured one'. He was said to be travelling with a prostitute from Tyre named Helen – in both Marlowe's and Goethe's plays Faust conjures up the shade of Helen of Troy. Simon Magus, possibly an insignificant Samaritan charlatan, became regarded as an arch-heretic in the early Christian Church, and was even associated with the Antichrist.

None of this was lost on Georg Faust's contemporaries. It may have been the legend of Simon Magus that moved Trithemius to dub Georg Sabellicus 'Faustus junior', following in the footsteps of his biblical predecessor. Luther himself is sometimes said to have been the first to pronounce Faust as an associate of the devil. A Lutheran sermon of the 1540s makes the connection explicit: 'Simon Magus tried to fly to heaven, but Peter prayed that he might fall . . . Faust also tried this in Venice. But he was sorely dashed to the ground.'[38] This most prosaic fall of course also echoes the casting down of Satan from heaven for challenging God's authority. No wonder, then, that both Paracelsus and Faust became heroes of the German Romantics. It was, after all, the quest for knowledge that emboldened man to place himself above God and to regard nature not as something to be feared and endured but to be understood and tamed. Jung believed that this heretical impulse was perceived unconsciously in the devoutly Christian Paracelsus, creating in him an inner conflict that accounts for much of his bizarre, bombastic and self-destructive behaviour.

FROM ICE TO DESERT

'For many years,' Paracelsus tells us:

> I studied at the universities of Germany, Italy, and France, seeking to discover the foundations of medicine. However, I did not content myself with their teachings and writings and books, but continued my travels to Granada and Lisbon, through Spain and England, through Brandenburg, Prussia, Lithuania, Poland, Hungary, Wallachia, Transylvania, Croatia, the Wendian Mark, and yet other countries which there is no need to mention here.[39]

These travels, we saw earlier, took him from Villach to Ferrara and then through Italy and southern France, down the

Mediterranean coast of Spain and to North Africa in the campaign against the Algerian Moors. Returning to Granada, he passed through Cordoba and Seville and up through Portugal to the sacred city of Santiago de Compostella, the final destination of the pilgrim's trail that passed through Einsiedeln. From there he toured the northern Spanish cities of León, Salamanca, Valladolid, Saragossa and Jaca, before crossing the Pyrenees to Toulouse.

In Paris he inevitably fell out with the conservative doctors of the Sorbonne. The Parisian physicians, he later said, 'despise all others and yet are nothing but utter ignoramuses themselves; they think that their long necks and high judgement reach right unto heaven.'[40] He reached England via Calais (which was then under English rule) and visited the chaotic patchwork of Tudor London shortly before Henry VIII's reformation left it disfigured by the ruins of churches and monasteries. It was not a welcoming city to foreigners: one Swiss medical student commented that 'the common people are still somewhat coarse and uncultured . . . and believe that the world beyond England is boarded off'.[41]

Paracelsus travelled all over the British Isles. He saw tin mines in Cornwall, lead mines in Cumbria, and the famous University of Oxford. He sailed to Ireland, he went to Scotland and Yorkshire. Then he recrossed the English Channel to Bruges in the Netherlands, and passed through Flanders to Louvain and the great commercial centre of Antwerp. He visited the Dutch university city of Deventer, where Erasmus was educated, and seems at some point to have crossed the sea to Zealand, between Denmark and Sweden.

This was in 1519, the year the Emperor Maximilian I died. The struggle for the Imperial throne between Francis I of France and Charles III of Spain and Burgundy sent ripples throughout Europe, and they were felt in the Netherlands. Formally, the region was a part of the Empire; but it effectively operated as an oligarchy in which many towns had their own constitutions and power rested largely in the hands of a few powerful merchant families. The Duke of Guelderland, the region of the east central Netherlands, had a long-standing enmity with the House of Burgundy, and refused to recognize Charles's authority. Little more than a robber-baron himself, the duke launched regular forays into Holland in the early sixteenth century, plundering the

Dutch cities that offered allegiance to the Emperor. Between 1517 and 1523 Guelderland and the Habsburg Netherlands were mired in a state of more or less continual, inconclusive warfare, abetted by the rebellious province of Friesland, which was also keen to resist Imperial authority and sometimes made common cause with the Guelderns.

Paracelsus had no quarrel with the Empire, but he was ready enough to offer his services as a military surgeon to the rebels. So when the Habsburgs gained the upper hand, he was forced to make a rapid exit from the Netherlands, whose people, he admitted, 'did not like me'.[42] He landed on the southern coast of the Baltic Sea, where he visited the mercantile cities of the Hanseatic League: Hamburg, Lübeck and Rostock. From there it was a short step to Copenhagen, where he arrived at the end of 1519.

His stock seems to have been high at that time. The king of Denmark and Norway, Christian II (1481–1559), welcomed the doctor to his court and appointed him as royal physician. In this capacity Paracelsus accompanied the Danish troops on an expedition to Sweden, where they overwhelmed and killed the ruler Sten Sture at the beginning of 1520, captured Stockholm and secured Christian's claim to the Swedish throne. Paracelsus was rewarded with both honours and gold.

He took the opportunity to explore far into the north, visiting the university at Uppsala and the copper mines at Falun. He is reputed to have passed beyond the Arctic Circle into the lands of ice and midnight sun, searching for hidden treasures – which makes for a fine tale, but for which there is sadly no firm evidence.

The situation changed, however, when the Swedes rebelled in November of 1520. Christian's conquest had been excessively brutal; when Sture's widow surrendered in the autumn, the Danish king beheaded two Swedish bishops before a witness of eighty of their parishioners, who were subsequently themselves slaughtered. Then Sture's rotting corpse was disinterred and burned, his young son was torn from his wife's arms and flung living into the flames, and the king's widow was forced to live as a prostitute. These public massacres and humiliations finally incited the Swedes to rebellion; the Danish invaders were forced out and Paracelsus' position in Sweden became untenable. He returned to Hanseatic territory, travelling by sea to Libau, which was ruled by the Order of the Teutonic Knights. From there he sailed to Stettin, the capital of Pomerania, and then to the port of Danzig, which was officially part of the kingdom of Poland.

In late 1520 a group of Teutonic Knights attempted to capture Danzig. They were repulsed, but Paracelsus decided to continue his travels with the retreating troops, whom he accompanied to Königsberg in Prussia, where the Grand Master of the Order held his court. In preparation for besieging Danzig, the Grand Master had sought an alliance with Grand Duke Vassily III of Moscow, the 'Ruler of all Russias'. Now Vassily had a proposal of his own.

He was keen to cultivate in the Russian court a level of sophistication comparable to that of Western Europeans, and to this end he aimed to woo some of the learned men of Europe to Moscow. Russia had for several centuries been carved up by Mongol invaders from the east, who arrived as Genghis Khan's Golden Horde in the thirteenth century. Muscovy survived that onslaught, remaining a principality and observing the Orthodox Christianity of Byzantium. But it was a precarious existence, continually threatened by the Mongolian 'Tartars' along its eastern borders, and Moscow's culture lagged behind Europe as it emerged from the Middle Ages. In 1462 Vassily's father Ivan III ('the Great') became Grand Duke of Muscovy and set about restoring his kingdom. He annexed several neighbouring principalities in the next two decades, and in 1472 he declared himself Tsar, sovereign of all Russia. The kingdom continued to expand into Lithuania after Vassily took the throne in 1502.

When Paracelsus decided to take up the Tsar's invitation in the winter of 1520–21, passing by sledge through Vilna and Smolensk to Moscow, the Tartar raids along the Russian borders were an increasing threat. The Tartars plundered towns and villages along the Oka and Volga rivers, even as far as the Moskva tributary and the suburbs of Moscow itself, carting off the inhabitants as slaves. These were not mere forays; the Khan of the Crimean Tartars, Mehmed-Girey, harboured plans of invading Muscovy, and in 1521 he laid siege to Moscow with an army of 100,000 men. The offensive was ultimately unsuccessful, since Mehmed-Girey was forced to withdraw to defend himself from an attack by the Astrakhan Tartars and he was killed in battle in 1523. But it was on one of these raids on Vassily's capital that Paracelsus and some of the other foreign visitors were captured in the outskirts of the city, where they were lodged.

To most men, this would have been a disastrous turn of events. The Tartars were seen as barbaric, cruel, almost subhuman. But Paracelsus' belief that all the world was an open book extended even to these uncharted places. As he was taken back though the Ukraine to the Crimea, he found a way into the confidences of his captors. It is said that the Tartars regarded a healer as a holy man, which is just how Paracelsus saw it too. He demonstrated his skills in medicine and surgery, and in return it seems that this Western doctor discovered marvellous things in the traditions of the Tartars. Their medicine was shamanic – rooted, Paracelsus says, in 'faith and imagination' – as well as drawing on herbal remedies. There are even rumours that Paracelsus underwent a shamanic initiation rite; the truth of that will never be known, but we can imagine that he would have welcomed the opportunity. 'Wondrous things happen in this world,' he said, when 'a man of Swabia brings physic and experiment right up to the Tartars'.[43]

There is a story that Paracelsus travelled with a Tartar prince to Constantinople, where a magus gave him the Philosopher's Stone. But it appears that, on the contrary, he left the company of the Tartars when a troop of Polish knights raided the borderlands and 'liberated' the European captives. Whether or not he welcomed this escape is not known, but it seems that he regarded his time with the Tartars as providential.

From Poland and Lithuania Paracelsus travelled to Hungary, Transylvania and Wallachia. Here and in the Ukraine, he complained of an affliction of lice, and devised a medicine against them. Likewise, when he had crossed the choppy English Channel he claimed to have invented a remedy for sea-sickness, which he called 'travellers' salt'. Who could fail to be astounded by this physician who created cures as fast as he encountered ailments?

Then came Slavonia, Croatia and Slovenia, where he visited the recently excavated mercury mine at Idrija. From there the natural course was to head for Venice across the upper Adriatic, and Paracelsus arrived in the Italian city some time in 1521.

Italy was still strife-torn, a battleground now for Charles V and his jealous rival Francis I. Paracelsus seems to have become a surgeon in the Venetian army in the campaign of September 1521, before travelling deeper into Italy through Ancona and Apulia until he reached Taranto on the southern coast. He set sail for the

Peloponnese peninsula and landed at the ancient port of Messene.

He was now moving along a Venetian trading route, for both Messene and his next stop, Crete, were ruled by Venice. The way led to Alexandria, cradle of alchemy on the north African coast. Here Paracelsus encountered at first hand the mystical traditions that informed his philosophy: Neoplatonism and Gnosticism. There is scarcely any record of what he found in Alexander's ancient city, although he writes of the Alexandrian physicians and says that in Egypt he received 'magical instructions'.

He continued to follow the Venetian trade route up the Nile to Cairo and the trading station of Syene (now Aswan) on the border of Ethiopia. For a man born and raised in the Swiss mountains, it was a journey into the realm of dreams. The climate and customs were strange enough, but the creatures of the Nile were like monsters from some dark corner of a Bosch painting: razor-jawed crocodiles, languid hippopotami, brightly coloured birds. 'In Africa', he later recounted, 'there are monsters, which forsooth are so dreadful of aspect that, instead of travelling further, you would want to creep again into your mother's lap.'[44]

At this point, it must be said, we are forced to work with the briefest of allusions in Paracelsus' own writings; there is some dispute as to whether he reached Africa at all. But his references to the Red Sea suggest that he sailed over it from the Egyptian coast to reach the Gulf of Aqaba, along a trading and pilgrim route that passed into Palestine and delivered the traveller to the Holy City of Jerusalem.

When the Ottoman Sultan Selim I drove the Egyptian Mameluke Muslims out of Jerusalem in 1516, he acquired a city dilapidated by centuries of neglect. Selim's successor, Suleiman the Magnificent, began an extensive rebuilding programme in the 1530s, but Paracelsus describes a city still in ruins. He left the Holy Land via Syria, probably through the port of Akka, a conduit for Christian pilgrims. This took him to Venetian-ruled Cyprus, and thence to Rhodes, the headquarters of the Knights of the Order of St John.

There is something uncanny about the way turmoil follows in Paracelsus' footsteps: war breaks out as he arrives in the Netherlands, in Denmark, in Moscow and Danzig. In Rhodes it was there waiting for him. The Knights Hospitaller of St John captured the island from the Muslims in 1307, and made it their base. The Order was formed at a Benedictine monastery established in Jerusalem in the eleventh century, where a hostel dedicated to St John the Almsgiver, patriarch of Alexandria in the sixth century, attended sick pilgrims in the Holy City. The Hospitallers eventually became independent of the Benedictines and developed into a powerful, wealthy and militarized brotherhood that played a central role in the later Crusades. By the end of the thirteenth century, however, they had been driven out of their strongholds in the Holy Land and were forced to relocate in Cyprus and Rhodes.

But that foothold in the Middle East became precarious as the Ottoman Empire expanded its borders. When Constantinople fell in 1453, the outlook was bleak. The knights withstood an attack on the island in 1480, but it was only a matter of time before they were overwhelmed. When Suleiman became the Ottoman Sultan in 1520, Rhodes was one of the first obstacles to his westward expansion. Two years later the Grand Master of the Order of St John, Philippe Villiers de L'Isle-Adam, received a message from the Sultan ordering him to relinquish Rhodes. Philippe had 6,000 troops; Suleiman amassed an awesome force of 700 ships and 200,000 men. The Grand Master must have known that his situation was hopeless, yet he resisted all the same. From July until December, Rhodes held out against the Turkish forces, despite a barrage of some 40,000 iron cannonballs. It was during this time that Paracelsus found his way by unknown means onto the beleaguered island. Here he attended to the wounded and the sick as the resistance of the knights slowly crumbled.

The Order held out until Christmas Day, but on 26 December the Grand Master accepted the Sultan's offer of free passage for all the remaining defenders. The evacuation began on New Year's Day in 1523: the Hospitallers were given a new home in Malta by Charles V, but Paracelsus, on one of the last boats to leave Rhodes, headed instead for the island of Cos, also under the rule of the Order of St John, where Hippocrates had been born.

Via Samos and the Cyclades, Paracelsus reached Athens in early 1523. The ancient Greek city was under Turkish rule, having been conquered in 1456. He seems to have visited the Temple of the Oracle at Delphi, where natural springs flowed from fissures in the seismically active hills overlooking the Bay of Corinth. Passing across the Aegean to Lesbos and Limnos, and through the straits of the Dardanelles, he arrived finally at Constantinople, renamed Istanbul by Suleiman and now the seat of his court. Here again various tales arose of how Paracelsus discovered the secret elixirs of the alchemists. He tells how, around this time, he met a German alchemist called Solomon Trismosin in Constantinople, who was learned in the Cabbala and Egyptian magic. Trismosin claimed to have discovered the secret of alchemical gold-making from a Jew in Italy, and he passed on his esoteric knowledge to Paracelsus.

It was not strange that a European and Christian could enter the heart of the Ottoman Empire unmolested. Christendom may have been at war with Islam, but it was at that time generally a dignified war that observed the courtly rules of the age – as Suleiman's conduct at Rhodes demonstrated. In most respects the Ottoman culture surpassed that of the Europeans: its arts and scholarship were supremely refined. Ogier Ghiselin de Busbecq, ambassador for the Emperor Ferdinand (Charles's successor) between 1554 and 1562, admitted that he was impressed by much of what he found in Istanbul. The Ottoman court was a meritocracy, he said, where there was no inherited nobility aside from the Sultan's family: 'each man is rewarded according to his deserts', and sloth and dishonesty 'never attain to distinction'.[45] But learning, especially scientific learning, was increasingly hampered in the Islamic world by religious constraints that frowned on printing (the holy books were too sacred to be reproduced in such a crude, mechanical manner) and forbade clocks (the muezzins were the timekeepers).

It was a long and dangerous journey back to Europe: Thrace, Bulgaria, Macedonia, Albania, Herzegovina, through lands wild and remote where a traveller could easily disappear for ever in dark woods and rugged mountains. But at last he found himself on the Adriatic coast in Dalmatia, a colonial outpost of Venice. He reached the city again in late 1523, where allegiances had shifted once again in the Italian wars and the Venetians now supported the Holy Roman Emperor. No wonder Paracelsus took such a dim view of the posturing of the European princes, after having survived all kinds of perils and seen astonishing sights only to find on his return that the same squabbles were being fought out between arbitrarily reorganized allies. He did not linger in Italy, but made his way back over the Alps in the winter of 1523–4 to return to his father's house in Villach.

When, rested and refreshed, he set out again in the late summer, it was the last time he would see his father alive. He could not have failed to perceive that the world was increasingly troubled, as southern Germany trembled with social and religious tensions. The signs were in the stars: a crisis was imminent, and some believed it would be the greatest upheaval since the Fall of Man.

6

A New Religion

The Trials of Reformation

> Now note, fellow boozers, that while Greatclod was saying his dry Mass, three church bell-ringers, each with a great bowl in his hand, were passing among the crowd, crying: 'Don't forget the happy people who have seen him face-to-face.' As we came out of the temple, they brought Greatclod their basins, which were quite full of Papimaniac money.
>
> Greatclod told us that this collection was for convivial purposes, and that, in accordance with a miraculous gloss, hidden in a certain corner of their holy Decretals, one half of the contribution would be spent on good drink, the other on good food. This was done, in a very fine tavern, which rather reminded me of Guillot's at Amiens. Believe me, the viands were copious and the drinks numerous.
>
> François Rabelais, *Pantagruel* (1532)

As a young student hurries anxiously towards Erfurt along the road past Stotternheim in Saxony, storm clouds darken the sky. Then down come the cataracts, till they have drenched the steeples and drowned the cocks; and the oak-cleaving thunderbolts begin to crash. A tempest like this is not a natural hazard but the work of demons, and the air is thick with their wild yells. Caught in the downpour, the terrified student stands in fear not just of his life but of his mortal soul.

He is flung to the ground by the forks of fire, and he smells the acrid aroma of his doom. Yet he cries out an appeal to the patron saint of his father's profession, St Anne, who watches over the miners of Saxony. If she will save him, his life shall henceforth be devoted to God.

This intense and febrile young man does not make oaths lightly. When the young Martin Luther reaches Erfurt, he graduates in law only to defy the wishes of his dour and violent father by entering the ascetic brotherhood of the Augustinian Eremites – an order dedicated to re-establishing the pious and humble ideals of the monastery in an age when monks are widely regarded with contempt and ridicule. It is 1505 and Luther is twenty-one years old, and he would be the last to suspect or to wish that his religious vocation is going to lead to war and misery and an end to a united Christendom.

As with most legends, it barely matters whether the *Sturm und Drang* of Luther's initiation into holy orders can be supported by historical evidence. The story captures the emotional truth of the narrative. That wild storm was soon to settle over much of Europe, and Luther himself was dispensing the thunderbolts, coarsely wrought and thrown with venom. Rage and terror marked his footsteps until he died in 1546. On the savage heath of Martin Luther's soul there were only primal absolutes: salvation or damnation, saint or villain, Christ or Antichrist.

If we were to see in this rough, irascible figure – determined to speak to the common man through the vernacular, pitting himself against a monumental institution that is defended by smooth-

Martin Luther, drawn by Lucas Cranach.

tongued academics, vilifying his enemies, brooking no compromise, accused (by Pope Leo X, no less) of being a drunken German – if we were to see in all of this the outline of another now-familiar profile, we would not be guilty of idle analogy. The contemporaries of Paracelsus remarked on the resemblance too, and they did not intend it as any sort of compliment. Paracelsus himself took offence not only at the censorious sentiment behind the comparison, but by what he perceived as its insufficiency. 'Do you think I am only Luther?' he fumes in the *Paragranum* (1530). 'Insofar as I am gifted beyond that which is required of a Christian, namely, with more of an office than an apostle (that is, with medicine, philosophy, astronomy, alchemy), and you think that I am only Luther? . . . Why do you do this? Because you hope that Luther will be burned and Theophrastus shall be burned also.'[1]

In the year these words were written, when attempts to reconcile the beliefs of the Lutheran Reformers with those of the Catholic Church finally collapsed at the Diet of Augsburg, there was every reason to suppose that Luther's enemies would have burned him if they could. Paracelsus never really stood in any such danger. His was a milder sentence: to be outcast, ridiculed and banned from publishing his theories. But it is very clear that he was conscious of the parallels with Luther throughout his life, and on occasion he encouraged the comparison explicitly. He was the *Lutherus medicorum*, come to reform medicine and to rescue it from corruption, decadence and ignorance.

But that is not the only reason to measure Paracelsus' life alongside Martin Luther's. For his combative style, his acrimonious disputes and his fluctuating fortunes become more comprehensible when viewed against the shifting social and intellectual context of the Reformation. There is no sharp distinction between the religious turmoil stirred up by the Lutherans and others who challenged the authority of the Church, and the metaphysical and scientific upheavals that were brought about by natural philosophers of the early sixteenth century.

The simple reason for this is that in both cases the arguments were at root theological. No one could divorce beliefs about nature from beliefs about God and the spiritual life of humankind. Paracelsus least of all recognized any such separation. Though he is regarded by some as a founder of positivistic science, his ideas are saturated in medieval theology. It can be argued that he was more of a religious thinker than a scientist – except that this distinction would be utterly meaningless to Paracelsus and his contemporaries.

The Reformation not only coloured Paracelsus' philosophy; it determined his immediate circumstances. A traveller could not take to the roads of Germany and Switzerland without experiencing the effects of religious strife and crisis. Luther's objectives were resolutely spiritual, but the fortunes of Lutheranism were shaped primarily by German power politics, in which local princes strove to strengthen and broaden their dominions in the face of attempts by Maximilian I and his successor Charles V to keep a tight hold on the Holy Roman Empire. Nor was Luther the only reformer; in Switzerland Ulrich Zwingli launched his own struggle against the corruptions of the Roman Church, with comparably bloody results. Anarchic Anabaptists wandered through central Europe preaching their fiery visions of a new Church reborn in apocalypse and ecstasy, while at the other extreme the humanists challenged the decadence of papal institutions with scholarly urbanity. Political allegiances often divided along religious lines, which made for some unlikely and uncomfortable bedfellows. Paracelsus, characteristically, belonged to no camp, but that did not prevent him from being regarded by some reformers as a convenient instrument for their own ends.

Luther and Paracelsus never met, and it is probably just as well. Neither man was at all conciliatory or diplomatic, and they would not for long have contained their differences. The closest they came to one another's orbit, it seems, is when Paracelsus wrote a dedicatory letter to Luther and his confederates at Wittenberg, prefacing a theological tract composed while he was in Salzburg in 1524. It looks as though Paracelsus was keen at that time to court favour with the Lutherans, but it was hardly an auspicious moment to do so: Paracelsus was, to his peril, shortly to become associated with the rebels of a peasant uprising, whom Luther condemned viciously. If Luther ever saw Paracelsus' overture, it is unlikely that he would have been impressed.

With both men, you were either for them or against them. There was no middle ground. Luther fell out violently with Erasmus, forcing the placid humanist to make an unusually vehement defence against his critic's harangues. Luther argued bitterly with Zwingli, although their theological views barely differed. It was not easy to like Martin Luther, not least because he could not care less whether you did or not.

Paracelsus and Luther both wrote mostly in the vernacular, not in Latin, because they wanted to reach a wide audience and to undermine the exclusiveness of scholastic and monastic learning. But this was not really anything new. There were German Bibles before Luther, and the English-language translations by John Wyclif and his followers, the Lollards, had circulated since the 1370s. Luther's versions of the scriptures were particularly effective, however, in part because he translated loosely, injecting idiomatic phrases and examples familiar to the common man. And Luther had the incomparable advantage (of which he vigorously availed himself) of the expansion in printing in the early sixteenth century, which made his translations more widely available and accessibly priced. Paracelsus' German writings on medicine, meanwhile, were prefigured by a tradition of books and almanacs collecting medical, herbal and surgical inventories, such as Ortolf von Baierland's *Artzneibuch* (1477) or Lorenz Fries's *Spiegel der Artznei* (1518). Such books gave rise to the eclectic vernacular craft manuals known as *Kunstbüchlein* from the 1530s. Few of Paracelsus' works, in contrast, appeared until after his death.

But Luther had another reason to write in German. His opposition to the Roman Church was strengthened by the implication that Germans were being ruled from abroad, beholden to a capricious and avaricious papal authority in a far-off land. The resentment against Rome and the rise of nationalistic feeling in Germany in the sixteenth century were not primarily of Luther's making, but he saw that they could help his cause.

Some have argued that Paracelsus, far from being to science what Luther was to religion, was his opposite. Luther, they say, was bigoted, superstitious, conservative and medieval; Paracelsus, open-minded, rational and radical, a man of the Renaissance. Although Luther would be hard pressed to evade those accusations, it takes a very selective reading to put Paracelsus at the opposite pole. He defies such easy categorization, not least because he is himself a morass of contradictions.

All the same, Paracelsus' theological philosophy, while hardly itself scientific, *enables* science whereas Luther's prohibits it. Indeed, it is not easy to separate Paracelsus' 'science' from his theology, and to the man himself there could never be any such division: his chemical cosmos was comprehensible only within the framework of his distinctive and idiosyncratic Christianity, for God's work was visible everywhere in nature.

In this sense at least, Paracelsus' outlook is sympathetic to the aspirations of humanism and the Renaissance. The Germanist scholar Andrew Weeks puts it succinctly: 'Paracelsus and Agrippa were men who would know all things; Luther was the man who would know the one necessary thing.'[2] For Paracelsus, God was everywhere in nature – a belief close to pantheism. God in his wisdom had designed nature so that humankind might understand it: one had only to learn the language of nature, how to read the signs and signatures. Although devout Christianity underpinned all of Paracelsus' speculative theory, nevertheless nature was accessible to reason. To Luther, on the other hand, 'reason is the devil's harlot'.[3] 'The Virgin birth', he admitted, 'was unreasonable; so was the Resurrection; so were the Gospels, the sacraments, the pontifical prerogatives, and the promise of life everlasting.'[4] That was why reason was terribly dangerous. It was sheer hubris and blasphemy to suppose that one could decode God's handiwork. Men should not understand, Luther thundered – they should only believe.

Thus, while Paracelsus' religious beliefs, which brought a kind of alchemical materialism to Christian philosophy, were certainly unorthodox – one might even regard them as perverse – they were in essence allied to the humanistic quest for understanding and order. His God was not Luther's ineffable Being. Luther's demand for incurious acquiescence before the Creator was anathema to Paracelsus, who felt that all of nature could be understood by anyone who was prepared to open his eyes. This progressive, optimistic view of human nature and capability was summed up in the remark of the great Florentine artist and architect Leon Battista Alberti: 'Man can make whatever he will of himself.'[5] Man, said the humanists, was not sinful and fallen from grace, but walked on earth with dignity. It was a vision expressed eloquently in the *Oration on the Dignity of Man* (1498) by the Italian Pico della Mirandola, in which God says to Adam:

> The nature of all other beings is limited and constrained within the bounds of laws prescribed by Us. Thou, constrained by no limits, in accordance with thy own free

> will . . . shalt have the power, out of thy judgement, to be reborn into the higher forms which are divine.[6]

Paracelsus' theology has been under-emphasized.* He began his religious writings during the period covered in the next chapter – his stay in Salzburg in 1524–5 – but most of them stemmed from the turbulent years between 1529 and 1535, when he wandered through Europe as a lay preacher. Forty of Paracelsus' theological monographs still survive, as well as sixteen Bible commentaries, twenty sermons, twenty works on the Eucharist and seven on the Virgin Mary. Half of these have never been properly edited, let alone printed in modern form. There is no question that Paracelsus thought long and hard about Christianity, and by styling himself a professor of theology (without, it seems, any official academic sanction) he implies that he regarded this component of his output to be the equal of his medical and chemical theories. That his role in the history of science and medicine has received far more attention than his theological oeuvre is, however, understandable and probably apt, for it cannot be said that he had much influence even on the religious debates of his day. In theology he never aspired to be a Luther, and that would in any case have been a futile aspiration for one so lacking in political acumen or the ability to foster disciples.

UNREST IN CHRISTENDOM

It was surely not so much the corruption of the Church that drove Luther to adopt a stance of extreme anti-rationalism, as some dark urging of his own soul. The humanists, faced with the same dismaying spectacle of a depraved papacy and a corrupt and

* The German medical historian Karl Sudhoff, who was largely responsible for Paracelsus' modern rehabilitation, accepted that an analysis of Paracelsian theology should be left to theologians. This aspect of his work was studied and collated first by William Matthiessen in the 1910s and 1920s, and later by Franz Strunz and Kurt Goldammer, and has never expanded far outside the German language. It is being continued by the Paracelsus Project at the University of Zurich; but even now there are religious works of Paracelsus that await a proper modern translation and analysis.

ignorant ecclesiastical rank and file, reached very different conclusions. Yet it was clear to anyone of free mind that something was very rotten in the state of the Holy See.

Christendom – the unified earthly kingdom that served Christ – was a medieval concept that, by the fifteenth century, had almost run its course. It had, for one thing, no supreme monarch. Throughout the Middle Ages the authority of the papacy was split in two. In all spiritual matters Rome had the final say. But the arm that wielded Christ's sword had, since the time of Charlemagne, belonged to the Holy Roman Emperor. In practice, neither of these figureheads was much more than a territorial prince. Nor were they always of one mind: popes were even imprisoned by Emperors in the fourteenth century, and Christendom had experienced the indignity of seeing rival popes contend for power. (In the late fourteenth century, two popes – each with a full house of cardinals and other officials – held court in Rome and Avignon for over thirty farcical years.) When the papacy was made formally subordinate to a collective of Christian monarchies at the Council of Constance in 1415, it could not pretend to be much more than a mere principality, albeit one that could exact tribute payments from all over Europe.

The popes of the late fifteenth century certainly behaved more like princes than spiritual leaders. They bribed, swindled and murdered their way into high office. When popes took mistresses, bishops felt licensed to follow suit. Rodrigo Borgia, succeeding his uncle Alfonso, became Pope Alexander VI in 1492 by buying off the opposition, and he went on to make his family name synonymous with nepotism, vice and debauchery. His son Cesare became the papal Captain-General, possibly after taking his own brother's life. Alexander was said to have shared the sexual favours of his daughter Lucrezia with her two brothers and several husbands.* When she bore a son, Alexander publicly decreed him to be Cesare's, born of an unidentified 'unmarried woman', but

* In Friedrich Maximilian von Klinger's 1791 play *Faustus*, which revitalized that legend in German Romantic culture, Alexander VI confronts Faust and Satan in hell, and is horrified to discover that Faust is a mere printer, for 'he gave himself out at my court for a gentleman, and slept with my daughter Lucretia!' Faust replies defiantly: 'My noble invention will sow more good, and will be more profitable to the human race, than all the popes from St. Peter down to thyself.'[7]

privately he claimed the boy as his own. During her pregnancy Lucrezia entered a nunnery, and by the time she left the morals of the whole place had been loosened by her example.

Popes not only fathered children; they feathered their offspring's beds. Where nepotism was concerned, they seem to have known no shame. Pope Leo X (Giovanni de' Medici of Florence), a relatively cultured man with humanist tendencies, made his bastard cousin Giulio de' Medici a cardinal and then helped to engineer his ascent to the papacy as Clement VII. He was at least frank about his intentions: 'God has given us the papacy,' he wrote to his brother. 'Let us enjoy it.'[8] And they did, gambling, eating, drinking and fornicating with abandon.

Avaricious and ostentatious popes made money-raising their first concern, levying a charge for every spiritual service. The official posts of the Church were filled not by appointment but by purchase. Some positions were created solely because they could be sold, making the papacy a flabby institution of small-minded and uninterested bureaucrats – many of whom were nobles lacking any theological training. By 1520 there were two thousand offices in the papal service that had essentially no purpose other than, through their purchase, to swell the papal coffers by two and a half million gold florins.

Among the most lucrative of holy bestowments were the documents known as indulgences. In 1343 the papacy declared that Christ's death on the cross had created an excess of merit, a kind of credit account of holy grace that could be dispensed – for a fee, naturally – to all good Christians. True, what you got for your money hardly looked impressive: a scribbled scrap of paper handed out by some itinerant cleric like the medicines of a quack doctor. But this was a credit note that could be redeemed in the next life. A papal indulgence reduced the holder's time in Purgatory, where they bided time in mild misery before admission to heaven, and since the average sinner could expect a typical wait of a thousand or more years, there was great demand for these papal dispensations. According to Thomas Gascoigne, chancellor of Oxford in 1450, some vendors neglected the financial motives of their trade by accepting payment in kind, giving out their letters of holy pardon 'for the hire of a harlot, or for carnal love'.[9] Yet on the whole it was a simple business transaction, in line with the view of Christian piety that was widespread among the common people and encouraged by the clergy: salvation was just a matter of good book-keeping to keep your spiritual account in credit.

If the papacy was marked by corruption and wantonness, the foot soldiers of God were afflicted with apathy, sloth and ignorance. The ordinary clergy were poorly trained and ill motivated. Monasteries were full of men who felt little religious vocation but rather saw a monk's life as (marginally) preferable to the drudgery and uncertainty that oppressed the laity. Since a monastery's well-being depended largely on donations, some tried to win favours with bribes or by being lenient in the penances it prescribed, or through shameless advertising of the healing powers of its shrines. Because senior clergy would buy their way into several offices to maximize their money-raising opportunities, many positions were held *in absentia*, so that the lesser members of the Church lacked effective leadership.

All this meant that monks were widely despised as lazy, immoral, gluttonous and foolish. François Rabelais, who spent some time as a Franciscan monk, was presumably speaking from experience when he describes a monk who drinks with the giant Gargantua as having 'a paved stomack as hollow as a But of malvoisie, or St Benedictus boot, and alwayes open like a lawyer's pouch'.[10] In his view, monks did no more than sponge off the guilty consciences of the people, which left them smelling of all the foulness they sucked up: 'they eat the ordure and excrement of the world, that is to say, the sins of the people, and like dung-chewers and excrementious eaters, they are cast into the privies and secessive places.'[11] The point was made in scatological style in a leaflet from 1551 titled *About the Origin of Monks; About the Origin of Antichrist*, which shows tonsured monks being defecated into a heap by scaly demons squatting atop a latrine. The abbot Trithemius of Sponheim complained of his own brethren that

> The whole day is spent in filthy talk; their whole time is given to play and gluttony . . . They neither fear nor love God; they have no thought of the life to come, preferring their fleshly lusts to the needs of the soul . . . The smoke of their filth ascends all around.[12]

At every level, then, the Church was in desperate need of reform and renewal by the early sixteenth century. And the religious unrest of these times was mingled with dissent about broader social issues and coloured by millenarian spiritual unease. Prophets of a new Christian vision began appearing across Europe: Luther's vision of reformation vied with those of men like Ulrich Zwingli in Zurich, Jean Calvin in Geneva and Martin Bucer in Strasbourg. And nothing could be further from the dour fury of Luther and the frenzied dreams of the Anabaptists than the measured, compassionate and warmly intelligent words of humanists such as Erasmus of Rotterdam. Where Luther wielded the clumsy broadsword of blunt invective, Erasmus waved the deft foil of satire. He gives the lie to the claim, often made in defence of both Luther and Paracelsus, that their violence of language was simply typical of their time. 'Whatever is ingenious, scholarly and wisely written', a later proverb went, 'is termed "erasmic", that is, unerring and perfect.'[13]

Desiderius Erasmus (Gerhard Gerhards) came from a tradition of religious integrity that stood in stark contrast to the excesses of Rome. Born the illegitimate son of a Dutch priest in Rotterdam in 1466, he was educated at the famous theological school in Deventer, run by the Brethren of the Common Lot. The Brethren rejected the dry and pedantic scholasticism of much academic theology, preferring instead to seek a humble spiritual union with God, as advocated in the words of the movement's founder Thomas à Kempis: 'Lofty words do not make a man just or holy; but a good life makes him dear to God.'[14]

Erasmus became Latin Secretary to the bishop of Cambrai, then a teacher in Paris and England, and subsequently a professor of divinity and Greek at Cambridge from 1510 to 1513. Here he met and befriended the English humanist scholars Thomas More and John Colet. In 1503 Erasmus published *The Handbook of the Christian Knight*, which extolled a Luther-like message of faith before reason. But his *Praise of Folly* (1511) addressed the abuses of the Church head on – albeit with Erasmus' characteristic wit. Churchmen, he said, are 'satraps, caring nothing for religion, its blessings and its ceremonies and holding it cowardly and shameful for a bishop to die otherwise than in battle. Their clergy valiantly follow their example and fight like warriors for their tithes.'[15]

He combined acute wit with profound scholarship, and if this meant that his words flew over the head of the common man (who could not read Latin anyway), even popes were amused by the jokes (at least until the reactionary Paul IV put all Erasmus'

Erasmus of Rotterdam (*c.* 1466–1536), painted by Hans Holbein in 1523.

books on the Index). In *Iulius exclusis*, Pope Julius II argues with St Peter for admission through the pearly gates. He is forced to admit that he has shown favouritism to one of his sons.

'What?' St Peter exclaims. 'Popes with wives and children?'

'No, not wives', says Julius, 'but why not children?'[16]

Erasmus and the humanists thus established a climate of criticism of the failings of the Christian Church, which could not but have been to Luther's advantage when he mounted his own more forceful assault. Some said 'Erasmus laid the egg and Luther hatched it.' But Erasmus did not see it like that at all. 'I laid a hen's egg,' he admitted, 'but what Luther hatched was a bird of a quite different sort.'[17]

BY FAITH ALONE

That hatching was an agony. Once he had devoted himself to God, young Martin Luther found himself in a state of perpetual and mortal terror. How poorly he felt capable of serving God, although his conduct was nothing if not devout. His efforts to atone for this perceived unworthiness verged on the psychotic. 'If I had kept on any longer,' he admitted, 'I should have killed myself with vigils, prayers, readings and other works.'[18] As it was, he suffered from morbid depression and attacks of acute anxiety – to which was added the misery of constipation, which plagued him all his life.* It was small wonder, then, when he confessed that in his depths 'I did not love, rather I hated this just God who punished sinners.'[19]

Luther might have ended as another mad monk mumbling red-eyed in the cloister had he not, some time between 1513 and 1515, come across a very different view of God from the vengeful Master of his imagination. In Psalm 22 God was revealed instead as Lord of Compassion. He sacrificed His Son on the cross, Luther decided, not out of ire or caprice but so that He might, through Christ's blameless suffering, become reconciled to the imperfect world. 'God hides His goodness in severity, . . . His mercy in anger,'[21] Luther concluded. Moreover, he discovered in St Paul's Epistle to the Romans that the harshly judgemental God described in Latin as *justitia dei* could be interpreted otherwise in the original Greek: *justitia* did not necessarily imply judgement, but the act of putting things right – 'justification'.

From this prescription Luther derived his vision of salvation. The just, he concluded, shall live by faith alone: '*Justus autem ex fide vivit*', as St Paul said. Far from being doomed to struggle for grace by endless acts of penance and self-denial, mankind could reach heaven by a sincere and self-effacing belief. There was no formula that guaranteed redemption; all depended on faith. With this realization, said Luther, 'I felt myself to be reborn and to have gone through open doors into Paradise.'[22]

* It is perhaps for this reason that defecation looms large in Luther's theology. The devil, he once said, threw shit at him, and he threw it back. Then Satan departed, leaving behind a great stool 'which left a foul stench in the chamber for several days afterwards'.[20] The devil would often appear to show Luther his backside; Luther threatened to shit in his face, or better still, in his trousers, which he would then hang around Satan's neck. Wishful thinking, perhaps. Luther even admitted that the revelation of the true meaning of God's justice came to him on the privy, where he presumably had time aplenty to contemplate theological matters.

At this time Luther was a professor of theology at the newly inaugurated (1502) university in Wittenberg, and he had no wish to upset the Church. But in the next few years his misgivings deepened. What disturbed Luther was not so much the academic technicalities of doctrine – the question of what was or was not endorsed by the Bible – and he could probably have turned a blind eye, in remote Saxony, to the decadence of Rome (which he'd seen at first hand on an errand for his order in 1510). But if the preaching of the priests was in error, then their flocks stood in danger of eternal damnation. Having struggled under the shadow of that awful prospect, Luther could not stand by to watch it engulf other souls.

In particular, if salvation was a matter of faith, then the people were being desperately misled by the idea that they could buy God off with 'good works': going to church and to confession, paying their tithes and their priests, and dutifully mumbling pious formulas. Matters came to a head with the arrival near Wittenberg of the Dominican friar Johann Tetzel, bearing papal indulgences for sale like a mountebank peddling miracle cures. These were no ordinary indulgences, but were charged by the Church with immense redemptive power. Not only would they secure forgiveness for all the buyer's own sins, but they would release from Purgatory the soul of a deceased beloved friend or relative. Any sin is excused, Tetzel promised. Even incest. Even violation of the Holy Virgin. You will even be granted pardon for sins not yet committed.

In fact the certificates were a way for Pope Leo X to raise money for rebuilding the basilica of St Peter in Rome. But when Pope Leo entrusted their sale to the Archbishop of Mainz, who was the Elector Albrecht of Brandenburg, he granted Albrecht a share of the proceeds – for Albrecht had bought his bishopric with a loan of 30,000 gold florins from the Fuggers, and was now struggling to repay the debt.

So when Luther, now priest of the Castle Church at Wittenberg, sent a complaint to the Archbishop about these credit notes for grace that the townsfolk presented to him, he could hardly have taken a more disastrous action. And he compounded the matter by accompanying the letter with a copy of his *Ninety-Five Theses*, a formal declaration of his arguments against indulgences. This is the document that Luther is said to have nailed to the church door in

Wittenberg on 31 October 1517. If this happened at all, it was not quite as dramatic a gesture as it sounds – this was not an uncommon way to distribute pamphlets and polemics, and the *Theses*, written in Latin, would not have been accessible to most of the lay townspeople. But the timing – on the eve of All Saints' Day – made the challenge auspicious, and the document was soon thereafter distributed in a German translation by a local printer. Before long, everyone knew of Luther's provocative message.

The rest is a familiar story. But it is critical to recognize that, once Albrecht informed Pope Leo of Luther's dissent, and this minor contention began to spiral into a dispute that threatened to open a rift in the Church, Luther could not have survived without the support of Frederick the Wise, the Elector of Saxony. The Pope regarded Frederick as the preferred successor to the ailing Maximilian I, and was keen to remain on good terms with the German prince. For his part, Frederick saw Albrecht of Brandenburg as his main rival and so he had good reason to endorse Luther's attacks. But his support was also motivated by nationalistic feelings of resentment at the tribute payments exacted by Rome. Luther's cause became caught up in the political machinations of the German princes, the Emperor, and the papacy – and the Wittenberg rebel was canny about exploiting these currents.

Following a formal disputation in 1519 in which Luther and his associate Andreas Karlstadt faced the formidable theologian Johann Eck in Leipzig, Luther's case was deemed heretical, and in 1520 Pope Leo issued the bull* of excommunication *Exsurge Domine*. When Luther got his hands on a copy, he burnt it publicly.

The affair culminated at the Diet of Worms in April 1521, to which Martin Luther was summoned to explain himself before the new Holy Roman Emperor Charles V, in front of the princes and representatives from all over the empire. Before Worms, Luther was merely an awkward dissenter. After it, he began to emerge as the leader of a new Christian Church, although it would be several more years before they became known as the Protestants. The offences with which Luther was charged could hardly have been more serious: it was said that:

* Papal decrees were called bulls after their official lead seal, called *bulla* in Latin.

> He utterly takes away obedience and authority, and writes nothing which does not have the effect of promoting sedition, discord, war, murder, robbery and arson, and which does not subserve the complete collapse of the Christian faith. He teaches a loose, self-willed kind of life, without any kind of law, utterly brutish.[23]

In other words, he wasn't merely a heretic but was a revolutionary. One cannot help but be impressed by Luther's famous response, framed in brave and defiant words that expose a minatory Rome under the spotlight of integrity:

> Unless I am convicted by Scripture and plain reason – I do not accept the authority of popes and councils, for they have contradicted each other – my conscience is captive to the Word of God. I cannot and will not recant anything, for to go against conscience is neither right nor safe. Here I stand. I cannot do otherwise. God help me. Amen.[24]

Luther was now beyond redemption. In the Edict of Worms, Charles V declared that thenceforth he was an outlaw: his writings were to be destroyed, and none should give food or shelter to him or his supporters. But Luther was gone from Worms before the edict was announced. Frederick arranged for a mock kidnap that spirited Luther into hiding at Wartburg Castle, near Eisenach in the dense Thuringian forest, the stronghold of the rebel Imperial knights Franz von Sickingen and Ulrich von Hutten.

The Imperial knights (*Reichsritter*) owed allegiance to the Emperor, but their restless, warlike nature tended to make them unwilling to bow to any authority. Many served as mercenaries when there was no Imperial war to fight, or else they turned to brigandage. Typically of a nationalistic and anti-papal persuasion, some of them saw Luther as a potential ally and his cause as an opportunity to take up arms. After sheltering Luther from the confrontation at Worms, Sickingen and Hutten launched an assault on the ecclesiastical principalities of the Rhineland in 1522 under the Lutheran banner. They were defeated at Landstuhl the following year, where Sickingen was mortally wounded. Hutten, once an elegant humanist poet, became a pitiful figure, a desperate fugitive from the Empire. Syphilitic and penniless, he wandered from Basle to Zurich, where he eventually died on an island in the lake west of the city, an old man of thirty-five.

Under Frederick's protection, Luther returned to Wittenberg in 1522. He was ill pleased with what he found there. During his absence, certain of his supporters had pursued his cause with excessive zeal, purging the churches of all images and unnecessary ritual, performing the mass partly in German, allowing priests to marry, and ignoring the pleas of Frederick (who never became a Protestant himself) to slow the speed of reform. Aware that his future depended on political support, Luther dismissed the more impetuous of his devotees, who had been incited to their excesses by the hot-headed Andreas Karlstadt. For Karlstadt, religious reform went hand in hand with radical social change, and he was impatient for both. Rejected by Luther, he became an agitator in the subsequent political unrest before joining Zwingli in Switzerland.

For Luther, it was now a matter of slowly spreading the new Word. To this end, he had translated the New Testament into German while in Wartburg Castle, and when his translation was published in September 1522 it sold out within a month. Luther was knowledgeable but never the most careful scholar; he relied on passion and fire rather than meticulous argument and patient persuasion. Fortunately for the progress of the Reformation, these latter qualities could be found among Luther's supporters, most prominently in Melanchthon, one of the key intellectual figures of the Renaissance. Melanchthon was a man of voracious intellectual appetite. As well as a theologian, he was well versed in the natural and occult sciences. He was a frequent correspondent of Erasmus and a gifted educator, and surrounded himself with learned men. By nature milder and more studious than Luther, Melanchthon helped to forge the Wittenberg message into a coherent doctrine that he hoped might prevent a schism in the Church. Luther recognized that this brilliant and moderate go-between was essential to his cause: 'I am rough, boisterous, stormy and altogether warlike, I am born to fight innumerable monsters and devils, to remove stems and stones, cut away thistles and thorns and clear away wild forests, but Master Philip comes along softly and gently with joy, according to the gifts which God has abundantly bestowed upon him.'[25]

Melanchthon's interest in medicine put him on Paracelsus' trail,

Philip Melanchthon (1497–1560), Luther's scholarly collaborator.

and he was not impressed with what he found. In 1531 he denounced 'medical empirics' who had no formal education but had merely got their remedies from barbers and apothecaries. This looks like a reference to Paracelsus, whose unorthodox works had at that time just been banned by the Nuremberg town council; and indeed Melanchthon's pupil Joachim Cureus tells how his master sometimes spoke harshly of Paracelsus.

PARACELSUS IN THE PULPIT

In the early stages of the Reformation, Luther would have seemed rather an attractive figure to Paracelsus: plain-spoken, unpretentious, opposed to the blind adherence to authority, and an underdog. But what would he have made of Luther's message?

One thing is certain: despite some claims to the contrary, Paracelsus never became a Protestant. It was not that he shrank from dispute and iconoclasm; rather, that his Christian beliefs were too idiosyncratic to be fitted into any particular system. Some have allied his religious convictions with the free-thinking 'spiritualists' like the Lutheran pastor Sebastian Franck, who are sometimes in turn (and rather unjustly) placed together with the anarchic Anabaptists. But in the end he would follow no one's creed but his own. 'Every fool', he said, 'praises his own club; he who stands on the Pope stands on a cushion; he who stands on Zwingli stands on emptiness; he who stands on Luther stands on a water-pipe.'[26]

Paracelsus repeatedly asserts his independence of religious thought, insisting that it is the duty of the true Christian to reject all schools, all leaders, all doctrines except the simple truth of the Bible. He had no time for the pomp and ceremony of the Church, which found no justification in Scripture:

> God requires from us our heart, and not ceremonials, for with these faith in Him perishes . . . If we seek God we must go forth, for in the Church we find Him not . . . Faith in God and in His only begotten Son Jesus Christ is enough for us. Our fasts, our masses, our vigils, and the like effect nothing for us.[27]

Rituals such as the blessing and scattering of holy water were superstitious impositions that degraded the glory of divine nature: 'God blessed the water to quench the thirst and to breed fish,' said Paracelsus, 'not to serve as a sprinkling against the devil.'[28] In the pugilistic tract *On Seven Points of Christian Idolatry* from his fiery Salzburg days (Chapter 7), he deplores senseless rites with a zeal that surpasses even Luther's. Church services are a sinful waste of time, and so are prayers and religious holidays. Fasting is idolatrous; almsgiving and pilgrimages should not be turned into institutionalized obligations; and the veneration of church bells, altars, cloisters and all the other paraphernalia of the Roman Church is the devil's work.

No wonder, then, that he initially courted the Lutherans, dedicating treatises to them in the 1520s and sending them friendly overtures. It may have been their rejection of these offerings – for they could barely understand a man whose religious views were so bizarre – that later led Paracelsus to denounce the reformers as vehemently as the opulent Roman Church. But in any event, Luther's notion that faith was all did not sit comfortably with Paracelsus' practical mission of healing. 'Faith without works is dead,'[29] he said.

Johannes Oporinus, Paracelsus' amanuensis in Basle (Chapter 10), confirms that by the late 1520s his master had little time for any formal creed:

> I never heard him pray or inquire after the Evangelical doctrine which was then practiced in our town. He not only despised our good preacher, but threatened that one day, as he had done to Hippocrates and Galen, he would set Luther's and the Pope's heads right. He also said that none so far who had written about the Holy Scriptures had grasped their right meaning.[30]

And indeed Paracelsus eventually swept them all aside:

> In the end, whether they be papists, Lutherans, baptists, Zwinglians, they are all of them ready to glory in themselves as alone possessing the Holy Spirit and alone justified in their construction of the Gospel: and each cries 'I am right, right is with *me*, *I* speak the word of God, Christ and His words are what *I* tell you: after *me*, all of you, it is *I* who bring you the Gospel. And yet it was just that which was the sin of the Pharisees . . . It is a sin against the Holy Ghost to say: the Pope, Luther, Zwingli etc. are the Word of God, or speak to us from Christ, or are they who represent Christ, are His prophets, are His apostles: he who holds and esteems their discourse as the Word of God sins against the Holy Ghost . . . Thou hearest not what Christ says, but only what they say.[31]

Or more pithily, and in one of his most inspired barbs, Luther and the Pope were 'two whores debating chastity'.[32]

Paracelsus' Christianity was a kind of mystical pantheism which would have been widely considered heretical. The Neoplatonic idea that God exists in all things, which can be traced back to the

Arabic scholar Salomon ibn Gabirol (Avicebron; 1020–70) and the nonconformist philosopher David of Dinant (died 1209), was popular with many of the radical sects of the fifteenth and sixteenth centuries, such as the Anabaptists and the Brethren of the Free Spirit. Paracelsus' theological works have long exerted an influence on Christian mystics, ranging from Jacob Boehme in the early seventeenth century to the German Romantics and *Naturphilosophes* Novalis, Goethe and Rudolf Steiner. Unfortunately, because his theory of the natural world is so intimately woven into his mystical religious views, many theosophical writers have found it necessary to ossify Paracelsus' 'scientific' framework into an absolute description of the world – without which their own theories collapse – instead of allowing it its proper place in the evolution of ideas.

For Paracelsus, one cannot help but suspect sometimes that it was the other way around: his beliefs about nature shaped his theology. And so we find him trying to tease out rather strange, literal interpretations of Scripture that soon lead to something redolent of heresy. On questions concerning the Trinity and the Conception, he is keen to establish an order of events, a chain of cause and effect, where others felt he should instead have been lowering his gaze in reverence.

How did God come to be 'three in one'? In the beginning, Paracelsus says, God was alone; there was 'no one with him'. 'And [He] remained so long alone until it pleased Him to wed and increase Himself', whereupon He 'caused Him to divide into three persons, into three kinds, into three beings, into three properties' – a kind of stepwise fission in which 'God has first become twofold.' This amoebic splitting was hardly concordant with traditional theology. In the same way God 'made from His person a woman': Mary, who is thus a goddess. And Jesus was 'born of two persons, namely from God and the goddess'.[33] There is some strange biology going on here, and it does not sound much like orthodox interpretations of the Holy Book.

The same materialistic attitude forced Paracelsus to confront one of the thorniest issues of Reformation Christianity: the nature of Christ's flesh, which underlay the doctrine of the Eucharist. It was this issue that estranged Luther both from Erasmus and from Zwingli and prevented the early Reformation from achieving any unity in the German lands. The authors of the Scriptures inadvertently ignited this dispute with vague wording that left subsequent theologians haggling over irreconcilable ambiguities. 'This is my body,' said Jesus at the feast of the Passover, 'this is my blood'.[34] What did he mean? The conventional interpretation was that the bread and wine of the Eucharist literally became the flesh and blood of Christ. But did this imply that one substance was transformed, as if by magic, into another? To Paracelsus, it sounded remarkably like alchemy.

To Zwingli, on the other hand, the Eucharist was merely symbolic: 'a sign and nothing more'. A rational humanist, he rejected this magical sleight of hand and felt that that the Church was bewitching the people into thinking the Mass 'is something strange and unusual'.[35] Transubstantiation, he maintained, was a fallacy. Luther, however, held the Eucharist in deep reverence as one of the only two genuine sacraments authenticated by the Bible, and he could not bring himself to relinquish this recurrent miracle. Instead he devised the somewhat clumsy doctrine of consubstantiation, which proposed that blood and wine, flesh and bread, could coexist in the same substance. The material elements of the Eucharist become sacred while remaining mundane. On this point, the obstinate Luther would not be moved.

To Paracelsus these disputes raised the question of what kind of flesh Christ inhabited in the first place. Was he a man like other men, descended from the flesh of Adam, which God had made from the *limbus* of the earth (page 266)? Paracelsus did not think this was so; rather, Christ received 'holy flesh' directly from God while in Mary's womb, absolving him of the original sin that tainted all the sons of Adam. This idea, derived from an old Gnostic belief, was endorsed in the early 1530s by the Anabaptist leader Melchior Hoffmann.

Man, however, was much more than just flesh. That was the least of a person's constitution. Paracelsus asserts that humans have both a 'visible' and an 'invisible' body, and it is the latter, he says, which interacts with the occult forces of nature, such as the influences of the stars, herbs and medicines. There are times, however, when he – never one to be troubled by consistency – implies that the human being is *three*fold, having an elemental (fleshy), sidereal (invisible) and spiritual aspect. The question for theologians was: which of these is eternal, and which not? The New Testament states that in the Last Judgement, man will be resurrected in body and soul. But Paracelsus tended to disregard the body, the raw meat of existence, and suggested that only the spiritual self would enter into Paradise. At least, that is what he usually said. But he also made the claim that human flesh may indeed be admitted to heaven, once it has been sanctified by regeneration after perishing in the ground – an alchemical viewpoint that invokes the characteristically Paracelsian idea of purification by putrefaction. Christians acquire this 'eternal flesh', Paracelsus said, through a spiritual rebirth in the sacrament of baptism. This flesh is then nourished by the Eucharist: holy food for a holy body. This hedging did not save him from later charges of heresy, but it does illustrate how his religious thinking merged seamlessly with his 'chemical philosophy' of the universe: a rational scheme (or so he hoped) of materials and their transformations.

THE REJECTION OF REASON

At first, the humanists mistook Martin Luther for one of their own: committed to a Christian creed that was rooted in reason and moderation, and opposed to the superstition and corruption infecting the Church. But Luther, as we have seen, distrusted reason, and by 1525 it was no longer possible for humanists to pretend otherwise. That was when Luther wrote *On the Bondage of the Will*, in which he argued that man has no free will and can be saved only by God's grace. And so there opened a rift between the Reformation and humanism, and humankind was not again to be allowed such dignity until the Enlightenment. To Erasmus, life had no meaning unless man could control his own destiny. What was the point of religion if men had no choice about their fate? No, he insisted, 'God indeed preserves the ship but the mariner conducts it into harbour.'[36]

But Luther could not be contradicted lightly. He turned against Erasmus with fury and venom, scorning his belief 'that all can be accomplished with civility and benevolence'.[37] It was folly, Luther said, to suppose that man is open to persuasion by reason; rather, he must be overawed by the exigencies of faith, or if necessary by force. The attacks prompted Erasmus to defend himself with words that were unusually blunt and heated:

> The whole world knows your nature, according to which you have guided your pen against no one more bitterly and, what is more odious, more maliciously, than against me . . . How do your scurrilous charges that I am an atheist help the argument? . . . Wish me any curse you will except your temper, unless the Lord change it for you.[38]

The Lord never did any such thing, and Luther ended his days an obstinate and spiteful old man, watching a bloody revolution unfold which he no longer had any power to direct even had he cared to. Lutherans began to renounce learning, books, and even reading and writing. In Calvin's Geneva there was no singing or dancing, life was joyless, and heretics were as likely to end their days at the stake as they were in Rome. Erasmus was dead by then, but his despairing words rang true: 'I see how much easier it is to start than to assuage a tumult.'[39]

7
Revolution under the Sign of the Shoe

Sedition in Salzburg

> I will not oppose a ruler who, even though he does not tolerate the Gospel, will smite and punish these peasants without offering to submit the case to judgement. For he is within his rights, since the peasants are not contending any longer for the Gospel, but have become faithless, perjured, disobedient, rebellious murderers, robbers and blasphemers, whom even heathen rulers have the right and power to punish; nay, it is their duty to punish them, for it is just for this purpose that they bear the sword, and are 'the ministers of God upon him that doeth evil'.
>
> Martin Luther, *Against the Robbing and Murdering Hordes of Peasants* (1525)

Paracelsus left Villach in the late summer of 1524 and travelled north to the city of Salzburg. He arrived there with a head full of alarming notions – all of them radical, whether they concerned religion, politics or medicine. He came as a doctor but behaved like a crazed preacher, spouting dangerous ideas from any makeshift pulpit he could climb. Theologically, he might be best described as reformist in spirit, Catholic by default, and wildly unorthodox in practice. It was not a happy mixture in the southern German territories in those turbulent days, when the radical sects known as Anabaptists were threatening a revolution that would overturn the social order.

Setting up a medical practice at number 11 in the narrow Pfeifergasse, where his friend Hans Rappl ran a public bathhouse, he also took to addressing unruly crowds in taverns like some charismatic evangelist. His head was filled with visions of apocalypse. Like the Anabaptists, he mixed his religion with explosive political polemic, decrying the injustices of the world and calling for social equality based on Christian principles:

> No good can happen to the poor with the rich being what they are. They are bound together as with a chain. Learn, you rich, to respect these chains. If you break your link, you will be cast aside.[1]

Paracelsus promoted the view that one should forsake worldly wealth and seek instead to live a devout and pious life before God:

> Because we know that our kingdom is not of this earth, we need no more than enough food and clothing, and we cannot take it with us [when we die] . . . We should enrich ourselves in works and virtues, which will follow us into the next world. Even if we have the favour of all princes and kings and are held in high esteem by them, if the Highest King does not approve, what are all the kings on earth worth? Therefore: do not seek to please men, seek to please God; do not be rich in earthly things but in heavenly things.[2]

All this sounds saintly and unremarkable enough, even if it was not likely to endear him to affluent clerics. But when Paracelsus pursued the social consequences of these Christian thoughts, as he did in *De honestis utrisque divitiis* (*On Virtue or Wealth*) (1533), a tract representative enough of his lifelong views, they took on the appearance of something more seditious:

> If you are one who does not accumulate riches but works for his needs, then you are happy and you will prosper. This is because you are not stealing. For theft occurs on account of riches, whenever one makes a living without working. Nor are you killing. For it is akin to killing when you procure the property of another or cause another to procure it for a mere wage . . . If you cut another's corn or use his hay, you are killing what is his. For you are eating and killing him on his field. It is no sin to kill what one has planted; it is a sin and murder when one has not grown it but cuts it down . . . you should feed yourself by the sweat of your brow.[3]

Yet although he sometimes quoted from the works of Anabaptists such as Michael Geismeyer and Thomas Müntzer, he was no anarchist wishing to see society go up in flames. Unlike those men, Paracelsus was essentially a pacifist and forswore persuasion by force. Neither did he wish to see an end to all hierarchy. In this respect he looked back rather than forward: back to an idealized vision of the medieval world where all had respected their place and had not sought to exploit or abuse their power. Time was when the common man was permitted to fish and hunt on common land, when trade did not produce indolent and offensively wealthy merchants, when peasants were not taxed to the edge of starvation. In that golden age, individuals recognized that they had a communal role to play, each according to his abilities, so that society arose from the cooperation of lords and peasants:

> To us on earth God has given gifts and virtues, which each may and should use in the service of others, not for himself. One should therefore consider how each gift is to be used with regard to one's neighbour, in order that the commandment of God be fulfilled . . . No one could say to the other, what do I care about you, I have my own gift for myself alone.[4]

To a man who believed so assertively in the hierarchy of nature, a hierarchical society seemed both inevitable and desirable: 'there is no monarchia without its master and servant'[5], for after all, even Christ needed his apostles. To the master fall the responsibilities of the more capable mind:

> But although they may have servants, the masters are not idle . . . each master works by ensuring that his servant does his job properly, forgets nothing, and leaves nothing incomplete . . . The worries of interest payments and securities are the work of the master . . . And as can be observed in any craft, masters have special gifts which servants do not possess . . . Therefore he should leave the common work to his servant and work on the subtle tasks.[6]

In *De ordine doni* Paracelsus sketches a benign arcadia in which there is neither subjection of the poor nor profiteering by the rich, and where those afflicted by bad harvests are succoured by donations of food from more fortunate regions. 'No one should think otherwise than: if my neighbour had more than I, a good order would result.' The role of the authorities was essential to maintain social order: they 'should help to establish this neighbourly love and ensure that everyone does not do as they like'.[7]

But he admits that things no longer work this way:

> Self-interest makes mendacious people out of you noblemen and authorities . . . In order to cover your needs you must buy corn, meat and salt for money and so you raise taxes . . . Buying increases, prices rise, and cereals and bread are dearer . . . How much milder and better you would govern if you were rid of money. For wherever there is money, there is worry. That is what causes murders; bread and meat do not cause them. You become profiteers, gamblers, debauchees, and whore from money, and your money does no one any good. You do not invest it in the country but give it to the whores and the Pharisees, and there is only bad feeling, dishonour, mischief, and evil, wealth, dancing, tournaments, and banqueting.[8]

Paracelsus' vitriol is, however, directed mostly not at the lords and princes but at grasping merchants and money-lenders. 'Lending money and earning interest destroys commonality,' he charges – it causes inflation and is the devil's own work. As for the merchants:

> there can be no greater deception of one's neighbour than the doings of businessmen. Their whole life is devilish. They seduce rich and poor . . . They steal the land and labour from princes and lords with cunning and polite deception. They seduce all estates and want to be the best, they are the greatest favourites of the princes and are held in high regard. They mix with the nobility and the princes only to swindle them hugely . . . What is their luxury, what is their belief, what is their whole Christian life? Nothing, for it is all deception and a devilish life . . . They are whores and rogues, thieves and scoundrels . . . Where does their wealth come from? From lies, deceptions, and unjust takings.[9]

It was the businessmen who were fast eroding the stable structures of the Middle Ages and giving rise to a capitalist society in which inequalities did not have the support of tradition. Paracelsus did not like this new world order.

How much of this he vented in the Salzburg taverns, we cannot know. In 1524, the year of an auspicious astronomical Great Conjunction of planets, it is likely that his preachings would have been occupied more with eschatology than with politics. In happier times he might have been dismissed as a harmless ranter. But at that moment Salzburg was a pot about to boil, and it was no time to mount the tavern bench and hold forth.

THE RADICAL REFORMATION

Three iron cages hang from the Gothic tower of St Lambert's church in Münster. Their function is now long obsolete, but citizens of Münster in the late sixteenth century knew what they signified: a warning not to challenge the authority of lords and priests.

In 1535 these man-sized metal baskets contained three corpses: the mutilated bodies of Anabaptist rebel leaders who for a heady year established one of the most bizarre sovereignties Germany has ever seen. The most notorious of them was Jan Bockelson of Leiden, a Dutch tailor and leader of a group of revolutionaries that seized control of the city. Defeated by the besieging forces of local princes,* the self-styled King Jan of Münster was bound in front of the cathedral, had his tongue ripped from his mouth, and was tortured to death before being placed in his iron cage and left to rot. Several thousand of his followers in Münster were rooted out and slaughtered mercilessly, like so many rats polluting the city.

Germany in the early sixteenth century was a bitter ferment, agitated by disgruntled groups of radical Anabaptists. By denying the validity of the medieval social hierarchy and the authority of the Church, these sects seemed to threaten revolt and destruction of the social order, and they were often persecuted mercilessly.

The social unrest associated with millenarianism – the antici-

* On the day the defence of Münster collapsed, one of those said to have been present was none other than Georg Sabellicus Faustus, up to some mysterious business of his own.

pation of the Last Days and the Final Judgement – was not merely the result of the millennium's ominous passage past its half-way point. There were much more concrete, mundane reasons why many common folk were dissatisfied with their lot. Population began to rise at the beginning of the sixteenth century for the first time since the Black Death had arrived in Europe. Cities were becoming overcrowded and people feared famine. It came sure enough: there were major crop failures in Germany every ten years or so. Tensions between peasants and landlords were endemic, and exacerbated by the enclosure of common land and the restriction of hunting rights. As the scale and frequency of war increased, lords and princes had to impose greater taxation to finance it, and the populace, if not ruined or killed by the armies of either side, had nothing to show from these games of power except a lighter purse and an emptier stomach. The Church added to these financial burdens by demanding tithes and payment for every priestly service. Sebastian Franck in Nuremberg wrote of oppressive tithes, death duties, forced labour, tributes, interest payments and other 'serious grievances' as causes of peasant dissent in the 1520s. 'Nobleman, may a cow shit on you' ran one of the peasants' blunt slogans.

The peasants and oppressed people lacked any real forum for voicing their complaints, and often saw no alternative but to seek recourse in violence. There were several peasant revolts in Europe in the fourteenth and fifteenth centuries. Food riots were frequent, sometimes triggered by the suspicion (which was not always unfounded) that merchants were withholding grain to push up the price. Society might have existed in a state of permanent revolution if the poor had been better organized and had not had it ground into them from birth that they must accept the fate that God had given them, however grim it might be. Many chose instead to drown their sorrows in beer and let out their frustration and fury in tavern brawls. To the increasing number of urban dwellers, rural folk and peasants were animals whose savagery was feared and could be tamed only with threats or violence. 'A peasant', according to a popular German saying, 'is just like an ox, only he has no horns.' The radicals of the Reformation, however, were considered a more dangerous kind of beast.

The Anabaptists were a loose grouping of radicals who were

neither Catholics nor Lutherans but had abandoned the orthodox Church entirely, for a variety of reasons. Luther's differences with Rome hinged on issues of academic theology; the Anabaptists did not necessarily recognize the authority of the Bible at all. Some, indeed, regarded it as the work of the devil. They were generically of that brand of mystic which believes God may be found through direct revelation; some claimed that man himself could become God, and that nothing was forbidden.

The term 'anabaptism' refers to the belief that the baptism of children is invalid and meaningless: only as an adult is one truly ready to receive the grace of God. And so Anabaptists were *re*baptized as adults in the sight of the Lord. Beyond that, the Anabaptists spanned a broad spectrum, their ideas often shaped by those of charismatic leaders like Jan Bockelson. Some were pacifists: the Schleitheim Confession of 1527, probably the work of the Anabaptist leader Michael Sattler, spoke of 'the unchristian, devilish weapons of force'.[10] Most were strict puritans, advocating a humble life of abstinence. Some were fanatics who thought that Christ would return to the world only once all non-believers were dead. Many were heretical Unitarians, denying the Trinity – it was largely his anti-trinitarian views that brought the Spanish physician Michael Servetus to the stake in Calvin's Geneva in 1553. The Bohemian Brethren rejected the pope, elected their own bishops (who could marry), and denied the doctrines of transubstantiation and confession and the existence of saints.

But even as they denounced both orthodox and reformed Christian faith, Anabaptists could be as intolerant, exclusive, dogmatic and judgemental as any Roman priest or strict Calvinist. The world, they had decided, was full of sin, and the only proper response was to turn one's back on it. What really ensured the perpetual persecution of the Anabaptists, however, was not their religious doctrines but their political views. In today's terms, they might be crudely regarded as communists and anarchists. They renounced private property, maintaining that all goods should be shared. They refused to pay taxes or to serve in armies, and would not swear oaths of allegiance. Many looked forward to the overthrow of the Church and the civic establishment, which would be replaced by the kingdom of God. To the rich and powerful, this was the worst kind of heresy. It was unacceptable enough to deny the Word, but another matter to refuse the levies and obedience due to one's betters.

And so the scattered and disparate groups of Anabaptists became as downtrodden and abused as the Jews. In the Netherlands and in remote Bohemia some were able to live relatively unmolested so long as they avoided confrontation: the Frieslander Menno Simons created a community that survives today as the Mennonites. But elsewhere, and particularly in Catholic states, they were executed in their thousands or banished. Even in remote Moravia, Jakob Hutter was not safe when he set up his communitarian society with a small band of several hundred in the 1520s: in Innsbruck in 1536 he met a death every bit as barbaric as Jan of Leiden's, being dipped in freezing water, lacerated and rubbed in brandy which was then ignited. Lutherans and other reformers were more tolerant, but they had their limits. Martin Bucer's Strasbourg became a haven for Anabaptists and visionaries, and by 1533 they constituted perhaps a tenth of the city's population. Melchior Hoffmann proclaimed that Strasbourg was about to become the New Jerusalem, from which Anabaptism would spread across the world. This was too much for Bucer and his colleague Wolfgang Capito, who felt compelled to expel the disruptive elements.

THE POT BOILS

The revolt that became known as the Peasants' War began in 1524. Trouble had been widely anticipated that year because of the Great Conjunction (see page 293) – although after several years of unbroken famine in Germany, one did not need to be an astrologer to see that a storm was brewing. It started in the town of Stühlingen, near Schaffhausen in Switzerland, where a group of malcontents associated with the Zwinglian preacher Balthasar Hübmaier issued a set of demands called the Sixty-Two Articles, calling for the radical restructuring of society and the revocation of exorbitant rents and duties imposed by landlords. The movement spread through southern Germany, becoming more radical in 1525 when the Swabian pastor Christoph Schappeler and a tanner named Sebastian Lotzer in Memmingen drew up twelve Articles demanding hunting, fishing and wood-gathering rights for the poor, fairer tithes and the abolition of death duties, freedom of congregations to elect their own pastor, and, most outrageously, that lordships be conferred only through consultation with the peasants.

All this was vexatious enough to the nobility; but some dissenters went further still. While Luther's arguments were strictly with the Roman Church, some of his supporters regarded these theological disputes as just one aspect of a broader social rebellion. In Basle, Luther's expelled associate Andreas Karlstadt allied himself with the Anabaptist Thomas Müntzer, another ex-Lutheran, who was influenced by the apocalyptic preachings of the twelfth-century Cistercian abbot Joachim of Fiore. After his extremist views earned him Luther's condemnation in 1521, Müntzer denounced the Wittenberg friar as a vain and pampered academic who toadied to princes. He continued to preach revolt at Allstedt in Saxony until even the tolerant Frederick the Wise had no choice but to exile him. Müntzer was a millenarian, convinced that his actions would hasten the advent of a new kingdom of God. He cast himself as the prophet of the new age, a latter-day Elijah.

In southern Germany, peasants flocked to join Müntzer's revolutionaries under a banner depicting a *Bundschuh*, a peasant's shoe. 'The people want to be free,' said Müntzer, 'and God wants to be their only master.'[11] He considered the very principle of lordship to be anti-Christian, and committed himself to an anarchic position that called for the massacre of the ruling classes. Likewise, in the Tyrol the rebel leader Michael Geismeyer, former secretary to the Bishop of Brixen, called for all 'godless men who persecute the common man' to be executed. In the spring of 1525, under the leadership of Müntzer, the peasants formed a violent mob – 'army' would be too generous a word – that began to terrorize, rob and depose the rich. Castles were ransacked, towns plundered, all without any real plan or authority or political agenda beyond causing chaos and destruction. This was a little too much for Karlstadt, who, fearing the inevitable retribution, extricated himself from Müntzer and fled.

It is understandable that Luther and Zwingli should be concerned that their case for religious reformation would be undermined by association with the social dissent being spread by Müntzer and Geismeyer. But that does not fully explain the vehemence of Luther's response to the rebels. He was not wholly unsympathetic to the hard lot of the poor, but he never particularly concerned himself with social change. In part this was because his gaze was fixed not on this world but on the next; but he also wished to see the status quo maintained because he considered that men needed to be governed by a strong authority. 'An earthly kingdom', he said, 'cannot exist without inequality of persons.'[12] His respect for medieval hierarchy had a pragmatic element: if Luther wanted to overthrow the clergy, he needed allies, and Luther sought them in sympathetic rulers. After all, if it had not been for the support of rebel Imperial knights and of Frederick, he would surely have gone to the stake.

So Luther denounced the Peasants' Revolt not because, like Zwingli and Erasmus, he deplored the violence and vulgarity of their methods but to avoid alienating the princes. In 1525 he published a pamphlet the very title of which – *Against the Robbing and Murdering Hordes of Peasants* – left no doubt of his position. These revolutionaries, Luther wrote, have broken their vow to be 'faithful, submissive and obedient' to their rulers and have thereby forfeited body and soul, 'as faithless, perjured, lying, disobedient knaves and scoundrels are wont to do'. What is more, they claim to find justification for their actions in the Bible, and thus 'become the greatest of blasphemers of God and slanderers of his holy Name'. In consequence, Luther was prepared to encourage the princes to any extreme of retaliation, and his words were shocking even to those who were called upon to discharge them:

> Any man against whom it can be proved that he is a maker of sedition is outside the law of God and Empire, so that the first who can slay him is doing right and well. For as when a fire starts, the first to put it out is the best man. For rebellion is not a simple murder, but is like a great fire, which attacks and lays waste a whole land . . . Therefore let anyone who can, smite, slay and stab, secretly or openly, remembering that nothing can be more poisonous, hurtful or devilish than a rebel. It is just as when one must kill a mad dog; if you do not strike him, he will strike you, and a whole land with you . . . Here, then, there is no time for sleeping; no place for patience or mercy. It is the time of the sword, not the day of grace. The rulers, then, should go on unconcerned, and with a good conscience lay about them as long as their hearts still beat.[13]

Which is just what they did. To those who viewed Luther's own cause as consonant with that of the Peasants' War, this condemnation of the rebels sounded like treachery. But to those nobles sympathetic to the Protestant cause, Luther gave assurance that God would endorse any degree of violence against the revolutionaries.

Faced with the well-trained and well-armed forces of the lords and barons, the peasant rebels never stood a chance. The revolt was crushed swiftly and brutally. In May of 1525 Müntzer and the rebels were defeated in battle at Frankenhausen by the army of Philip of Hesse, the leading defender of Lutheranism after Frederick of Saxony died that year. Tens of thousands of peasants were killed, and their leaders, including Müntzer, were captured and tortured to death, some of them tied to horses and torn apart. By late 1525 the Peasants' War was over, and the poor and dispossessed were no better off for it: it had only made clear to the princes and lords how important and effective it could be to stamp quickly and firmly on any sign of dissent. This meant it went badly not just for Anabaptists but for anyone who would not readily align themselves with either the Roman Church or Luther.

A PREACHER IN SALZBURG

When Paracelsus came to Salzburg, the city was brewing a peasants' revolt of its own. Its focus was a churchman who ironically might have drawn Paracelsus to the city. He was Cardinal Matthäus Lang, who as both a humanist sympathizer and an 'adoptive' Carinthian (he was born in Augsburg) might have given Paracelsus reason to anticipate a favourable reception.

But the Cardinal was also one of those acquisitive and decadent lords against whom Paracelsus would later inveigh. He cultivated ties with the Fugger family, enjoyed influence in the Church and in Imperial politics, and had an appetite for luxury that was bankrolled by taxes imposed on the people of Salzburg. One tax in the 1520s was levied to cover his expenses for a lavish appearance at the coronation of Charles V in 1519. This provoked much complaint, especially as the nobility were sometimes exempted from duties that the towns and markets had to pay. But in 1523 Lang marched with a mercenary force into the centre of Salzburg and compelled the burghers to cede all right of protest.

In early 1525 Lang dismissed any illusion of tolerance by imprisoning a preacher from the Tyrol on charges of heresy. In response, some of the local people rebelled and freed the preacher from captivity, leading Lang to order the arrest and execution of one of their leaders. The fuse was lit. A mob of peasants, miners and townspeople clasping scythes and pitchforks took over the city and blockaded Lang and his troops and nobles in the towering Hohensalzburg castle, impregnable on its pillar of rock. Throughout the summer the rebels ruled in Salzburg, turning back a relief force at Schmalding.

But by then, Paracelsus had already been compelled to leave the city in haste. Exactly what happened is not clear, but it seems that his outspoken views had eventually been deemed too disruptive. In any event, he was arrested on suspicion of fomenting insurrection, and was probably lucky to escape the noose. It has been claimed that he offered his services as a physician to the rebel peasants and that after the rebellion was suppressed he was accused of having encouraged it. According to this story, Paracelsus' sentence was commuted from execution to exile only because of his status as a doctor and because he did not bear arms in the revolt. But that seems untenable, given that the rebels were still in command of the city when he left. Rather, it appears likely that Paracelsus succeeded in taking a position sufficiently eccentric to alienate himself from all parties.

Paracelsus spent much of his time in Salzburg writing religious polemics in pamphlets that he distributed to the townspeople. Although at least one of them was dedicated to Luther, Melanchthon and their colleague Johannes Bugenhagen, their contents would have elicited little interest in Wittenberg, being concerned mostly with obscure theological questions about the Virgin and the saints. What got him into real trouble, however, was not his religious views but his reaction when the authorities took issue with them. It had not been his intention to say anything that ran contrary to the established position of the Church – he was indeed intent on defending the divinity of the mother of Christ against suggestions that she was mortal. Some churchmen, however, viewed his statements about the Holy Trinity as heretical. At first he countered these accusations mildly, but in such a convoluted and confused way that he only made matters worse. When the criticism increased, he lost his temper and began to rave against the Church's authority and practices in provocative language. Here, as later, we can see Paracelsus responding to the threat of calamity by hastening it upon himself.

His tract *On the Seven Points of Christian Idolatry*, written in late 1524 or early 1525, reveals the state to which matters degenerated. Now, it seems, he is being not only contradicted but ridiculed – and that is something he always found intolerable. He paints a vivid picture of how he had spread his message, and in doing so he gave ammunition for later accusations of inebriation:

> Your daily cavilling and inciting against me on account of the truth which I have spoken, occasionally and several times, in taverns, drinking places, and inns, against the senseless attendance of church, luxurious ceremonies, fruitless praying and fasting, alms giving, making of sacrifices, tithing, . . . confessing, taking of the sacraments, and all other priestly commandments and [priestly self-]support; moreover, [your having alleged] against me, this was in drunkenness, because it happened in taverns, and taverns prove to be unseemly places for truth, and because of which [you] call me a mountebank preacher.[14]

If he really said these things in public, he was undermining more or less everything the Church represented. It amounted to a rejection of all ecclesiastical authority.

But Paracelsus seems outraged as much by criticism of his methods as of his message:

> Why do you do this to me now at this time, inasmuch as you were silent and well pleased when I spoke in the taverns [saying that] people should make sacrifices to you and follow you and not say anything against you? If that was proper in the taverns and a service for you, then accept now the truth that is spoken in the taverns. For in the taverns I was faithful in you, but now I am faithful in Christ and no longer in you.[15]

Even during the rebellion, such an attack on the Church could not be made with impunity. The Salzburg rebels, after all, were not led by Müntzer – they did not question the Cardinal's religious authority, but demanded only social and political reform. When he was forced to flee, Paracelsus had no time even to pack his clothing and possessions; he left some medicines, some fine clothes, a compass and a portrait of his father with his host Wolfgang Büchel. He took to the road with barely an overcoat to his name.

The Salzburg rebellion outlasted Thomas Müntzer's revolt, but not by long. Its suppression was equally brutal, and Cardinal Lang was soon once again in control of the city. The revolution feared by the lords and forecast in the stars by millenarian rebels was in the end no more than a brief cry of anger from the poor and oppressed, who in their thousands would continue to pay for it with their lives.

8
Transmutation at Ingolstadt

Making Gold

> For what, in the end, do we know about the causes and motives that prompted man, for more than a thousand years, to believe in that 'absurdity' the transmutation of metals?
>
> Carl Jung, 'Paracelsus as a Spiritual Phenomenon' (1942)

So up the broad valley of the Danube comes Paracelsus, while all around him southern Germany is on fire with revolution. It is not a good moment to be travelling, with the nobles fearful of rebellion and the peasants driven into a frenzy of chiliastic terror. A wise man would keep his head down; but for this one, controversy is as unshakeable as his shadow.

He presents himself merely as a humble doctor, but legends stick to him like dust. Riding north from Salzburg, he passes through Munich and reaches the Danube at Ingolstadt, where he takes lodging at an inn owned by an alderman. The man's 23-year-old daughter has been paralysed since birth, but the reputation of the travelling doctor has reached miraculous proportions, and he is brought to her bed. He prepares a cure with the colourful alchemical name Azoth of the Red Lion. This concoction could have been almost anything: Azoth indicates that it is an 'alchemical mercury', a universal medicine, and Red Lion implies a residue, the matter remaining at the bottom of a flask after volatile vapours have been sublimed away. Some metal salt, then, made potent in the fire.

Take a pinch of this remedy with a spoonful of wine after every meal, the doctor commands. It will induce copious sweating, but that is just a sign of healing and no cause for alarm. Yet the young woman has been taking the medicine for just a single day when she – who has previously been unable to take a step unaided – rises, walks into the chamber where her parents are consulting with the doctor and throws herself in tearful gratitude at Paracelsus' feet.

So they say. The man is a wonder-worker; stories like this attend his passing wherever he goes. In this fashion the itinerant physician makes his way west along the great artery of central Europe.

A few miles from Ingolstadt he stops at Neuburg, where the Duke of Bavaria owns a castle. Here a man named Hans Kilian keeps the duke's books and records, but the impecunious duke also employs Kilian as an alchemist, hoping that he will manufacture gold. We would hardly expect Paracelsus to pass through unnoticed, and it is not long before he and Kilian are labouring with bellows and retorts in the castle basement.

Gold-making – chrysopoeia – is not, in Paracelsus' alchemy, the goal that it was for Isaac Newton and Robert Boyle; but neither is he about to dismiss it, because after all a man has to live. Stories abound of how he could manufacture the precious metal. One of the more credible tales comes from his assistant Franz of Meissen, whom he engaged shortly after his stay at Neuburg. Franz explained the process in a letter published in 1586, in which he claimed it was his master's way of escaping poverty.

'Franz, we have no money,' the alchemist said one day, giving his assistant a guilder to buy a pound of mercury from the apothecary. Paracelsus poured the mercury into a crucible, which he placed on bricks over burning coals. Time passed, until he said to Franz, 'See, take hold of the mass with the pincers and cover the crucible. Put the fire out and let it stand.' Another half hour went by; alchemy cannot to be rushed. Then: 'We must now see what God has given us,' said Paracelsus. 'Take off the lid.'

'The fire was quite out,' Franz recalled, 'and in the crucible all was solid.' 'It looks yellow like gold,' he said. 'Yes,' Paracelsus replied, 'it will be gold.' And so it was, as the goldsmith testified to Franz: barely an ounce less than a pound, which the smith exchanged for a purse 'full of Rhenish guilders'.[1]

How was it done? There was, needless to say, a secret ingredient, kept by Paracelsus in a roll 'about the size of a big hazelnut done up with red sealing wax'. 'What was inside,' Franz confessed, 'I did not dare, being a young man, to ask, but I think that had I asked him, he would have told me, for he always showed me liking.'[2] What else could it be, but the Philosopher's Stone?

For every alchemist who has become a legend is said to have possessed, at some time in his life, this fabulous agent.

THE CHEMICAL PHILOSOPHY

In 1704, a Fellow of the Royal Society named John Harris passed a judgement on alchemy which pretty much reflected the view of scientists for the next two centuries. It is, he said, 'an art which begins with lying, is continued with toil and labour, and at last ends in beggary'.[3]

But a few later commentators were more perceptive about the scientific value of alchemy. The nineteenth-century German chemist Justus von Liebig wrote:

> the most lively imagination is not capable of devising a thought which could have acted more powerfully and constantly on the minds and faculties of men, than that very idea of the Philosopher's Stone. Without this idea, chemistry would not now stand in its present perfection . . . In order to know that the Philosopher's Stone did not really exist, it was indispensable that every substance accessible . . . should be observed and examined.[4]

That is to say, if you want men to leave no stone unturned in their enquiry into nature, say that there is gold to be found under one of them.

This is to assert that the means justify the end: that chemical knowledge benefited from a futile quest. Paracelsus' own contributions to alchemy had a profound impact on the development of the science of chemistry, but we should never forget that this was not his intention. Scientists today take the generally modest view that their efforts might, if they are diligent, smart and lucky enough, add a brick to the great temple that science is constructing. Not so Paracelsus. He wanted to explain everything, and to explain it in a way that left scant room for further improvement. Describing how Paracelsus' alchemy assisted the development of chemistry tells us little about what alchemy meant to the man himself.

For Paracelsus, all of nature was a form of alchemy. Humankind was a chemist's crucible, and the whole of creation likewise. For the central teleological pillar of Paracelsus' world was a belief in the cosmic design that made man a mirror of the universe: 'heaven is man, and man is heaven.'[5] 'In natural philosophy heaven and earth, air and water are a man, and man is a world with heaven, earth, air and water . . . We must understand, therefore, that when we administer medicine, we administer the whole world.'[6]

This dictum – 'as above, so below' – was familiar to the Hellenistic alchemists of Alexandria. They believed that the processes that occurred in living bodies (the microcosm) were reflected by those in the cosmos (the macrocosm), and vice versa. Thus, for example, just as it was thought that a 'seed' (whether in the ground or in the womb) must die before it could grow and bloom, so the alchemist had to 'kill' his materials in order to begin the process by which they grew into the perfection of gold. This notion can be traced to Aristotle, who claimed that things come into existence via the passing away of something else: corruption must precede generation. That was very much how Paracelsus saw it too.

It is common today to delight in the poetic metaphors of the Renaissance, as for example when Leonardo da Vinci speaks of geological water as the blood of the earth. But for Leonardo and his contemporaries, this was no metaphor; the correspondence between the greater and the lesser worlds was to be taken literally. As Leonardo said:

> Man has been called by the ancients a lesser world, and indeed the term is rightly applied, seeing that if man is compounded of earth, water, air and fire, this body of the earth is the same; and as man has within himself bones as a stay and framework for the flesh, so the world has the rocks which are the supports of the earth; as man has within him a pool of blood wherein the lungs as he breathes expand and contract, so the body of the earth has its ocean, which also rises and falls every six hours with the breathing of the world; as from the said pool of blood proceed the veins which spread their branches through the human body, in just the same manner the ocean fills the body of the earth with an infinite number of veins of water.[7]

The idea of macrocosm and microcosm makes sense only in a universe designed by a Creator. Fundamentally it is a way of saying that the universe exists *for us*. There is nothing that happens which does not relate in some way to what happens to us. The religious belief in a god who created the universe in order to put people in it, and who sits in judgement over their actions, stems from the same impulse. It is of course the impulse of the infant who cannot imagine any action or mind or event that is independent from itself. But it is also a way of rendering significance to human existence, and perhaps every culture since the beginning of the world has needed to do that.

And this belief in cosmic correspondences was by no means necessarily incongruent with science. For Isaac Newton, who had no doubts about the existence of a Creator, it suggested that the gravitational force that he perceived to be acting in the celestial sphere operated also in the human realm, drawing an apple to the ground. Indeed, he even speculated that this force governs the interactions between invisible atoms. Looked at this way, 'as above, so below' is not a magical formula but an expression of one of science's fundamental tenets: knowledge is generalizable, so that by looking in one place we can find out about phenomena in other places too.

For Paracelsus the principle of analogy implied by the correspondence of macro- and microcosm was the key to all knowledge, particularly in medicine and physiology. It meant that the processes one could see and study in the outer world (including those of the alchemical laboratory) were a sort of cipher that revealed the way the human body worked. God created man in his own image, but he also created the world in the image of man: in *Eleven Treatises*, one of Paracelsus' earliest medical books, written around 1520, he says, 'The one who put the winds and seas, sun, moon etc. in the heavens, he put it in the human being as well . . . For what is the entire world but a sign that it is of God and has been made by God?'[8] Moreover, God intended that man should recognize this handiwork and its relation to the human body, and therefore He left His signature everywhere. That is to say, the properties and virtues of things in the wide world can be discerned from their visible appearance – a clue to their value for the physician. Heart-shaped plants may be used to prepare medicines for the heart; those with leaves that look like the liver (or plants with yellow, 'jaundiced' colours) administer to that organ; orchids cure diseases of the testicles. 'Just as a carved image is a sign of its mason or carver, so too, with all other things, God has made these works and signs that they are his production.'[9]

> Behold the Satyrion root, is it not formed like the male privy parts? Accordingly, magic discovered it and revealed that it can restore a man's virility and passion . . . The *Siegwurz* [gladiolus] root is wrapped in an envelope like armour, and this is a magic sign showing that like armour it gives protection against weapons. And the *Syderica* [vervain] bears the image and form of a snake on each of its leaves, and thus, according to magic, it gives protection against any kind of poisoning . . . also the thistle – do not its leaves prick like needles? Hence there is no better remedy against internal stitches.[10]

By the same token, cures can be deduced from the corresponding processes in nature. In *Eleven Treatises* Paracelsus explains how dropsy is a swelling caused by an excess of water, 'like a man in a flood that overtakes him'.[11] Floods are remedied by the sun's heat, and so, likewise, dropsy should be treated with 'sun-like' medicines such as the yellow 'crocus' of Mars (iron oxide) and sulphur.

These are not mere guessing games set by God. There is a mechanistic principle at work: outward similarities signal a correspondence between inner virtues: sulphur has in it something of the sun. It is because of these innate properties that seemingly bizarre treatments were supposed to exert their effects. Thus a tortoise's paw cures gout in the feet, and the genitals of birds such as turtle doves and sparrows, plucked from the poor creatures 'at the time of rut' (and, one presumes, first cooked to tenderness), produce feelings of love when eaten. Thus the art of physiognomy: 'the shape of a man is formed in accordance with the manner of his heart.'[12]

Paracelsus' chemical philosophy, in which alchemy held the key to understanding man and nature, provided the compass points by which many natural scientists of the seventeenth century oriented themselves. It seems almost unjust that the whole scheme was simply wrong, since it is so attractive a description of the world that one is tempted to feel it *ought* to be right. If God had truly planned the Creation, He could not have devised a better scheme. The miner and the baker, the doctor and the painter, are, in this view, all engaged in the same activity: the great art of transformation. And this is no art of human devising, but simply a reflection of the natural art that makes a flower grow, that stores up metals in the earth, and brings wind and rain. By taking alchemy out of the smoky laboratory and setting it free in wild nature, Paracelsus stakes his claim to genius.

DUPES AND DECEIVERS

There has never been a time when alchemy was not tainted with infamy and associated with charlatans. The Roman Emperor Diocletian is said (perhaps apocryphally) to have outlawed it in the third century AD for fear that it was too efficacious – that the artificial manufacture of gold would undermine his currency – but for the most part, alchemists were more prone to the kind of dismissive mockery meted out by Chaucer in the *Canterbury Tales*. His party of pilgrims is joined by a Canon and his yeoman or assistant, who proceeds to tell his tale – to the Canon's great discomfort. The Canon is not at all lordly in his apparel; indeed, he is a shabby sight. Ah well, explains the yeoman, that is because he is an alchemist, who has laboured unsuccessfully these past seven years to achieve a 'projection' of base metals into gold. This has reduced the two of them to poverty:

> All that I ever had I've lost thereby,
> And so, God knows, have many more than I . . .
> That slippery science stripped me down so bare,
> That I'm worth nothing, here or anywhere.

His experiences have convinced the yeoman that all alchemy is but a useless chasing of shadows which reduces a man to a stinking tramp.

This pathetic figure is the archetype that Paracelsus, during his ragged wanderings, seemed to many of his contemporaries to fit. Familiar with such caricatures, townspeople would have immediately been on their guard against someone claiming to know the secrets of alchemy. But itinerant alchemists of this ilk might represent a yet more unwholesome breed. Chaucer's Canon's yeoman goes on to relate how he once worked for another alchemist – not a deluded puffer, but a calculating mountebank who tricked a priest into believing he could project mercury and copper into silver. This transmutation was apparently effected by some powder which was in truth 'not worth a fly', the recipe for which the Canon sold to the priest for forty pounds. The yeoman advises the party of pilgrims that they had best not waste their time with the Philosopher's Stone: 'my advice would be to let it go.'[13] Sound advice, but seldom heeded.

By the fourteenth century, not only were the alchemical arts open to charges of obscurantism, plagiarism and charlatanism but they were also of questionable legality. In 1317 Pope John XXII, warned of a plot to kill him using black magic, decided to take pre-emptive action against such infernal practices, and anything that smelled of the occult was tarred with the same brush. The pope issued the bull *Spondent pariter*, which forbade alchemy and accused the alchemists of deception. 'To such an extent', it said, 'does their damned and damnable temerity go that they stamp upon the base metal the characters of public money for believing eyes, and it is only in this way that they deceive the ignorant populace as to the alchemical fire of the furnace.'[14]

Those caught manufacturing alchemical gold, the bull decreed, would be compelled to give to the poor an amount of true gold equal to that which they had made. If they could not pay, they would be imprisoned. And if they minted coinage from their false gold, they would spend the rest of their life in the dungeon. In this climate, practising alchemy could be a dangerous business: the Italian alchemist Cecco d'Ascoli was burned at the stake by the Inquisition in Florence in 1327.

The Renaissance alchemist did not generally run that kind of risk, although there was still much mockery and distrust. The fifteenth-century Italian writer Francesco Filelfo wrote:

> Those who think that, by spoiling and corrupting copper, silver or gold can be made, seem to me stupid fools. But triflers of that type devise certain names of herbs and other such things, known to themselves alone and which produce nothing but wild fancies and frauds.[15]

Chaucer's mendacious alchemist is the prototype for Ben Jonson's wily Subtle, the swindler of *The Alchemist* (*c.*1610) whose confidence tricks are hidden behind a veil of arcane terminology. Yet Jonson, however mocking of alchemy's charlatanism, had clearly taken the trouble to verse himself well in its traditions: his parody is such a good one because it is so well informed. He puts his own thoughts on the lips of the perspicacious 'gamester' Surly, who proclaims, 'Rather than I'll be brayed, sir, I'll believe that Alchemy is a pretty kind of game, somewhat like tricks o' the cards, to cheat a man with charming.'[16] The character of Subtle fits both Simon Forman, a wily London quack of the late sixteenth century, and Edward Kelley, the trickster who persuaded the Elizabethan magus John Dee that he could converse with angels. But in Jonson's day there was no shortage of models for the stereotypical sham alchemist.

HERMETICA RESURRECTED

Yet although Paracelsus once lamented that 'alchemy has now fallen into contempt and is even considered a thing of the past',[17] in fact the advent of Renaissance humanism helped to revive it. In 1460 the wealthy banker and humanist Cosimo de' Medici, de facto ruler of the Florentine republic, engaged the scholar and priest Marsilio Ficino to translate into Latin the famed collection of some seventeen or so Greek works on the occult known as the *Corpus hermeticum* or Hermetica. They were a curious mixture, including works on alchemy, astrology, magic, medicine, botany, occultism, theology and philosophy. In the fifteenth century it was widely believed that these books represented ancient wisdom, although it seems that most of the Hermetica was actually written in the second and third centuries AD, and that the collection was assembled around the year 500.

Ficino (whom Paracelsus considered 'the best of the Italian physicians'[18]) finished the job in 1463, and his translation, published in Treviso in 1471, sent ripples through the burgeoning field of Renaissance magic. The *Corpus hermeticum* was the central text of the school of mystical philosophy now known as Neoplatonism. Conceived in Alexandria, Neoplatonism was, like the city itself, something of a melting pot for occult and spiritual ideas. Although rooted in Plato's philosophy, it drew on the works of Pythagoras and was influenced by Egyptian, Jewish and Persian and perhaps even Buddhist mysticism. Beyond the superficial chaos of mundane existence, the Neoplatonists insisted, there lies an insensible and in some sense divine web of forces and correspondences that accounts for it all.

The core tenets of Neoplatonism are generally ascribed to the Greek philosopher Plotinus (AD 204–70), born at Lykopolis in Egypt. Plotinus was not a Christian, but to humanist scholars he served a vital role in uniting classical Greek thought with Christian theology. Not only does he invoke an omnipotent supreme being, but he makes Him into a Trinity: the One, the Spirit (*nous*), and the Soul. Central to Plotinus' philosophy was the belief that we can know the One, the supreme God, by knowing ourselves – by searching our own soul. Such knowledge may come in the form of a revelation to those who are 'divinely possessed and inspired', whereupon they 'have at least the knowledge that they hold some greater thing within them, though they cannot tell what it is'.[19] Those who acquire this insight are overcome with ecstasy – literally, a feeling of standing outside oneself. Plotinus says he experienced this state himself many times.

To know the universe, then, you must look within. To this extent, Neoplatonism overlaps with Gnostic ideas developed in the second and third centuries AD. Gnosis is the direct, revealed knowledge of God, something that demands a leap of faith beyond the logic and rationalism of Aristotle. It has been called a 'Hellenization of Christianity': a version of the Word blended into the heady stew of Middle Eastern esotericism.

To Plotinus, the true magician does not use occult forces to contrive events, but is instead a collaborator, if not indeed a servant, of nature. Pico della Mirandola, Ficino's friend and protégé, echoed this view: 'As the farmer weds his elms to vines, even so does the *magus* wed earth to heaven, that is, he weds lower things to the endowments and powers of higher things.'[20]

Woven into this basically Graeco-Egyptian design was another strand of ancient wisdom that, in the eyes of advocates such as Pico and Ficino, had a strong link with the roots of Christianity: the Cabbala. The word means 'reception' in Hebrew, for this was knowledge that had for generations been passed on only by word of mouth, ensuring that it became known only to trustworthy adepts and was kept hidden from the undeserving. It was sacred wisdom that came directly from God. 'Moses on the mount', said Pico, 'received from God not only the Law, which he left to posterity written down in five books, but also a true and more occult explanation of the Law.'[21] The Commandments were divine edicts for proper moral conduct, intended more or less as a recipe for maintaining social order; but the other knowledge that God gave Moses – the Cabbala – provides genuine enlightenment because it revealed the how and the why of the world. Pico became convinced that Pythagoras and Plato were essentially echoing this ancient wisdom of the Hebrews, which was the word of God and therefore synonymous with the Christian faith.

The Cabbala played almost no part in Jewish religious life until the fourteenth century, when it emerged among Jewish scholars familiar with alchemy, who discerned many parallels between the two arts. Indeed, according to the Paracelsian writer Gerard Dorn in the late sixteenth century, Hermes Trismegistus, the mythical father of alchemy, 'was taught by the Genesis of the Hebrews'.[22] Agrippa was thoroughly familiar with the doctrines of the Cabbala; Paracelsus says that he learnt them in Constantinople. Johann Reuchlin, who knew Hebrew fluently, was considered an authority on Cabbalism in Pico's day.

Numerology lay at the heart of Cabbalism. One of its key techniques, called gematria, consisted of interpreting biblical and other words from the 'numerical' values of the letters (that is, the consonants, since Hebrew has no written vowels) they contained. The Hebrew alphabet has twenty-two letters, each of which is assigned a numerical value: for example, *qof* has the value 100, and *shin* is 300. By counting up these numbers, Cabbalists could find hidden messages in the Bible. Trithemius of Sponheim claimed that angels could be summoned and commanded using the Cabbalistic numbers associated with their names: 4,400 invokes the angel Samael, for instance.

Owing in part to the status awarded to Neoplatonic philosophy by Ficino's translation (he founded the Platonic Academy in Florence in the 1460s), some knowledge of alchemy was considered desirable by most educated men in the sixteenth century, among them artists such as Leonardo, Dürer, Cranach, Giorgione and Jan van Eyck. Luther spoke of alchemy's 'beauty' and delighted in its spiritual metaphors, while not blind to its practical benefits:

> The science of alchymy I like very well . . . I like it not only for the profits it brings in melting metals, . . . I like it also for

> the sake of the allegory and secret signification, which is exceedingly fine, touching the resurrection of the dead at the last day.[23]

Even Pope Leo X, coming from the Florentine house of the Medici humanists, had a strong interest in this and other occult arts.

THE THEORY OF GOLD-MAKING

What gave alchemists the idea that one metal could be transformed to another? Was this sorcery, or was it merely a manipulation of natural forces – a 'science', in other words? In ancient Egypt, chemistry was essentially a practical craft, albeit an important one. Artisans made glass, ceramic glazes, cosmetics, pigments, dyes, unguents and metals, and there was no real perception of a deeper meaning behind their manipulations of matter. The art of *khemeia* itself, however – which is what chemistry was called in Alexandrian Egypt, a word that Pliny links etymologically to the 'black soil' of the Nile delta – was deemed to come from the deity Thoth, physician to the gods, whom the Greeks identified with their own Hermes. Chemistry was the Hermetic art.

The practical, experimental tradition of Egypt contrasts with the strong theoretical bias of classical Greece, whose philosophers were eager to rationalize the transformations that they saw both in 'art' and in nature. The Greeks, beginning with the Milesian school of Thales (*c.*620–*c.*555 BC), aimed to reduce all the world's substances to just a handful of basic *elements*. For Thales there was but a single element – water – but his successors postulated a variety of elemental schemes. The most famous today is that promulgated by Empedocles (*c.*490–*c.*430 BC), a polymath from Acragas who seems to have been as much a wizard as a philosopher. His quartet of earth, air, fire and water was adopted a hundred years later by Plato and his most brilliant pupil, Aristotle.

It would have seemed entirely natural, even inevitable, that the Empedoclean (ultimately to be Aristotelian) elements could be transformed from one to another. Wasn't the drying up of a puddle a case of water being transformed to air, for instance? According to Aristotle, these interconversions were effected by altering the *qualities* that the elements possessed: hotness, coldness, dryness and wetness. Cold, wet water became cold, dry air by turning wetness into dryness. Thus the substances of the universe were fluid things, shape-changers that dallied for a while in this form or that. If solid earth could be transmuted to intangible air, it must surely be a comparatively minor matter to convert one dense, shiny, cold metal into another?

In Alexandria, practical chemistry collided with the mystical strands of Greek thought represented by Plato and Pythagoras, and from that union alchemy was born. One of the key texts in this tradition, much plagiarized by later alchemists, was *Physica et mystica* by the Egyptian sage Bolos of Mendes, who flourished around 200 BC. His book is credited with launching the sect of Neopythagoreanism, which held that occult forces govern man and nature. By the first century BC Bolos had become famous, the equal of Aristotle.

Despite its title, *Physica et mystica* is a relatively straightforward account of chemical practices. Bolos devotes an entire section to gold-making, reeling off almost casually a list of recipes for this most elusive of operations. For instance:

> Treat pyrites until they become incombustible, after having lost the black colour. Treat with brine, or uncorrupted brine, or with sea water, or oxymel [vinegar and honey], or what you will, and heat until they become like particles of gold which have not been submitted to the action of fire. This done, mix in native sulphur, or yellow alum, or Attic ochre, or what you will. Then add the silver in order to have gold, and gold to have the coral of gold.[24]

Perhaps the only real hint of *mystica* in Bolos' great work comes in the cryptic inscriptions that unite triplets of recipe-book passages: 'Nature triumphs over nature. Nature rejoices in nature. Nature dominates nature.' This fleeting reference shows that behind these simple instructions there is now a theory of how nature works.

When the Arabic scholars encountered Alexandrian alchemy, they lent it a rational aspect based on quantification and theory. The alchemical philosophy of the Arabs is much indebted to a writer of the first century AD known as Balinus, who was probably a Syrian named Apollonius of Tyana. Figures like Balinus throng among the shadows of alchemy's origins: he was a magus and wonder-worker, and was even apparently the inspiration for a pseudo-Christian cult in Rome in the third century. Balinus is nominally the author of a work called the *Secret of Creation*, although this was more probably written between AD 600 and 750 by Arabic authors who may or may not have had access to the genuine works of Balinus.

The *Secret of Creation* introduced the core text of medieval alchemy: the *Emerald Tablet*, which pseudo-Balinus attributes to Hermes Trismegistus (the 'thrice-great'). This Hermes was thought by some to be a great magician who lived at the time of Moses; but he merges beyond the edge of history with the deity Hermes/ Thoth of ancient Egypt, as much a god as a man. Some say his *Emerald Tablet* was discovered by Sarah, the wife of Abraham; others, by Alexander the Great. Translated into Latin around 1200, it became the subject of extensive commentary in the Middle Ages. The short text consists of a series of cryptic proclamations that allow for an infinite variety of interpretations: 'Whatever is below is like that which is above, and whatever is above is like that which is below, to accomplish the miracles of one thing.'[25]

'As above, so below': this Neoplatonic precept became a strong theme in Arabic alchemy. It was one of the central principles of the Isma'iliya sect, which became powerful in the Muslim world in the tenth to the twelfth centuries. The Isma'iliya believed that the true leaders of Islam were descended from Ali, the son-in-law of Mohammad. Although later Muslim tradition apocryphally ascribed the inception of Islamic alchemy to a seventh-century prince named Khalid ibn Yazid, its roots really lie in the late ninth- and early tenth-century works of Jabir ibn Hayyan, one of the many 'great alchemists' who may be wholly fictitious.

Alchemical, Hermetic and Neoplatonic writers often masked their identity by attributing their works to someone else. These writers did not expect their books to bring fame and fortune; they were more concerned about whether anyone would read them at all, and realized that the chances were improved if the author was someone more illustrious than themselves. Often it did not matter if the pseudo-author had been dead for centuries – in the Middle Ages, new translations of classical authors abounded, and so a 'new' work by 'Pliny' or 'Aristotle' might be accepted quite readily as the genuine article. This meant that, once any scholar or philosopher became famous, he might posthumously be awarded a whole catalogue of works that he had never written at all. It makes the interpretation of medieval literature a complex business, since one can never take it for granted that a book or pamphlet bearing the name of 'Raymond Lull', 'Albertus Magnus' or indeed 'Paracelsus' was written by that person. Thus, while Jabir is credited with a substantial corpus, it might all be the work of a collection of anonymous Isma'ili alchemists. And many of the medieval works attributed to Jabir, which were published under the Latinized name of Geber, seem to have had an entirely separate genesis, originating in southern Italy in the thirteenth century.

The Jabirian alchemical literature is concerned mostly with metals: what they are, how they are formed, and how they might be transformed. Pseudo-Balinus' *Secret of Creation* says that sulphur and mercury are the 'parents' of all metals: that is to say, they are all mixtures of these two substances. It was natural to suspect some generic relationship between the metals, for they have much in common – high strength, high density, malleability, shininess, fusibility. Surely, then, they are but different varieties of the same basic substance, like different types of cheese? Yet the Arabic alchemists phrased this as the proposition not that there was one fundamental Ur-metal, but that they were all composed of these two fundamental components, blended in different ratios.

What is special about sulphur and mercury, that they should be assigned this role? The first is a pungent yellow mineral – not at all like a metal. The second is already a metal itself, and a most peculiar one at that: a metal that flows, known to medieval alchemists as quicksilver ('living silver'). Hardly obvious candidates, then, for the father and mother of all metals. But we can hazard a few guesses as to why they were so distinguished. They both have remarkable and quite singular properties. Sulphur burns avidly – it is the critical ingredient of gunpowder, and most probably a component of the incendiary mixture used by the Byzantines, known as Greek fire. And it is the colour of the sun. So sulphur seemed a kind of mundane representative of the ancient Aristotelian element of fire itself. Mercury, meanwhile, remains unearthly even today: a fluid, shimmering mirror, a contradiction, a metal that embodies the qualities of fire's contrary element water. Its silvery sheen evokes the complement of sulphur's hot sun: the cold light of the moon. So sulphur and mercury could be linked, in a mind attuned to cosmic correspondences, to fire and water, sun and moon, hot and cold. And the Great Work of alchemy was intrinsically bound up with this union of opposites, this Chemical Marriage.

But we must recognize that alchemical Sulphur and Mercury were not the two substances that alchemists could extract from the earth and hold in their hands. They were ethereal materials, beyond our power to make or see or touch. One could only ever encounter alchemical Sulphur and Mercury in impure forms, admixed with one another.

The sulphur–mercury theory, expounded vigorously in the Jabirian corpus, underpinned all notions of transmutation and the synthesis of gold. If metals differed from one another only by the amount of each principle they contained, it should be possible to transmute one to another by alchemical manipulations that readjusted the balance. This notion was fused seamlessly with the Greek conception of matter, since Jabir says (that is, the Jabirian writers say) that each metal may be attributed some of the four Aristotelian qualities: hot, cold, wet, dry. To change one to another, the alchemist must alter these qualities. This, said Jabir, requires the intervention of an agent of transformation: an elixir, which became the *lapis philosophorum*, the Philosopher's Stone.

It is not known when gold-making was first deemed to require this catalyst; the notion of the Stone may go all the way back to early Christian Alexandria. It was well established by the thirteenth century, when the Spanish alchemist Arnald of Villanova said, 'There abides in Nature a certain pure matter, which, being discovered and brought by Art to perfection, converts to itself proportionally all imperfect bodies that it touches.'[26] His contemporary Roger Bacon claimed that the Stone could transform a million times its weight of base metal into pure gold. The alchemist's quest became a search for this miraculous substance.

Jabir lent support to the notion of correspondences between the microcosm and macrocosm – between heaven and earth – by associating the genesis of metals with the stars. Each of the seven known metals was correlated with one of the heavenly spheres; Jabir said that the formation of metals in the deep earth from the

Albertus Magnus (*c.*1235–80) (left) and Roger Bacon (*c.*1214–92) (right) were the greatest 'scientific' scholars of the thirteenth century.

union of sulphur and mercury happens under the influence of their respective planets. This 'fermentation', he says, relies on the fact that mercury is hermaphroditic, and so, in combination with sulphur, can give rise to both the 'male' metals gold, iron and lead, and the 'female' metals silver, copper and tin.

ALCHEMY SEES REASON

The Arab conquest of southern Spain spread Islamic scholarship to Europe. Maslama al-Majriti's book *Rutbat al-Hakim* (*The Sage's Step*) was particularly instrumental in introducing alchemy and practical chemistry to the West in the late tenth century. By the start of the twelfth century, thanks to the work of translators such as Robert of Chester and Gerard of Cremona, the major works of Arabic alchemy were widely known in Europe.

The scholars who read these books were men like Albertus Magnus and Roger Bacon. Albertus and Bacon are commonly regarded as the greatest of the 'medieval scientists' – and if that is not a terribly meaningful description, nevertheless it is fair to say that they display modes of thought that the modern scientist can recognize. They were both products of the new university system which arose out of cathedral schools and theological institutions. A Dominican friar from Swabia, Albertus (Magnus is a Latinization of the designation 'the Great') was educated at Padua, later the alma mater of Vesalius and Galileo. He taught in Paris and Cologne, and his pupils included Thomas Aquinas, who shared Albertus' reverence of Aristotle. Bacon belonged to the Franciscan order; at Oxford he studied under Robert Grosseteste, an important translator of classical works and an avid experimentalist who deduced the optical principles of the rainbow centuries before Descartes and Newton. A proficient astronomer, Bacon proposed the reforms to the Julian calendar that were finally implemented by Pope Gregory XIII 300 years later; and his writings contain those technological prophesies so beloved of occultists today: boats powered without the aid of rowers, diving equipment, flying machines, suspension bridges and so forth.

Albertus and Bacon became famed throughout Europe, known respectively as Doctor Universalis and Doctor Mirabilis. Although they were both reputed to be magicians, it is easier to see continuity between them and the scholars of the early Enlightenment than it is to find room for the intervening centuries of magic and wonderment that Paracelsus typifies. Albertus and Bacon attempted not just to understand Aristotle but to extend and broaden his natural philosophy. On chemistry, Aristotle had next to nothing to say (which earned him Paracelsus' censure, and Bacon's too); but Albertus and Bacon were fully conversant with the Hermetic art, and attempted to reconcile it with the philosophy of the ancient Greeks. They were, in the words of the science historian Robert Multhauf, 'the first Latin philosophers to know both Aristotle and alchemy, and the last to consider them compatible'.[27]

Albertus Magnus and Roger Bacon actually wrote rather little that was *directly* concerned with alchemy, though several later alchemical works were falsely attributed to Bacon. Albertus believed in transmutation but doubted that the gold it produced was of a quality comparable to the natural metal. Bacon attempted to distinguish two kinds of alchemy, according to what we might now regard as the distinction between pure and applied science. On the one hand, there was gold-making – or more generally, the practical business of transforming matter. This 'operative alchemy', said Bacon, 'teaches how to make the noble metals and colours and many other things better and more copiously by art than they can be made by nature and is more important than other kinds of science since it produces the most useful products.'[28] On the other hand, Bacon's 'speculative' alchemy is concerned with theories about the nature of matter, and the distinctions that he drew between 'spiritual' and mundane, changeable matter helped Thomas Aquinas to separate theology from science and thus to argue that one could study God's creation without trespassing into His domain. That was a division which helped to make science permissible – indeed, even to make it conceivable.

Every natural philosopher of the Renaissance paid heed to Doctor Universalis and Doctor Mirabilis. But the alchemy of Paracelsus was indebted also to Doctor Illuminatus, who was an altogether more Paracelsian character. This was the nickname of the Catalonian Raymond (Ramon) Lull, who spent a privileged youth as a philandering courtier in Majorca. When he discovered that one of the married women he pursued was being consumed by cancer, Lull was shocked into abandoning his life of dissipation and vowed to devote himself to God. Inspired by ecstatic visions, he developed the missionary zeal of the convert. His determination to carry the Word to the Muslims took him on some hair-raising adventures in Tunis and Algiers, where he was imprisoned several times. In 1315 he returned to Tunis with religious books he had composed in Arabic, and made no secret of his arrival despite having been barred from the city. He was duly stoned to death by a mob.

Once again, much of Lull's posthumous repute as an alchemist stems from the tendency of other authors to attach his name to their writings. Among the important works attributed to him was the *Clavicula* (*Little Key*) which purported to explain the deepest secrets of alchemy. Lull's followers believed they could uncover all knowledge through the manipulation of letters of the alphabet that symbolized esoteric concepts, using charts, disks and geometrical patterns. This was the Lullian Art, a system based on the idea of correspondences in nature. Another Lullian treatise explains how to make *aqua regia*, 'water of kings', from nitric acid and sal ammoniac, and how this potent solvent will dissolve gold. And his prestige was enhanced when he became wrongly identified as the author of one of the most important pre-Paracelsian texts on medicinal alchemy, *De consideratione quinta essentia*, which was in fact the work of another Catalonian, John of Rupescissa (page 180).

The sixteenth-century Italian mystic Giordano Bruno regarded Lull as a key influence on Paracelsus, who reaped what Lull had sown: he cut the Majorcan cloth into Swiss short trousers, Bruno said. Paracelsus, however, expressed a rather low opinion of the Lullian oeuvre – but then he was never keen to acknowledge his predecessors.

ART AND EXPERIMENT

Roger Bacon evidently preferred the 'operative' over the 'speculative' in his science; in *Opus maius* (1268), he says that science should be based on 'experience' rather than 'argument'. Here he anticipates the viewpoint of his seventeenth-century namesake Francis Bacon, who considers that science should be directed towards 'art': the fabrication of devices, materials and other contrivances (including new living things). It is a perspective that has long been out of fashion, since today science is generally portrayed as a theoretical system for understanding the world, and applied science or technology is its soiled and disreputable cousin. The conventional histories of science look now for ideas, not 'works'.

Pedro Garcia, the bishop of Ussellus in Sardinia, saw things differently from Doctor Mirabilis. In his assault on the defence of magic published by Pico della Mirandola in 1486, Garcia argued that science has nothing to do with anything practical, which is to say, with 'experiment':

> To assert that such experimental knowledge is science or a part of natural science is ridiculous, wherefore such magicians are called experimenters rather than scientists. Besides magic, according to those of that opinion, is practical knowledge, whereas natural science in itself and all its parts is purely speculative knowledge.[29]

Now, we need to be careful about what we understand as 'experiment' here. Scientists today engage in a delicate interplay between theory and experiment. They use the latter to deduce general patterns of behaviour, from which theories might be formulated, while theories are themselves tested against experiments. A well-designed experiment might falsify a theory, or enable selection between different theoretical interpretations.

None of this was done by 'experimenters' in or before the sixteenth century. Paracelsus repeatedly stresses the importance of 'experience', as opposed to rote learning of ancient books, in developing an understanding of the world. But it is unlikely that he ever once conducted an experiment to *test* his ideas, even though in some cases the most rudimentary of trials would have disproved them. This was no different to the position adopted by his contemporaries – whether your belief was based on the works of Aristotle or on your own iconoclasm, an experiment was simply a way of demonstrating that you were right.

Even so, there was a difference between actually doing the experiment and just talking about it. This was one of the strengths of Paracelsus, a man who positively boasted about his readiness to get his hands dirty, unencumbered as they were with the gold rings of the wealthy doctors. But it was also a characteristic of the Arabic alchemists of the early Middle Ages, who appreciated that experimental science must inevitably be a quantitative science. In contrast to the qualitative theories of Aristotle, reliable knowledge is gathered by weighing and measuring. That requires instrumentation: balances, rulers and the like. Not until such apparatus is available does experimental science become possible.

Roger Bacon's taste for experiment was doubtless whetted by his interest in alchemy, and the significance of practical observation for that art was emphasized by Petrus Bonus of Ferrara in the fourteenth century:

> If you wish to know that pepper is hot and that vinegar is cooling, that colocynth [bitter apple] and absinthe are bitter, that honey is sweet, and that aconite [the herb monkshood] is poison; that the magnet attracts steel, that arsenic whitens brass, and that tutia [zinc oxide] turns it of an orange colour, you will, in every one of these cases, have to verify the assertion by experience. It is the same in Geometry, Astronomy, Music, Perspective, and other sciences with a practical aim and scope. A like rule applies with double force in alchemy, which undertakes to transmute the base metals into gold and silver . . . The truth and justice of this claim, like all other propositions of a practical nature, has to be

> demonstrated by a practical experiment, and in no other way can it satisfactorily be shown.[30]

But experiment and measurement are not always the same thing. The entry of quantification into science was assisted in the fifteenth century by another of Paracelsus' intellectual forebears, Nicholas of Cusa. Born in Cues on the Moselle, Nicholas was an astronomer and a mathematician with an aptitude for turning his knowledge to practical tasks such as mapmaking. He was himself influenced by Archimedes, arguably the father of experimental science. Like Bacon, Nicholas regarded experimentation as the handmaid of useful arts. He employed fine balances and timing instruments such as sand glasses (the contemporary mechanical clocks were hardly up to the task). He even proposed Galileo's epochal experiment for observing the rate of fall of objects from a tall tower, and realized that one would have to take air resistance into account. Tellingly, however, Nicholas seems never to have actually performed the experiment, showing that even he was prone to the tendency to let description suffice as demonstration.

Nicholas suggested that one might forecast the weather based on measurements of air humidity, by weighing wool exposed to the air: it increases in weight as dampness accumulates. This seems a long way indeed from the view of his contemporaries that the weather was ordained in, and could be read from, the positions of the stars. Nonetheless, an essentially Neoplatonic astrological scheme underpinned all of Nicholas's studies. He mocked the excessive claims of the astrologers, but was far from denying astrology itself. When he weighed herbs and roots, it was with the intention of uncovering their occult virtues, which provided clues about their medicinal properties. When he proposed that one could detect the debasement of gold coins by using their displacement of water to measure density – the 'Eureka' experiment of Archimedes – he did not think in terms of density as such but of the relative 'virtue' of pure and adulterated gold.

It's no surprise, then, that Pedro Garcia lumped such apparently innocuous experiments together with magic. Science was the knowledge handed down by approved ancients such as Aristotle. It spoke about natural phenomena in abstract and in strictly qualitative terms. It was to be learnt and not questioned. The impiety of experimenters and magicians was that they dared to interrogate God's creation, to peer under stones, to look for answers where humankind was meant simply to accept what they found on the surface. For the thirteenth-century French philosopher William of Auvergne, magic was a 'passion for knowing unnecessary things'.[31] It was in such an intellectual climate that 'curiosity' became considered a sin, a 'lust of the eyes'[32] in St Augustine's words. That was precisely what led Adam astray in Eden.

And we should remember that this kind of tampering was abjured by scholars at the universities too: magic, like alchemy and surgery, was for the unlettered. It is for just this reason that the relationship between magic and humanism was, at best, uneasy. There is no doubt that it was the humanist enthusiasm for ancient texts that led to Ficino's translation of the *Corpus hermeticum* and to the flourishing of Neoplatonism in the fifteenth century. But humanism also threatened to create a kind of Renaissance-style 'two cultures', in which some intellectuals revelled in displaying their subtle mastery of the classics while disdaining anything to do with natural science, let alone the dirty business of experiment. There was an emphasis on style and show; for every shirt-sleeved Leonardo, there was a haughty Michelangelo. At best, humanism had rather little to say to or about science; it was concerned not with nature but with human nature.

In this respect, the magicians of the fifteenth and sixteenth centuries – Ficino, Pico, Paracelsus, Agrippa and Trithemius – had much more in common with medieval scholars like Albertus Magnus and Roger Bacon than they did with some of their contemporaries among the humanists. One could argue that Paracelsus' challenge to orthodox Galenic medicine pays little heed to the fresh vision of the humanists, but is rather a *revival* of ancient Hermetic ideas and attitudes that had declined after the medieval golden age of the thirteenth century. When Paracelsus befriended humanists, as he frequently did, it was because of a shared independence of thought; but they were seldom thinking about the same things. Erasmus and Thomas More, Agricola and Melanchthon, lived in the world of men; but the world of Paracelsus was nature itself, and forever enchanted.

A CHANGE OF COLOUR

The alchemists are traditionally mocked for pursuing a fool's quest, since no fire, philtre or mumbled formula will turn lead to gold. But it seems they did not always fail. There were many stories of successful transmutations, as the account by Franz of Meissen illustrates. All the most famous alchemists were said to know the secret. It would surely be absurd to imagine that alchemy could have persisted for hundreds of years if all it could offer was a history of unmitigated failure. So what were the alchemists really doing?

The fact is that the distinction between transmutation and fakery was hazy even in the minds of those who attempted either one of them. Chemical recipe books of the Middle Ages sometimes openly admit that a particular 'alchemical' process merely tinges a metal like tin with the *colour* of gold: a yellow lacquer called *doratura*, made with saffron, was used to make 'golden tin' as a cheap stand-in for gold. But so long as the underlying theory allowed the possibility of transmutation, there would remain some confusion as to whether a colouring process like this might not be truly a step along the way to making real gold. Even though smiths and metal-workers knew well enough how to assess the identity and purity of a metal by measuring its density, nevertheless the much more superficial quality of colour was deemed to be an important clue too. When Theophrastus of Alexandria, describing how to make a kind of silver from tin and 'white Galatian copper', writes, 'It becomes prime silver that will deceive even skilled workmen who will not suppose it to be made by such treatment',[33] he implies that the deception is simply about concealing the *origin* of the metal, for it is now silver all the same. Likewise, gold made by alchemy might be considered inferior to 'natural' gold (as Albertus Magnus claimed), but it was gold nonetheless.

So when alchemical texts refer to the 'dyeing' of metals, they may be talking not merely about changing their colour but about a genuine transformation. This confusion of colour with chemical identity is apparent in a description by the twelfth-century Benedictine monk Theophilus of how to make a substance called Spanish gold from a remarkable blend of copper, human blood (from 'a red-headed man'), vinegar and the ground ashes of basilisks – the fabulous creature that is part cock, part serpent.* The Florentine craftsman Cennini Cennino is less fanciful in his *Libro dell'Arte* (*c.*1390), but he provides a recipe for 'mosaic gold' that sounds suspiciously like the prototype for Paracelsus' demonstration to young Franz: 'Take sal ammoniac, tin, sulphur, quicksilver, in equal parts; except less of the quicksilver. Put these ingredients into a flask of iron, copper or glass. Melt it all on the fire, and it is done.'[34]

The key point is that both authors are happy to regard these substances as 'gold' of some description – after all, if a material looked and behaved like gold, what else could it be? It is not hard to understand how ready some practical craftsmen were to believe that gold (even if of inferior quality) could be manufactured, once we recognize the alchemical notion that all metals had basically the same ingredients in different states of purity. But this also implies that transmutation, imitation and deception were barely distinguishable, even in principle.

Such tricks of the trade do sometimes blur into explicit fraud. In a third-century collection of Egyptian craft and metallurgical recipes known as the Leiden papyrus, there is a description of how to fake gold from copper, complete with an assurance that it is hard for smiths to detect the difference. Theophilus warns how the prized red colour of 'Arabian gold' is sometimes counterfeited by alloying a little copper with 'pale gold', by means of which many unwary people are deceived.

All of this makes it a little easier to comprehend what Paracelsus was doing in his chrysopoeian exploits. The truth is perhaps hinted at in a tale recounted by the editors of a volume of Paracelsus' works called the *Philosophischer Phoenix* (1682):

> Once Theophrastus had cured a farmer's wife in Ambras with one of his wonder drugs. After many years, he came along the same road again, and inadvertently entered the

* Theophilus' basilisks are created by letting toads hatch the eggs of a cock that, in defiance of nature, has laid them after mating with another cock. This uncharacteristic foray into magic in Theophilus' text appears to represent a literal account of a chemical process he must have come across disguised in the outlandish garb of alchemical allegory. One possibility is that it is a description of how to make brass.

> same farm. The woman recognized him and showed her gratitude in such moving ways that the master was much affected by such goodness in a simple woman. He took a kitchen fork and smeared it with his yellow ointment. The fork changed into solid gold.[35]

This suggests that Paracelsus was simply applying one of the common metal-tinting processes.

Biographers like Anne Stoddart, mercifully free from the kind of credulity that allowed the theosophist Franz Hartmann in 1887 to believe in the literal truth of any magic ever attributed to Paracelsus, have struggled with the uncomfortable possibility that their hero was no better than the countless other self-deluding fools who thought they knew alchemy's deepest secret. So Stoddart suggests that Paracelsus was simply playing a trick on the likes of Franz, gently mocking their wide-eyed ignorance. It is hard, however, to credit that he would have intentionally deceived a simple peasant woman in the same manner by rubbing some yellow tincture on her iron kitchenware and calling it gold. More probably, he considered in both cases that he was producing something of value from baser stuff, even if he was aware that this 'chemical gold' might not be as virtuous or as precious as the natural metal.

It has been claimed that Paracelsus revealed his lack of illusions about these 'tingeing' processes in his *Greater Surgery*:

> In chemical research very wonderful medicines have been made which prolong life . . . but after these come the gold-making tinctures transmuting metals. Thus one tincture colours metal. These discoveries have given rise to the idea that one substance can be transformed into another, so that a rough, coarse, and filthy substance can be transmuted into one that is pure, refined, and sound. Such results I have attained in various ways, always in connection with attempts to change metals into gold and silver.[36]

But if anything, these remarks seem to indicate that Paracelsus believed that transmutation *was* possible, and that it was indeed linked to the tingeing of metals. In fact he was quite explicit about it in *De natura rerum* (*On the Nature of Things*; 1537):

> The transmutation of metals, then, is the great secret in Nature, and can only be produced with difficulty, on account of the many hindrances and difficulties. Yet it is not contrary to Nature or the will of God, as many falsely say. But in order to transmute the five lower and baser metals, Venus, Jupiter, Saturn, Mars, and Mercury, into the two perfect metals, Sol and Luna, you must have the Philosophers' Stone. But since we have already . . . sufficiently unveiled and described the secrets of the Tinctures, it is not necessary to labour further about this, but rather rest satisfied with what we have written in other books on the Transmutations of Metals.[37]

It would be odd indeed if Paracelsus were *not* to believe in transmutation, since it accords with – is perhaps required by – his chemical philosophy of nature. Yet exactly what he thought of the Philosopher's Stone is hard to say – his remarks are characteristically contradictory. 'I am neither the author nor the executor of that Philosopher's Stone, which is differently described by others,' he wrote in 1526, even as he worked alongside Kilian in their alchemical laboratory. 'Still less am I a searcher into it so that I should speak of it by hearsay, or from having read about it.'[38]

All of which makes us wonder how far he and Kilian progressed towards their goal in the bowels of Neuburg Castle. But there Paracelsus cooked up something of more lasting value: his first great book of medical alchemy, the *Archidoxa* or 'Arch-Wisdom'.

9
Elixir and Quintessence

A Chemical Medicine

> Strive to preserve your health, and in this you will the better succeed in proportion as you keep clear of the physicians, for their drugs are a kind of alchemy concerning which there are no fewer books than there are medicines.
>
> Leonardo da Vinci, *Notebooks* (*c.*1451)

In the *Archidoxa* Paracelsus offers a fresh vision for alchemy: one concerned not with gold-making but with medicines. The book's central premise is that, rather than trying to cure disease and illness by a spurious readjustment of the body's humours through Galenic treatments such as blood-letting, the doctor can heal by harnessing the natural 'virtues' of chemical remedies. This is why Paracelsus is often credited with originating the concept of chemotherapy – the treatment of illness with chemical drugs, which are the mainstay of modern medicine. He was not by any means the first person to think this way, but he was the most influential proponent of the idea, and in the seventeenth century his methods redirected the course of medical science.

Like most of Paracelsus' writings, the *Archidoxa* was not published until after his death – the first (Latin) edition appeared in Cracow in 1569, followed a year later by publications in Basle, Strasbourg, Munich and Cologne. The book begins with an account of humanity's plight that provides as good an indication as any of Paracelsus' own circumstances during his days in the Danube valley:

> If, my dearest sons, we consider the misery by which we are detained in a gross and gloomy dwelling, exposed to hunger and to many and various accidents, by which we are overwhelmed and surrounded, we see that we could scarcely flourish, or even live, so long as we followed the medicine

> prescribed by the ancients. For we were continually hedged in by calamities and bitter conditions, and were bound with terrible chains.[1]

Paracelsus' alternative to the 'medicine of the ancients' is one that he claims to have discovered by experience, not from books: 'we have drawn our medicine by experiment, wherein it is made clear to the eye that things are so.'[2] The key is to understand how to liberate the great powers and mysteries that lie in nature, which in the normal course of events are 'hindered by the bodily structure, just as if one were bound in a prison with chains and fetters.' 'For in its operation', he goes on, 'this mystery is like fire in green wood, which seeks to burn, but cannot on account of the moisture.'[3]

The liberation of virtues, says Paracelsus, is achieved by the alchemical process of separation: a parting of the detritus and waste of mundane reality from the vital, healing forces of nature 'which are too wonderful to be ever thoroughly investigated'.[4] This separation yields rarefied, pure essences of nature such as the 'quintessence' – 'nature fortified beyond its grade'[5] – of which there is but an ounce in every twenty pounds of ordinary matter.

The regular doctors, he says, do not know these secrets, but instead foist upon the sick their worthless, bookish remedies for which they charge extortionately. 'Many teachers', Paracelsus claimed:

> by following the ancient methods have acquired for themselves much wealth, credit, and renown, though they did not deserve it, but got together such great resources by simple lies . . . And this we say not in mere arrogance, but on account of the great frauds practised by apothecaries and physicians. Wherefore, not undeservedly, we call them darkness, or caves of robbers and impostors, since in them many persons are treated for gain by ignorant men; persons who, if they were not rich, would at once be pronounced healthy, since the practitioners know that there is no remedy or help for these people in their consultations.[6]

Paracelsus is convinced that the doctors are not simply wrong-headed fools but wilful deceivers who know that their medicine is useless yet persist with it nonetheless because it will bring them profit. He is unwilling to believe that anyone would oppose his ideas unless they are villains and cheats. We shall need to bear that in mind.

The *Archidoxa* is not intended for the lay reader. Paracelsus has been widely celebrated as the champion of the common man; but that does not mean he always sought to convey his knowledge and wisdom to all and sundry. He professes the intention to speak his thoughts 'so far openly that we may be understood by our disciples, but not by the common people, for whom we do not wish these matters to be made too clear . . . we shall endeavour to shut off our secrets from them by a strong wall and a key.'[7] This, it is true, derives not from a general perception that the masses are unworthy, but from a belief that among them are 'idiots who are enemies of all true arts' – such as the 'followers of Galen and Avicenna', who would seize on his recipes, if they could, to gain 'honour, fame, and riches which they thence would obtain to the writings of Galen only and to themselves, out of envy ignoring my name and glorious achievements'.[8] Nonetheless, secrecy is a familiar habit in esoteric writings, and by indulging in it Paracelsus reveals the extent to which he inherited the view that some knowledge is too powerful to be revealed to the uninitiated.

This notion goes back to the earliest alchemical texts, which were cloaked in a cryptic style and terminology that only an adept could decipher. The same guarded attitude persists throughout the Middle Ages, even in the bastardized recipe books that collected together bits and pieces from all manner of primary sources and which mutated eventually into the so-called Books of Secrets. The image of a lock and key appears repeatedly. A ninth-century Italian work that provided the inspiration for several later compendia was called *Mappae clavicula,* 'The Little Key of Knowledge', and in his tenth-century tract *On the Colours and Arts of the Romans*, the monk Heraclius asks, 'Who is now able to show us what these artificers, powerful by their immense intellect, discovered for themselves? He who, by his powerful virtue, holds the keys of the mind, divides the pious hearts of men among various arts.'[9]

These 'keys' were not exactly like mechanical ciphers (though some writers employed those too); rather, they had the potential to unlock the mind to a kind of Gnostic experience of the most profound mysteries of God's creation. By professing to hide his message in this manner, Paracelsus shows himself to be no plain-speaking craftsman like Theophilus, but a magus immersed in the Neoplatonic tradition. He admits this openly in the final book of the *Archidoxa*, the 'key' in which he supplies the crucial secrets to the preparation of his remedies: 'we have concealed our doctrine', he says, 'according to ancient philosophic method and cabbalistic practice.'[10]

Even so, the *Archidoxa* is not so obscure and baffling as many a work of alchemy. It is a kind of physician's cookbook of remedies in which the ingredients and recipes are typically listed rather clearly. Although underpinned by the author's mystical view of nature, the text is taken up with mechanical instructions that anyone could follow once familiar with the instruments, materials and processes of the alchemist:

> Take gems, margarites, or pearls, pound them into somewhat large fragments, not into powder, put them into a glass, and pour on them so much radicated vinegar as will exceed the breadth of four or five fingers. Let them be digested for an entire month in a dung-heap, and when this is over the whole substance will appear as a liquid.[11]

This practicality is one of the book's strengths. Neither the style nor the content of the *Archidoxa*, however, marks out Paracelsus as a revolutionary thinker. Only in the books that followed it do we start to glimpse the imaginative scope of his thought, and the extent to which his medicine was not mere empiricism but part of an immense, self-consistent view of the natural world.

DRUG STORES

As a manifesto for medicinal drugs, there was nothing especially remarkable about the *Archidoxa*. Drugs were widely used in classical times. Dioscorides compiled a pharmacopoeia, *Materia medica*, in the first century AD, and the Galenic chemical remedies known as Galenicals were the mainstay of the apothecaries in every big town. Most Galenicals were plant extracts, whereas Dioscorides' drugs are mainly minerals – and there are some fearsome ones among them. The Hippocratic texts, for example, suggest arsenic sulphide as an ingredient of a drug for internal use.

The Arabic physicians perpetuated and added to the pharmacies of the ancients. Constantine of Africa translated Avicenna and the Jabirian corpus into Latin, and extolled the use of ingredients such as mercury, antimony and vitriol – all standard materials in the Paracelsian medicine cabinet. A treatise called *Al-Tasrif* (translated in the West in 1471 as *Liber servatoris*, 'Book of the Preserver') by the tenth-century physician Abu al-Qasim Muhammad ibn Ahmad al-'Iraqi (known as Abulcasis) emphasized the preparation of chemical medicines using alchemical techniques such as distillation. The *Liber servatoris* was immensely popular, as was the Galenic pharmacopoeia *Grabadin*, attributed to 'Mesuë Jr', which appeared at the same time and went through about thirty editions by the seventeenth century.

The fact was that, while the 'physic' that students of medicine learnt at university was rooted in philosophy and based on the humoral theory of Hippocrates and Galen, the classical texts were often so arcane, complex, subtly nuanced and far removed from reality that they had little bearing on the doctors' day-to-day routine. For minor ailments the doctor might prescribe a range of pills and potions intended to restore humoral equilibrium. These tended to be grouped under exotic classifications such as confortatives, maturatives, mitigatives, repercussives and stupefactives. Among the most popular medicines were laxatives, often prescribed to purge the body pre-emptively, for constipation was widespread and greatly feared.

These drugs came from the apothecary shops. But if the reputation of the doctors among common folk was somewhat tarnished, that of the apothecaries was worse. According to a fifteenth-century text:

> The apothecary's craft is the most full of deceit of all crafts in the world, for the apothecaries lack no deceit in weighing their spice, for either the balance is not right or else the beam is not equal or else they will hold the tongue of the balance still in the hollow with their finger when they are weighing. They care nothing for the wealth of their soul in order that they may be rich.[12]

Cornelius Agrippa echoed this judgement in the 1520s, saying that the apothecaries 'minister one thing for another, or else make medicine of rotten, stale and mouldy drugs, do oftentimes give a deadly drink instead of a wholesome medicine'.[13]

The apothecary shop, where medicines are being prepared according to the prescription of the doctor on the left, who is examining a urine sample.

Imagine it, after all: you are the seller of rare imported powders costing prodigious sums; no one has the means or the knowledge to check what they contain; and the people are desperate for them. What temptation there is, then, to mix in a little flour, a little brick dust or powdered marble, a few dried leaves from the local hedgerow. The jar that has sat on the shelves for the past three summers: who can tell that it is out of date? Who will know? In his youth, Rabelais's Gargantua goes walking in the market-place with his tutor on rainy days, noting casually their deceptive ways: 'They visited the druggists' shops, the herbalists, and the apothecaries, and carefully examined the fruits, roots, leaves, gums, seeds, and foreign ointments, also the way in which they were adulterated.'[14]

Sugar was one of the most highly prized of the rare 'spices' stocked by apothecaries. Its sweet taste was assumed to be a signature of its medicinal virtue, and it is true that a shot of pure sugar in those under-nourished times could have worked temporary wonders. It was also used to make other medicines palatable: in the late thirteenth century the ailing English king Edward I spent the royal sum of £164 on making over 2,000 lb of medicinal syrups.

Better still was the remedy known as theriac, the root of the English word treacle, which was kept in ornate ceramic jars on the shelves of every self-respecting apothecary shop. The name comes from the Greek *therion*, meaning venomous animal, for theriac was supposed in classical times to counteract all venoms and poisons. Derived from the *mithridate* that King Mithridates VI, Eupator of Pontus, and his physician devised in the first century BC, it was brought to Rome by the general Pompey, who defeated Mithridates around 66 BC. Mithridate had fifty-four ingredients, but Nero's physician Andromachus improved it by adding more, and the Theriac of Andromachus that Galen used was a compound of sixty-four substances. The emperor Marcus Aurelius swallowed a dose every day.

A jar of the fabulous cure-all theriac, from eighteenth-century France.

The influential Arabic doctor Averroës (Abu'l-Walid Muhammad ibn Rushd; d.1198) devoted an entire treatise to theriac, which was translated into Latin in the thirteenth century as *Tractatus de tyriaca*. The reputation of the wondrous medicine grew to the point where it was regarded as a universal panacea, a cure for all ills. According to one fifteenth-century source, it could prevent swellings, unblock intestinal blockages, remove skin blemishes and sores, cure fevers, heart trouble, dropsy, epilepsy and palsy, induce sleep, improve digestion, restore lost speech, convey strength and heal wounds. This miracle drug was supposed to contain many exotic ingredients, including the flesh of skinned and roasted vipers (a reflection of its original purpose). It was said to take forty days to prepare and up to twelve years to mature. For this reason, town authorities tried to monitor the manufacture and supply of theriac to ensure that the citizens were not hoodwinked by substandard or counterfeit forms.

The medical profession also called for apothecary shops to be licensed and regularly inspected; Paracelsus' attempt to enforce such measures in Basle in 1527 (see Chapter 11) was by no means unusual. But it was very difficult to impose such standards, and the public entertained dark suspicions that the doctors were often in collusion with the apothecaries, rather as GPs and drug companies are sometimes accused of collaborating today. The doctors, they said, prescribe some expensive drug, the apothecaries sell it at an exorbitant price, and they split the proceeds. Chaucer hints at the intimate relationship between the two when he says of his physician that 'All his apothecaries in a tribe, were ready with the drugs he would prescribe.'[15] Chaucer's friend John Gower was more explicit:

> A crooked apothecary can deceive folk well enough on his own at home, but once he's teamed up with a physician then he can trick them a hundred times over! One writes out the prescription and the other makes it up, yet it costs a florin to buy what's not worth a button . . . The physician and the crooked apothecary really know how to scratch each other's backs: one empties your stomach as often as he can, and the other is an expert at cleaning out your purse, which simply melts away! . . . The healthiest digestion anyone could wish for is not proof against medicines, and no purse is so long that it cannot be drained by an apothecary.[16]

There was undoubtedly some truth in this. Some doctors loftily defended the high price of drugs, saying that encouraged the patient to place more trust in them. This early and lucrative placebo effect was disingenuously explained by the fifteenth-century Italian physician Anthonius Guainerius, who averred that the doctor should 'have a trustworthy apothecary who will affirm that this syrup has been made from conjugal substances, and see to it that he sells it at a high price so that it commands greater faith'.[17]

THE ESSENCE OF THINGS

Even if the alchemical medicines of the *Archidoxa* are not wholly original, no one wrote about alchemy like Paracelsus. He was both a chemical philosopher and a 'sooty empiric', ready to soil his hands and robes at the bench. Alchemy becomes so powerful and so beautiful in Paracelsus' hands because it is a part of a greater system: a magical vision of the universe distilled in the overheated alembic of his feverishly imaginative mind.

The *Archidoxa* is divided into ten books, most of which describe various categories of chemical remedy and their means of preparation from minerals, metals, herbs, leaves and other substances. A quintessence, says Paracelsus, is:

> so to say, a nature, a force, a virtue, and a medicine, once, indeed, shut up within things, but now free from any domicile and from all outward incorporation . . . It is a spirit like the spirit of life, but with this difference, that the life-spirit of a thing is permanent, but that of man is mortal.[18]

So one cannot extract a quintessence from living flesh or from 'things endowed with sensation', because in those cases the spirit resides in the soul, not in the material substance.

Each quintessence has its own particular virtues and so is suited to treating specific ailments: 'some quintessences are styptic [staunching bleeding], others narcotic, others attractive, others again somniferous, bitter, sweet, sharp, stupefactive, and some able to renew the body to youth, others to preserve it in health, some purgative, others causing constipation, and so on.'[19] These medicines are, unlike the catch-all Galenicals, to be targeted at certain conditions and diseases. Appropriately enough, Cornelius

Celsus himself had categorized remedies in a similar fashion, identifying emetics, narcotics and so forth. The physician Michael Savonarola, whose influence Paracelsus must have felt at Ferrara, also made lists of such specifics. Even Avicenna had to some extent sifted the classical pharmacopoeia to find selective remedies.

Paracelsus stipulates that there are various alchemical procedures for preparing a quintessence: 'by sublimation, by calcination, by strong waters, by corrosives, by sweet things, by sour . . .'[20] But most of his recipes utilize a process that came to hold a central position in the Arabic works of medical chemistry: distillation, the separation of volatile components of a substance by heating and condensing.

The influential Alexandrian alchemist Zosimos of Panopolis wrote about this art in the third century AD – a technique, he said, that uses equipment devised by the mysterious Maria the Jewess. This is the origin of the *bain-marie*, a water bath still used today to distil volatile compounds or to gently warm cooking ingredients. Maria is almost unique as a female alchemist, although Zosimos tells us next to nothing about who she was or when she lived. She became a legendary figure, and was said by the bishop of Salamis in the fourth century to have had a vision of Christ.

By means of distillation, oils and essences could be extracted from plants, and it was common belief that the same technique liberated essences from just about any substance: Abulcasis describes a recipe for making 'oil of bricks', and metals too were subjected to distillation. Wine yielded a distillate with particularly miraculous properties, a clear liquid that defied all expectations of a 'water' by burning avidly. Recipes for making alcohol start to appear in the twelfth century – Abulcasis does not mention it – and by the thirteenth century this inflammable liquid became known as *aqua ardens* or *aqua vitae*, water of life.

Alcohol is an Arabic word: *al-kohl*, derived from *kuhl*, the word for the black mineral stibnite or antimony sulphide, which itself stems from the Assyrian *guhlu*, meaning 'eye paint' – in ancient Assyria and Egypt it was used as a cosmetic. But it also cured eye infections, and Dioscorides lists it as a drug. It is a long stretch from this black solid to the clear, subtle liquid of *aqua vitae* – at first 'alcohol' designated powdered antimony sulphide, then a powder of any kind, then the 'essence' of any substance. Alcohol was not unambiguously identified with the distillate of wine until at least the mid sixteenth century, and some attribute this designation to Paracelsus himself, who spoke of the essence of wine as 'alcool vini'.

These volatile, preservative extracts were also known as 'spirits', for there was considered to be a direct analogy with the 'incorruptible' and incorporeal part of humankind. The connection is particularly explicit in the works of the two men to whom the *Archidoxa* is heavily indebted: Arnald of Villanova and John of Rupescissa.

It is no coincidence that Arnald and John, like the renowned alchemist Raymond Lull, hailed from Catalonia (then part of Aragon), for eastern Spain was a meeting place for Christian and Arabic thought. Arnald, like Paracelsus, was a restless, itinerant physician who was drawn into disreputable circles: he associated with Franciscan mystics and chiliastic Joachimites, and learnt folk remedies from empirics and local healers. The value he placed on personal experience over rote learning of old books, his diatribes against those who cure with sorcery and conjuring, his characterization of medicine as a divine calling and not a route to easy profit, and his description of Christ as the 'highest physician' – all are echoed by Paracelsus. So too was the notion that medicine must suit its time, so that the prescriptions of Galen might no longer be appropriate:

> The world has become senescent, both macrocosm and microcosm have grown old, and with this human nature progressively become weak, exhausted and prostrate. Thus the experience of the ancients must be adapted and converted according to the requirements of the present state of human intellect and personal experience.[21]

Although not afraid to make trouble, Arnald was never exactly an outcast. He became a professor of medicine at Montpellier, his services were sought by kings and popes, and at root his medicine was conventionally Galenic. On the other hand, Arnald suffered the inevitable accusation of being a sorcerer – some say he tried to create a man from a pumpkin stuffed with chemical drugs, and in 1317 the Inquisition of Tarragona ordered that his books be burned.

It has been said that Arnald in fact practised two types of medicine: one wholly conventional, for the benefit of his learned

The Catalonian alchemist Arnald of Villanova (*c.*1234–1311).

colleagues and for presenting to royal clients, and the other more radical and experimental, used in Arnald's daily routine where the strictures of scholarship need not apply. Through this latter art he anticipates Paracelsus by directing alchemical knowledge towards medicine. He recounted how distillation separates human blood into its constituent elements, which have potent medicinal properties. He claimed to have once seen a dying nobleman temporarily restored by the 'fire' extracted from blood so that the man might make his last confession. One of Arnald's key works is the *Rosarius* (*Rosary*), which is concerned partly with gold-making and lends a medical slant to the sulphur–mercury theory by describing mercury as the 'medicine of metals'. Arnald explains how to make various elixirs by the separation and recombination of 'elements'; one such elixir, a red powder, is apparently the

Philosopher's Stone. He perfected the method of distilling spirit of wine so that he could make almost pure alcohol. Refracted through the alchemical lens, alcohol took on a mystical aspect and was seen as a kind of condensed sunlight (thus linked with gold) which the grapes had accumulated and then released into the wine.

Chemistry, pharmacy, alchemy and religion were adeptly blended in the fourteenth century by John of Rupescissa, the spiritual and intellectual heir to Arnald and Lull. John was probably every bit as troublesome as Paracelsus. A Franciscan priest, he was frequently jailed for fiery preachings full of apocalyptic and heretical prophecies compounded by vilification of the pope. He believed that his ecstatic revelations came straight from God, and he forecast the overthrow of the wealthy clergy and nobles and a return to pious poverty for those who sought the spiritual life.

He associated the spirit or essence of mundane materials with the incorruptible aether from which Aristotle believed the heavens were made. This stuff was Aristotle's fifth element,* and likewise John believed that the spirits obtainable by distillation were the 'quintessence' of earthly matter: the result of separating the pure from the impure. Quintessences could be extracted from anything, he said in his treatise *On the Consideration of Quintessences*, and they had marvellous medical properties. Alcohol, the quintessence of wine, was a great preservative. Administered to wounds, it prevented infection: John recounts how, chafed raw by his fetters in prison, he acquired some *aqua ardens*, which cured his sores 'in the twinkling of an eye'.[22]

The preserving property of alcohol made eminent sense in theoretical terms, for Aristotle's fifth essence was supposed to be

* It is a wonder that Aristotle did not discover alcohol himself – or maybe this is a testament to his lack of interest in careful experimentation. He did not altogether disdain empiricism, for he explains how he discovered that salt water can be freshened by distillation. But when he extended his experimental studies to wine, the character of Aristotelian practical science, conducted to fulfil expectation and not to uncover new facts, shines through: 'wine and other liquids which can be evaporated and subsequently condensed to liquid again become water on condensation.'[23] Perhaps he did not even bother to check that this was so.

immune to change and decay. The idea of quintessence became linked to the Arabic elixir (see below), a substance that could prolong life. Some recipes for alchemical elixirs claimed that it was a form of liquid or 'potable' gold, and John of Rupescissa said that both potable gold and the spirit of wine could be made by quenching hot gold in the liquid, which he called 'fixing the sun in the sky'.

Rupescissa's distillations isolated some useful chemicals, among them the 'spirit' of urine, which is the potent alkali ammonia. One of the quintessences he most esteemed was that which could be made from antimony, 'which surpasses the sweetness of honey':[24] dissolve antimony sulphide (kohl again!) in vinegar and, as you distil it, the wonder cure appears as 'blood red drops'. John established antimony alongside mercury and gold as a central ingredient of alchemical medicine.

Hieronymus Brunschwick, a surgeon of Strasbourg, was influenced by John of Rupescissa's use of distillation as a means of preparing chemical remedies, and like John he extracted 'spirits' or quintessences from all manner of substances. Brunschwick described the procedure in transparent terms, with admirable woodcuts to assist the reader, in his *Little Book on the Art of Distillation* (1500). The book explained how medicinal plant oils and essences could be prepared using alembics and water baths. As a practical manual it contrasts with the mixture of experiment and metaphysical speculation in Paracelsus' works, where the theological undertones are never far from the surface. Brunschwick's book was widely known, and there can be little doubt that Paracelsus saw it and drew on it. The book was first published not in Latin but in German (*Buch zu Distillieren*) – another innovation not lost on Paracelsus.

If all this inclines us to view the *Archidoxa*'s recipes for quintessences as decidedly derivative, Paracelsus was keen to suggest otherwise. In his later *Great Surgery* he acknowledges that both Arnald and John of Rupescissa wrote on quintessences before him, only to claim rather scandalously that their books 'contain nothing of value' but are 'all mere boasting, in which there is no mention of philosophy, medicine, or astronomy'[25] – by which he means that they did not share his own overwrought metaphysical interpretations.

The title page of Hieronymous Brunschwick's *Little Book on the Art of Distillation* (1500).

POTIONS OF LIFE

While Paracelsus insists on the distinct virtues of medicines such as the quintessences, the *Archidoxa* goes on to describe other classes of preparation and extract whose properties and medical virtues are all but indistinguishable from one another. There are arcana, magisteria and elixirs, all capable of transforming disease to health. It is, Paracelsus insists, the physician's duty to know of these substances and the distinctions between them; but frankly, that is asking rather a lot of the physician.

Paracelsus admits that the arcana are easily confused with quintessences:

> That is called an arcanum, then, which is incorporeal, immortal, of perpetual life, intelligible above all Nature and of knowledge more than human . . . [Arcana] have the power of transmuting, altering, and restoring us, as the arcana of God, according to their own induction.[26]

Whereas ordinary medicines combat disease by confronting it with 'opposite' qualities, Paracelsus says in his later *Paragranum* (Chapter 13) that arcana fight like with like: 'Now, the difference between arcana and medicines is this, that arcana operate in their own nature, or essence, but medicine in contrary elements.' Thus, 'Medicines are those things where it is understood that cold is to be removed by heat and superfluity by purgation', whereas with arcana the physician is required 'to act not otherwise than as if two enemies opposed one another, who were equally cold or equally glowing with heat, and are both armed with similar weapons'.[27]

Because of comments like this, the idea arose that Paracelsus initiated the homoeopathic principle of *similia similibus curentur*: like cures like. The resemblance is no more than superficial, however, for homoeopathic remedies are too dilute for any 'like' substance to be present at all. Someone as fastidious about dosage as Paracelsus would scarcely have had much time for the idea that the most effective dose is no dose at all (for despite the modern linguistic currency of the phrase 'homoeopathic quantities', these quantities are in fact zero). And in any event, Paracelsus does not appear to be setting up 'like cures like' as a universal principle, but as one that applies only to arcana.

It is hard to know what anyone made of Paracelsus' absurdly fine paring of alchemical concepts into differentiated kinds of medical remedy. If it were not enough to cope with quintessences and arcana, the *Archidoxa* goes on to list further categories of potion such as the magisteria and the elixirs. Magisteria, a canonical Paracelsian concept, can be extracted from virtually any material, and seem to be identified more by their means of preparation than by their nature: they are 'that which, apart from separation or any putrefaction of the elements, can be extracted out of things'.[28] They are capable of spreading their influence throughout some other substance: 'If vinegar be poured into wine it renders the whole vinegar. Now, this is magistery.'[29]

Paracelsus' elixirs, like quintessences (indeed, rather like arcana too), are preservatives 'implanted in Nature herself':[30] an elixir may preserve a living body for ever without end. We might hope to distinguish elixirs from arcana according to the idea that the former preserve and the latter transform – except that Paracelsus then tells us that elixirs too can alter the corruptible nature of matter, taking away diseases 'due to the subtlety which they possess'.[31]

Here, as elsewhere, Paracelsus has taken a common alchemical concept and given it a twist, transforming it into a substance that at the same time can be bottled and administered to a patient while retaining a hint of mystical, spiritual significance. The elixir holds a similar position in the alchemical tradition to the Philosopher's Stone; indeed, the two are sometimes interchangeable. Zosimos of Panopolis wrote a 28-volume encyclopaedia of alchemy, the *Cheirokmeta*, in which he mentioned a potent alchemical preparation called the Xerion, a word derived from the Greek for 'dry'. It seems to allude to a dry powder, but in its Arabic form of *al-iksir* it became later identified as a miraculous potion.

The association of the elixir with longevity arose from Chinese alchemical tradition. Trade between the Middle East and China had led to an exchange of technologies and ideas since ancient times: travellers were traversing the Silk Road by 1000 BC, and once Chinese nomads settled in Bactria, the easternmost province of the Persian empire, around 106 BC, Alexandrian chemical philosophy found its way eastwards. Alchemy was well established in China by 144 BC, when the Chinese emperor anticipated Diocletian by issuing an edict against the counterfeiting of gold.

On the whole, however, the Chinese alchemists made gold to benefit not their wealth, but their health. They wanted to consume it. Because gold was apparently incorruptible – it does not tarnish – the alchemists believed that by imbibing its virtues they could prevent their own bodily decay and live for ever. In his book *Pao P'u Tzu* (loosely translated as *The Master Who Preserves Pristine Simplicity*, or *The Solemn Seeming Philosopher*), the foremost Chinese alchemist, Ko Hung (*c.* AD 260–340), says, 'As to the True Man, he makes gold because he wishes by its medicinal use to become an Immortal . . . the object is not to get rich.'[32]

Thus Chinese alchemy had from an early stage a different character from the Alexandrian and Western tradition. The latter was fixated on metals; the Chinese sages, on the other hand, wanted to make medicines. Paracelsus was pivotal in introducing such a change in emphasis in the West: echoing Ko Hung, he entreated, 'stop making gold; instead, find medicines.'[33]

The Elixir was, to the Chinese alchemists, the equivalent of the Philosopher's Stone: it was the goal of all their chemical experiments, a substance just as fabulous and equally irresistible. The two are sometimes hard to distinguish, for the Elixir of immortality could also transmute base metals to gold. Ko Hung relates how lead could be turned to silver and gold using 'medicines', and the alchemist Ma Hsiang was said to pay for his wine by using the Elixir to turn the wine shop's iron goblet into gold. Conversely, the metallurgical Philosopher's Stone became conflated with a life-giving potion in the West; by this means, Albertus Magnus allegedly brought to life a bronze statue owned by Thomas Aquinas. (The animated statue proved to be an unruly servant and had to be kept in order with a hammer.) In Chinese tradition, the Elixir could raise the dead. When emperors dined off golden plates and drank from golden cups, it was not just a vulgar display of wealth but reflected the hope that some of gold's imperishability would literally rub off on them.

Chinese alchemy thus often sounds far closer to Paracelsus' vision than does the gold-making (chrysopoeian) version of the Hermetic art pursued by many of his Western predecessors and contemporaries. The alchemical Elixir was often called the 'medicine', and in fact Chinese sages recognized a whole pharmacopoeia of elixirs that were prepared for specific ailments, such as epilepsy, blindness and melancholia. One Chinese elixir acted as a kind of food, which was equally sustaining if eaten 'a hundred times a day, or only once in a hundred days'.[34] Paracelsus clearly shared this vision of elixirs as a *category* of medicines; there was no single, marvellous Elixir that can cure all diseases.

SEPARATING MEDICINE FROM POISON

Paracelsian chemical medicine can sound fearsome, and no doubt sometimes it was: among his favoured remedies are antimony, mercury and sulphuric acid. But Paracelsus knew very well that such substances contained potential poisons. Indeed, that was the whole point: to treat 'like with like' was to recognize that the cause of disease could also be the remedy. That was why 'tartar' (potassium tartrate, which was deposited as a white crust on the insides of wine casks) also figured in his pharmacopoeia, despite his insistence that tartar was the agent of diseases such as gout. In part, Paracelsus sought to alleviate the hazardous effects of his ingredients by carefully prescribing the dose: this was why he railed against the administering of 'horse doses' of mercury to treat syphilis (page 277). But he also ascribed great importance to the way the remedies were prepared. Separation is the fundamental process in his chemical philosophy, and that is what the alchemical physician seeks to achieve as he makes his medicines:

> What the eyes perceive in herbs or stones or trees is not yet a remedy; the eyes see only the dross. But inside, under the dross, there the remedy lies hidden. First it must be cleansed from the dross, then it is there. This is alchemy, and this is the office of Vulcan; he is the apothecary and chemist of medicine.[35]

The search for ways of making harsh materials palatable led Paracelsus into chemical experiments that occasionally revealed something new: he is said to have been the first, for example, to prepare 'butter of antimony', which is antimony trichloride. He and his followers looked for compounds of mercury that mitigated the unpleasant side-effects of the metal itself, such as mercuric oxide – probably Paracelsus' preferred mercury cure. The deceptively sweet-sounding (and sweet-tasting) calomel, or mercurous chloride, was possibly discovered by the Paracelsian chemist Jean Béguin in the seventeenth century.

The recommendation of 'vitriol' (sulphuric acid) is particularly disconcerting: Paracelsus apparently used it against epilepsy, syphilis, dropsy, gout and miners' diseases. But it may not have been the acid itself that his patients were given, and certainly not in pure form. He mixed it with rosemary oil or 'good wine', and claimed that through one mode of preparation it took on 'the sweetness of honey' (echoing John of Rupescissa). This 'sweet oil of sulphur' was made by mixing sulphuric acid with wine and distilling it. 'It is so sweet', said Paracelsus, 'that chickens eat it and then fall asleep, but wake up again after some time without any bad effect . . . it extinguishes pain and soothes the heat and painful diseases.'[36] How can he have made a narcotic from something so caustic? It seems likely that Paracelsus discovered how to prepare the anaesthetic compound diethyl ether, the same 'ether' that doctors in the nineteenth century used to knock their patients out before surgery (a quintessence indeed!). Sulphuric acid induces a chemical reaction that transforms alcohol to ether – a volatile substance that can be easily extracted by distillation. Andreas Libavius describes the process more clearly in his great chemical textbook *Alchemia* (1597).

Sometimes Paracelsus does make use of the fearsomely corrosive properties of the pure acid – but only for external treatments. He recommends 'oil of vitriol' for skin complaints or 'gnawing leprous diseases', which it will burn away. That might be painful, of course, but he warned that in medicine you can avoid pain as little as can a woman in childbirth.

One of Paracelsus' most renowned chemical remedies is also one of his most controversial. He refers often to *laudanum*, as though it is a kind of panacea; some regard this as his greatest secret, a veritable Elixir in its universality, precious enough that his assistant Oporinus is alleged (probably falsely) to have risked stealing some when they parted company in Colmar in 1529. In an account of his tumultuous apprenticeship, Oporinus himself mentions laudanum with awe, although he was apparently never initiated into the secret of how it was made:

> [Paracelsus] had pills which he called laudanum and which had the form of mice excrements, but he used them only in cases of extreme emergency. He boasted that with these pills he could wake up the dead, and indeed he proved that patients who seemed to be dead suddenly arose.[37]

What was the laudanum of Paracelsus? In contemporary accounts of the history of medicine, it is often described as an opiate: an extract of poppy. That would certainly account for its seemingly marvellous effects when administered to patients suffering from some agonizing illness. But this identification doesn't in fact seem likely. Paracelsus certainly used opium regularly, but when he does so he refers to it explicitly. So it is hard to believe that he would also have sought to conceal it as a secret ingredient in some other wonder drug. The association of Paracelsian laudanum with opium is often traced to the seventeenth-century English physician Thomas Sydenham, who recorded recipes under this name that were composed of a tincture of opium in alcohol (he recommended good sherry), seasoned with saffron, powdered cloves and cinnamon. Such recipes are older, however – although they are not found in Paracelsus' writings themselves, the Paracelsian chemist Oswald Croll lists various 'laudanums' in his *Basilica chymica* (1609), including opiates. The description given by Jacques Bongars, a Huguenot humanist scholar serving as German ambassador for Henri IV of France (who had a fondness for Paracelsian medicine) in the late sixteenth century, certainly shows the hallmarks of an opiate:

> Out of numberless medicines which have been poured into me, none has had any good effect except the *laudanum* of Paracelsus, which my doctors somehow allowed themselves to be persuaded to use. It never failed. It relieved my pains for six or seven hours and induced a restful sleep for two or three hours. In a word, when I was nearly dead with pain, it restored intervals of life.[38]

With benefits like that, who cared what it contained?

In typically perplexing fashion, laudanum means several things to Paracelsus: he pulls together associations and allusions so as to shape a traditional term to new ends. One kind of laudanum is mundane: it is nothing more than the *ladanum* (or *ledanum*) known since antiquity, a kind of gum. This astringent resin is secreted from the leaves of varieties of the Mediterranean *Cistus* shrub (rock-rose), and is mentioned by Dioscorides and Galen, who recommends it for diseases of the uterus and as a preventative for balding. In sixteenth-century Venice it was widely used by perfumers; Paracelsus classifies it alongside myrrh and other gums such as mastic.

Elsewhere, however, Paracelsus' laudanum is clearly something more exotic: a kind of arcanum, prepared from precious ingredients, including gold and pearls. Obviously this is no ordinary drug intended for curing baldness. Gerard Dorn, one of the key editors and popularizers of Paracelsus' works in the late sixteenth century, takes pains to distinguish this remedy from mere ladanum. It surely did not escape Paracelsus' attention that *laudare* is Latin for 'to praise', and his mysterious laudanum seems to be not so much a substance as a characteristically Paracelsian metaphysical concept – a kind of healing principle. There was no consensus among the later followers of Paracelsus on how to make this laudanum, although opium does not appear to have been involved until the seventeenth century. All the same, the Paracelsians were happy to promote the notion that their mentor could prepare a wonder drug: the Strasbourg physician and publisher Michael Schütz (Toxites), another of Paracelsus' late-sixteenth-century editors, speaks of a Laudanum Theophrasti which can cure every disease except leprosy, and can even (a rumour initiated by Oporinus, perhaps) raise the dead. It sounds like the Paracelsians' own version of the fabulous theriac – a medicine that had to be dreamed into existence in an age at the mercy of nature's terrible scourges.

We have to wonder whether Paracelsus was confused in his own mind about the distinctions between his various remedies, his arcana and elixirs and quintessences and mysteria and laudanums and so on. Or are we, in reading the *Archidoxa*, simply struggling to fit a sixteenth-century beast into a twenty-first-century harness? Maybe this is not even the right question. Paracelsus is by no means the most obscure writer of his age, but neither are his hair-splitting yet still nebulous definitions the inevitable product of a natural philosophy struggling to find its feet. It is not so much a matter of what he wrote, but of what he intended.

We've seen that the *Archidoxa* is part of the Hermetic tradition that, afraid of placing profound wisdom in unworthy hands, seeks to cloak and to conceal the meaning of its texts. Yet Paracelsus' medical chemistry is not coded in terms of the kings and queens, the eggs and dragons of conventional alchemy: the chemical instructions are written in plain language, provided that one knows how to sublimate, to resolve and to coagulate elements. Rather, the book strikes the reader as somewhat akin to the words of an obsessive recounting his great passion. At face value, he wants desperately to communicate these marvellous thoughts and insights. But the piling up of details and terminology, of almost invisible distinctions and subtle ideas, tells a different story: at some level there is a desire to keep us at a distance so that we may not challenge or threaten this intricate vision. Some have called Paracelsus' writings unintelligible ravings; others have argued that he was admirably clear. Neither is correct, although there are times when both suggestions are warranted. It is perhaps more useful to look at what the texts tell us about their author. In part, the trials of reading Paracelsus arise because his vision of nature was so grand and weaves so fine and so dense a net that it is hard to pull on one strand without awakening all the others. But surely it is also the case that there are contradictory motivations beneath the words: the wish for our admiration and respect, the frustration and fury at his circumstances, and above all the fear. Fear that we will not understand him, fear that we will, that we may mock or criticize or disprove or remain indifferent. Paracelsus wants to hide himself behind a barrage of neologisms, categories and lists, but instead they merely highlight his internal conflicts.

THE SUMMONS

With the manuscript of the *Archidoxa* in his saddlebag, Paracelsus rode from Neuburg, leaving Hans Kilian to his quest for gold. He went upriver to Ulm, and then he took to wandering in the south-west corner of the German lands between the Rhine and the Danube. At Rottweil he cured the abbess; he passed through Freiburg and Tübingen, as in his student days. He treated lowly people on the road, but the rich and mighty also clamoured for his services. In Baden he was summoned to cure the Margrave Philip of diarrhoea.

Here things did not go well. The Margrave was already surrounded by physicians, who did not take kindly to this outsider arriving to treat what they could not. And Paracelsus had, as ever, not the slightest intention of ingratiating himself. Clearly these court doctors were more of the idiots who persisted in the futile traditions of the ancients, and they had nothing to teach *him*.

So when Paracelsus settled down to work his cure, he was surrounded by enemies already. Unfortunately, there was no miraculous recovery this time – the Margrave's return to health was slow and tentative. The doctors seized the opportunity to discredit the arrogant newcomer, and they told the Margrave that the medicine Paracelsus had prescribed was theirs, which he had perniciously appropriated. The Margrave believed the story and dismissed Paracelsus without his fee.

His response was characteristic: he fumed and spat insults at his enemies, calling the Margrave a cheat and saying of the court doctors, 'Just wait till you see what I wrote and what they wrote; then it will be clear who learned from whom.'[39]

At Tübingen it was a similar story. He set up there as a town doctor and began to attract students eager to learn his new medicine, or just to bask in his reputation. But the university physicians soon denounced his unorthodox methods and forced him out. The physicians of Freiburg also saw him off promptly. He found some small solace by studying the mineral waters at springs and spas of the region: at Göppingen, Wildbad, Liebenzell and Baden-Baden, where people came to take the health-giving elixir and to convalesce.

Where could he settle without incurring slander and threats? Strasbourg looked attractive: a large and prosperous city with reformist and humanist sympathies and an inclination to tolerate new ideas. Indeed, it was one of the few towns in which the Anabaptists could still express their views openly, until the puritanical reformer Martin Bucer became a dominant influence there in 1527.

The lack of a university was perhaps no disincentive to Paracelsus, for it meant there were no scholastic Galenists to stir up trouble. And there was a fine printing press in the town, which might be persuaded to publish his manuscripts. So in late 1526 Paracelsus went to Strasbourg; he purchased the rights of citizenship at the end of November and entered onto the citizen-roll on 5 December. He was required to join a guild, and he became a member of the guild of lucerns – the millers and corn merchants, who, curiously, admitted surgeons and physicians too. He was starting to look respectable.

At the beginning of the sixteenth century the school of surgery in Strasbourg had been run by Hieronymus Brunschwick. Paracelsus may have hoped that, given the congruence of his views with those of this famous chemist, the school would readily admit him. But it was not to prove so easy. Inevitably, the local practitioners were alarmed by his notoriety, and he was challenged to a public disputation by the surgeon Wendolin Hock (Vendelinus).

These scholarly debates worked to the advantage of those with a quick wit and a fluid tongue, and although Paracelsus could deal out robust invective on the written page, in person he was prone to stuttering and ill-suited to verbal fencing. His opponent, Hock, was no reactionary: he lectured in German, and like the other doctors of Strasbourg he had none of the contempt that most academic physicians showed for surgery. As a skilled dissector, Hock had a thorough practical knowledge of anatomy, and he chose this as the topic of the debate. That was a most unhappy choice for Paracelsus, whose understanding of anatomy had always been weak, and Hock inflicted a humiliating defeat.

Smarting from this embarrassment, Paracelsus would no doubt have erupted before long into behaviour that would have precipitated a swift departure from Strasbourg. But on this occasion the outburst was postponed, for he received a summons from elsewhere. A good seventy miles down the Rhine Johann Froben, a famous publisher and humanist of Basle, lay stricken beyond the power of any doctor to cure. Some accounts say he had suffered a fall, others a stroke, others an infection; in any event, one leg was badly afflicted and the pain was getting so unbearable that amputation was contemplated. Desperate to avoid such a drastic and dangerous measure, Froben sought the aid of the now legendary Doctor Paracelsus.

10
Bitter Medicine

Paracelsus among the Humanists

And when many masters and physicians are assembled before the patient or sick man, they ought not there to argue and dispute one against another. But they ought to make good and simple coalition together, in such wise as they be not seen in their disputing one against another, for to encroach and get more glory of the world to themselves than to treat the well-being and health of the sick man.

William Caxton, *The Game and Playe of the Chesse* (1474)

A sketch of Basle made by an unknown artist in the early sixteenth century shows us what Paracelsus would have seen as he approached the city from the bank of the Rhine: a confident, bustling town, dominated by the twin-spired *Münster*, a city of some 10,000 inhabitants, a centre for printing and for the apothecary's trade, and a gathering place for reformers.

In Basle the reformers came from all parties. Some followed Luther, some adhered to the new doctrines of Zwingli. Some were liberal humanist intellectuals, some were grubbing merchants, others wild-eyed Anabaptists and millenarians. At the university, the scholars were generally loyal to Rome and hostile to all these camps, as were the conservative town magistrates. One was well advised to seek allies by pinning one's flag to one mast or another.

Johann Froben aligned himself with the Zwinglians. Holbein's portrait of the publisher* shows the kind of mild, thoughtful man you might expect to befriend the humanists, whose works he published. When Paracelsus arrived in the city, there was already another lodger at Froben's house: Erasmus of Rotterdam, who

* Holbein was an illustrator for Froben, and he painted the portraits of several of the Basle humanists. His sketch known as *Young Man with a Broad-Brimmed Hat* was once thought, wrongly, to be a portrait of Paracelsus.

Basle around 1510–20, by an unknown artist.

The Basle publisher Johann Froben (1460–1527).

was working as the publisher's editor. While Paracelsus wrought his cure, the two men lived under the same roof.

Paracelsus established his reputation in the city by speedily curing Froben's leg (the treatment is not recorded), no doubt to the chagrin of the Basle physicians. Erasmus, who had his share of the ailments that afflicted any well-to-do man of that time – gout and complaints of the liver and kidneys – seized his chance to benefit from the new doctor's attentions. 'I have had a long life,' he later wrote in 1535, when he was sixty-nine years old, 'counting in years; but were I to calculate the time wrestling with fever, the stone and the gout, I have not lived long.'[1]

Some say that Erasmus' complaints stemmed largely from a hypochondriac nature. But he was pleased enough with Paracelsus' diagnosis, as he indicated in a formal letter that typifies his scrupulous discourse:

> To Theophrastus, the highly expert doctor of Medicine, Erasmus sends his greetings. Perhaps it is not out of place to wish perpetual health of the soul to the doctor who heals our bodily ailments. I am amazed to see how well you know me to the marrow although you saw me only once. I know that the riddles [of Paracelsus' diagnosis] are verily true, not from any knowledge of medicine, which art I never studied, but from my sensations of the disease. In fact I did feel pain in the liver although I could not guess the seat of the evil. I could see the liver troubles in my urine for several years. As for the third item in your diagnosis, I cannot understand it, but it seems likely to me. At present I have no time for a cure, indeed I have no time either to be sick or to die, for I am engaged in exacting studies. However, if you know something that might give me relief, please let me know it; and I wish you would explain to me briefly that which you have alluded to and prescribe a medicine for me. I cannot offer any compensation adequate to your art, but I promise to bear gratitude toward you. You have brought back from hell Froben, who is my other half. If you restore my health too, you will give us back to each other. Let Fortune retain you in Basle, and I hope you will be able to read this hastily written letter.[2]

It is hard to imagine, however, that this cultured man would have thought there was anything to be gained from the Swiss doctor besides a convenient remedy. Paracelsus lacked the ability to bridge the gap between his extravagant Neoplatonism and the measured intellectualism of men like Erasmus. Were he ever to have outlined his alchemical philosophy, we can imagine Erasmus listening with the polite and patient silence of a modern physicist receiving a lecture from a starry-eyed advocate of ley lines. Erasmus no doubt sensed as much, and kept Paracelsus at a tactful distance. For his part, Paracelsus probably cared little for Erasmus' good opinion.

Comfortably supported by a respected patron and welcomed into an influential circle of humanists, Paracelsus found the greatest opportunities of his career in Basle. But this was a time of tremendous political upheaval in the city, and once again Paracelsus was to lack the tact and prudence needed to negotiate the currents.

THE OTHER LUTHER

The Swiss Reformation began in Paracelsus' birthplace of Einsiedeln. Here in 1516 a priest named Huldrych (or Ulrich) Zwingli sat in the monastery library studying the Greek edition of the New Testament published that year by Erasmus, and brooding on the same corruptions of Christian doctrine that were starting to make Martin Luther doubt the authority of Rome. Erasmus' new translation persuaded Zwingli that the official interpretation of the scriptures was wrong on several counts. As *Pfaffer* (parish priest) of Einsiedeln from 1516 to 1518, he began to preach against the selling of indulgences.

But Zwingli was not Luther. His rejection of papal theology was measured and he had little appetite for confrontation. Zwingli was no gloomy, tortured dissident; he was an optimistic, candid humanist in the mould of Erasmus himself. By degrees, however, his position hardened as he came to regard the Scriptures as the arbiter in all holy matters, so that the word of the Church was subordinate to the Word of God. Salvation was not a matter of appeasing the clergy – but neither was it to be obtained through Luther's prescription of total and unquestioning faith. Rather, the true course was to be found in the Bible. The Holy Book made no mention of saints, fasts, pilgrimages or celibacy. If it was not in there, it was of no account; if it was, it lay beyond questioning.

All this would have been of little consequence had it come from a simple parish preacher of Einsiedeln. But when Zwingli was appointed priest of the Great Minster at Zurich in 1518, his influence grew. Zurich was the most powerful of the Swiss cantons, an independent city-state and a centre of progressive thinking. It proved receptive to Zwingli's reformist ideas, and soon the city council openly supported this 'people's priest'.

The early stages of the Swiss Reformation give us a hint of how things might have unfolded in Germany if religion and politics had not been so inextricably entangled. Zwingli's demands for reform, the Sixty-Seven Articles, were no less radical Luther's Ninety-Five Theses, but they were passed and adopted without controversy by the council of Zurich in 1523. The following year Zwingli quietly disregarded the principle of clerical celibacy by taking a wife. In 1525, the citizens of Zurich began to attend a reformed communion – not a Mass but a 'memorial service' of the Last Supper. Zwingli and his colleague Leo Judae began to translate the Bible into Swiss German, which they published in 1530.

Zwingli was inevitably labelled a Lutheran, but he insisted that 'I will not bear Luther's name for I have read little of his teaching and have often refrained from reading his books.'[3] He said that, when he came to Zurich, the people there knew almost nothing about Luther, and so his reforms were certainly not a response to the situation in Germany. His colleague Wolfgang Capito, a friend of Paracelsus in Strasbourg, claimed (albeit perhaps with some exaggeration) that he and Zwingli were taking 'counsel how to cast down the pope . . . while Luther was in the hermitage and had not yet emerged'.[4]

All the same, it seemed natural for the two leaders of the Reformation to make common cause. But for Luther there was no common cause – there was only his own. He and Zwingli agreed on almost everything, but the subtle differences that remained became ground for bitter dispute (page 124). Martin Bucer in Strasbourg sought in vain to mediate between the two men, and Philip of Hesse, seeing the political expediency of a united opposition to Charles V, persuaded them to debate their differences at his castle in Marburg in 1529. But it was a hopeless enterprise. Luther departed from the meeting with the barbed comment that 'we would not call them [the Zwinglians] brothers or members of Christ, although we wish them well and desire to remain at peace'.[5] Zwingli, meanwhile, was convinced that he had won the debate, although 'even so, Luther kept on exclaiming that he hadn't been beaten'.[6]

The Reformation could have belonged as much to Zwingli and humanist tolerance as it did to Luther, Calvin and belligerent dogma, had it not been for two considerations. First, Luther's coarse, blunt language had populist appeal; Zwingli was too calm, too reasonable. Moreover, Luther fully exploited, as Zwingli did not, the printing press as a propaganda tool. And second, Luther was not killed.

Zwingli never wished to start a civil war in Switzerland, but he did so all the same. The souls of the people were at stake, since Zwingli, like Luther, believed that they risked damnation while they continued to be misled by the lies of the clergy. He would have been content to spread his reformation peacefully, but while Berne and Basle accepted it, the other Swiss cantons resisted the message of the Zwinglian preachers. Deciding that armed conflict was inevitable, Zwingli mobilized Zurich, in alliance with Berne and Basle, to make a pre-emptive strike against the rest of the Swiss Confederation in 1529.

The First War of Kappel was not pursued with great vigour on either side. The armies held back from battle, and eventually there was an attempt to break the impasse by negotiating a peace settlement. It failed, however, and so the Second War of Kappel began in 1531. The Zwinglians were stronger, but the Catholic forces out-manoeuvred them at the Battle of Kappel. Zwingli was knocked from his horse and run through by a lance. His troops were scattered, his body was quartered and burnt, and the Swiss Reformation foundered until Calvin began to disseminate his own austere and inflexible form of Protestantism from Geneva in the second half of the century.

A POLITICAL APPOINTMENT

Erasmus' presence in Basle had a moderating effect on the reforms of the Zwinglians, who were keen to eradicate Catholicism altogether. Zwingli's leading acolyte in the city was an evangelizing priest named Johannes Heusgen, known as Oecolampadius.

Oecolampadius (1482–1531), Paracelsus' sometime advocate in Basle.

When he persuaded the town council to remove the municipal physician, a Dr Wonecker, from his post for expressing criticisms of the reforms, Erasmus and other liberal humanists spoke out against this curtailment of religious liberty.

Oecolampadius was introduced to Paracelsus through the circle of their mutual friend Capito, and he saw how he might settle the matter of Wonecker's dismissal by taking advantage of Paracelsus' reputation. Why not appoint him in Wonecker's place? He had surely proved himself a worthy doctor; and he seemed to Oecolampadius also to have reformist sympathies. The appointment carried with it the entitlement to lecture at the university, where Oecolampadius was keen to counterbalance the dominant papist tendency (most German universities were still dominated by the Roman Church). The town council approved the appointment, and Paracelsus found himself in March 1527 suddenly in a position of authority and power in Basle.

The university's medical faculty reacted to this blatantly political manoeuvre entirely as one might expect. They had no intention of accepting the new member without his first facing the customary disputation to prove he had the suitable medical knowledge. And who else should the faculty appoint to cross-examine Paracelsus than his nemesis from Strasbourg, Wendolin Hock.

Paracelsus knew all too well how this encounter would go, but he was spared the ordeal when the council insisted that, as the appointment was theirs and not the university's, the usual procedures did not apply. The medical faculty were forced to admit him. They did everything they could to resist, refusing to acknowledge Paracelsus as a faculty member and extending him none of the usual academic privileges. In this tense and awkward situation, a man like Erasmus might have sought to build bridges and soothe ruffled feathers. But we would expect no such thing from Paracelsus.

ACADEMIC IMPOSTURES

It is the beginning of the summer term of 1527 at the University of Basle. There is an anticipatory buzz in the lecture hall where the medical students have gathered to hear the fabled Dr Theophrastus. He is rumoured to be unorthodox, and he does not disappoint his audience. There is controversy from the outset, for this is not even a regular lecture hall of the medical faculty: he has not been granted that platform, since the faculty does not recognize his position. Neither has he formally registered with the university, nor graced his new faculty colleagues with a visit, as courtesy and tradition demand. He has high-handedly met their hostility with disdain.

Nor has he sought approval from the faculty for his intended course of lectures – and he would not have had the slightest chance of receiving it. His intentions, summarized in an 'intimation' to the new students, are a litany of medical heresy and implied insult:

> Theophrastus Bombast von Hohenheim, doctor of both medicines and professor, greetings to the students of medicine. Of all disciplines medicine alone, through the grace of God and according to the opinion of authors divine and profane, is recognized as a sacred art. Yet few doctors today practise it with success and therefore the time has come to bring it back

to its former dignity, to cleanse it from the leaven of the barbarians, and to purge their errors. We shall do so not by strictly adhering to the rules of the ancients, but exclusively by studying nature and using the experience which we have gained in long years of practice. Who does not know that most contemporary doctors fail because they slavishly abide by the precepts of Avicenna, Galen, and Hippocrates, as though these were Apollo's oracles from which it is not allowed to digress by a finger's breadth. If it pleases God, this way may lead to splendid titles, but does not make a true doctor. What a doctor needs is not eloquence or knowledge of language and of books, illustrious though they be, but profound knowledge of Nature and her works. The task of a rhetorician is to bring the judge over to his opinion. The doctor must know the causes and symptoms of the disease and use his judgement to prescribe the right medicine.

Thanks to the liberal allowance the gentlemen of Basle have granted for that purpose, I shall explain the textbooks which I have written on surgery and pathology, every day for two hours, for the greatest benefit of the audience, as an introduction to my healing methods. I do not compile them from excerpts of Hippocrates or Galen. In ceaseless toil I created them anew upon the foundation of experience, the supreme teacher of all things. If I want to prove anything I shall not do so by quoting authorities, but by experiment and by reasoning thereupon. If, therefore, dear reader, you should feel the impulse to enter into those divine mysteries, if within a brief lapse of time you should want to fathom the depths of medicine, then come to me at Basle and you will find much more than I can say here in a few words.

To express myself more plainly, let me say here, by way of example, that I do not believe in the ancient doctrine of complexions and humours which has been falsely supposed to account for all diseases. It is because of these doctrines that so few physicians have correct views of disease, its origins and its course. I bid you, do not pass a premature judgement on Theophrastus until you have heard him. Farewell, and come with a good will to study our attempt to reform medicine.

Basle, June 5, 1527[7]

It is a perfect précis of Paracelsus' lifelong mission, a demonstration of just how clear and concise he could be when the mood took him. And it had the intended effect: the faculty was furious, while students flocked to hear him. And not just students, for Paracelsus opened up his lectures to surgeon-barbers and other non-academics. For the benefit of the untutored, he delivered his surgical discourses in German rather than the customary Latin – another transgression that outraged the conservative faculty. In 1526, five new students enrolled to study medicine at Basle; in 1527 there were thirty-one, and while one cannot prove that this was a consequence of Dr Theophrastus' fame and notoriety, it is not hard to believe.

And when he appeared in the lecture hall, it was not in the gown of a distinguished professor but in the plain smock of an artisan, stained and smeared with the residues of the chemical laboratory. Throughout the summer (ignoring the usual academic holiday) and the winter of 1527 he spoke on the diagnosis of illnesses, the preparation and prescription of medicines, the treatment of wounds and injuries, and on surgery and dissection.

The medical faculty quickly tried to put an end to the outrage by prohibiting Paracelsus from lecturing anywhere. He appealed to the town magistrates: 'They think that I have neither right nor power to lecture in the college without their knowledge and consent; and they note that I explain my art of medicine in a manner not yet usual and so as to instruct every one.'[8] The ban was lifted: the university could do nothing to stop him, even when he staged his most overt and dramatic affront to the traditionalists.

St John's Day, 24 June, was an occasion for the kind of celebrations and revelries typical of student life. In Basle they would build a great bonfire in the market place, and it was not unknown for this to become a pyre for effigies of unpopular figures. In the summer of 1527, however, a different fuel fed the flames: the great *Summa* of Avicenna, his collected writings on medicine. The book was allegedly cast into the fire by the rebel Doctor Theophrastus, surrounded by baying students, so that, as he put it, 'all that misery might go up in the air with the smoke'.[9]

Once again, this legend is open to interpretation. Paracelsus himself recounts this episode, but he refers to the destroyed book only as the *Summe der Bücher* – which does not clearly identify it as Avicenna's classic. Those keen to play down his (self-) mythologizing have suggested it may have instead been some minor tract, although it is hard to understand why Paracelsus should have been so half-hearted in his gesture, nor does there seem to be any reason to doubt his willingness to consign his illustrious enemies to the flames. After all, this 'Luther of medicine' surely wished his act to allude to the Wittenberg theologian's rejection of the Church authorities by burning a papal bull six years earlier. Given the highest platform in his life, Paracelsus was enough of a performer to know how to use it – if not ultimately to his best advantage.

Paracelsus seemed eager to ingratiate himself with the medical students, knowing how youth loves rebellion and craves a leader. While in Zurich in the autumn of 1527 (page 217) he joined them on vacation, staying in the roisterous Stork Inn on the bank of the Limmat. Thereafter he referred to his young followers as his *combibones optimi*, his best drinking companions. To his later critics, this was tantamount to a confession of habitual drunkenness.

Paracelsus engaged several assistants in Basle, generally poor students whom he would provide with food and clothing. This, it seems, was scant reward from a demanding, unpredictable and often frightening master, and few of his protégés stayed with him for long. Paracelsus acknowledged as much in his later life, although he denied responsibility for it and argued that he was sometimes poorly treated himself. 'There is a further complaint against me, with regard in part to the servants who have left me, in part also to my pupils, that none of them could stay with me because of my strange ways.'[10] But he explained that fully twenty-one of his servants died, while others had cheated him, demanded half of the fees he was paid, and pretended to have acquired his secrets.

Franz of Meissen was one of his better students: he worked studiously and eventually became a respected physician. He had nothing but good to say of his mentor: 'He cured lepers, dropsicals, epileptics, syphilitics, and gout patients, besides innumerable other diseases. The Galenic doctors could not do likewise, and envied the honours he earned.'[11]* The relationship between Paracelsus and his assistant and secretary at Basle, Johannes Herbst, known as Oporinus, has often been cast as far less amiable. Oporinus later

* The authenticity of Franz's letter has been questioned, however.

Paracelsus' assistant Johannes Oporinus (1507–68).

became a professor of Greek and a publisher of Basle, and it was he who published Vesalius' *De fabrica*, the central text of Renaissance anatomy, in 1543. But for all this apparent respectability, Oporinus has incurred untold opprobrium and defamation from Paracelsus' hagiographers for having the temerity to make critical remarks about his former master. Stoddart says of him:

> In order to have money for study, he married an elderly widow, who embittered his life with her scolding. He went to Paracelsus in the hope that he might surprise the secrets with the possession of which he accredited him, and so might make fame and money on his own account.[12]

Poor Oporinus has been painted as a scoundrel, sneak, traitor with a 'dirty little soul', although there is no evidence for any of it. To Walter Pagel he was merely a 'timid scholar'[13] who was frightened and exhausted during his days as Paracelsus' assistant – although he could not have been as timid as all that, for he later published a Latin translation of the Koran, suffering a jail sentence for his pains. There is every reason to suppose that he was a competent and diligent humanist, and he prepared Paracelsus' medical lectures at Basle, which, unlike the surgical lectures, were delivered in Latin. As far as we can tell, Paracelsus always thought well of him: he praised the faithfulness of his assistant on one occasion, and indeed Oporinus stayed at his side after his downfall in Basle.

Perhaps Oporinus' worst fault was that he was a little too honest. In 1555 he wrote a letter to Johann Weyer, physician to the duke of Cleves, who was writing a book on magic and demonology and had requested some information on the famous Paracelsus from his former disciple. Oporinus sent Weyer the best first-hand account we have of Paracelsus at work in his Basle days:

> As to Paracelsus, he has been dead for a long time and I should hate to speak against the spirit of his death (as the saying goes). While he was living I knew him so well that I should not desire to live again with such a man. Apart from his miraculous and fortunate cures in all kinds of sickness, I have noticed in him neither scholarship nor piety of any kind. It makes me wonder to see all the publications which, they say, were written by him or left by him but which I would not have dreamed of ascribing to him. The two years I passed in his company he spent in drinking and gluttony, day and night. He could not be found sober, an hour or two together, in particular after his departure from Basle. In Alsace, noblemen, peasants and their womenfolk adulated him like a second Aesclapius. Nevertheless, when he was most drunk and came home to dictate to me, he was so consistent and logical that a sober man could not have improved upon his manuscripts. I had to translate them into Latin and there are several books which I and others thus translated. All night, as long as I stayed with him, he never undressed,

> which I attributed to his drunkenness. Often he would come home tipsy, after midnight, throw himself on his bed in his clothes wearing his sword which he said he had obtained from a hangman. He had hardly time to fall asleep when he rose, drew his sword like a madman, threw it on the ground or against the wall, so that sometimes I was afraid he would kill me. I would need many days to tell what I had to put up with. His kitchen blazed with constant fire; his *alcali*, *oleum*, *sublimati*, *rex praecipiti*, arsenic oil, *crocus martis*, or his miraculous *opoldeltoch* or God knows what concoction. Once he nearly killed me. He told me to look at the spirit in his alembic and pushed my nose close to it so that the smoke came into my mouth and my nose. I fainted from the virulent vapour . . . Only when he sprinkled water on me did I come to . . . He pretended that he could prophesy great things and knew great secrets and mysteries. So I never dared to peep into his affairs, for I was scared.
>
> He was a spendthrift, so that sometimes he had not a penny left, yet the next day would show me a full purse. I often wondered where he got it. Every month he had a new coat made for him, and gave away his old one to the first comer; but usually it was so dirty that I never wanted one . . .
>
> In the beginning he was very modest, so that up to his twenty-fifth year, I believe he never touched wine. Later on he learned how to drink and even challenged an inn full of peasants to drink with him and drank them under the table, now and then putting his finger in his mouth like a swine.[14]

There is no obvious reason why Oporinus should have fabricated all of this. But he was naïve to send it to the anti-Paracelsian Weyer, who duly published it – which was never Oporinus' intention. In later life Oporinus expressed regret to the publisher Michael Toxites that he had spoken so openly and so ill of his former master. Paracelsus' reputation was waxing at that time, so Oporinus' retraction may have contained an element of expediency. But since Toxites, an avid Paracelsian, was said to have 'loved Oporinus like a brother', it seems unlikely that Oporinus was generally regarded as a critic of his former mentor.

Besides, other sources confirm that Paracelsus was no stranger to wine. 'It is true', said Aegidius von der Wiese, 'that he enjoyed drinking. But on the other hand, when he had undertaken anything he scarcely ate or drank until he had finished his work, and then, when he felt free, he became mightily merry.'[15] His drunkenness soon became a part of Paracelsus' legend, inducing Giordano Bruno to acknowledge his debt to Paracelsus with this double-edged tribute: 'Who after Hippocrates was similar to Paracelsus as a wonder-working doctor? And, seeing how much this inebriate knew, what should I think he might have discovered had he been sober?'[16] The sixteenth-century English Paracelsian Richard Bostocke took a commendably pragmatic stance in response to the litany of accusations assembled by his contemporary Thomas Erastus (Chapter 19):

> If Paracelsus some tyme woulde be dronke after his Countrey maner I can not excuse hym no more then I can excuse in some nations glottenie, in others pride, and contempt of all others in comparison of themselves, in others breach of promise and fidelitie, in others dissimulation, triflying and much babbling, but lett the doctrine bee tried by the worke and successe, not by their faultes in their lives.[17]

Bostocke, the first prominent Paracelsian voice in England, was supported by a mysterious advocate who signed himself only as 'I.W.' and devised the ingenious defence that Paracelsus could hardly be blamed for taking a drink or two when, after labouring for hours at the hot furnace, he needed to cool off.

Paracelsus clearly did not win the adulation of all the students at Basle. One of them poked fun at this iconoclast by parodying the 'intimation' with which he announced his course of lectures, showing that little has changed in the deadpan nature of student wit:

> Valentinus ab Riso greets the reader; this is about Theophrastum and his writings. He was born at Einsiedeln in Switzerland and the *Athenienses* call him the Great Paracelsus. For he has written 230 books on philosophy, 40 on medicine, 12 on government, 7 on mathematics and astrology, and 66 about secret and magic arts. He has also combined three works in a book he calls Theophrastia. The first is *archidoxa*

in which he teaches how to separate the pure from the impure; the other *Parasarchum*, in which he treats *de summo bono in aeternitae*; the third *carboantes*, in which he deals with transformations in substance and essence. Gellius Zemeus wrote of this German Theophrastus as follows: 'There is a young man now in Germany the like of whom cannot be found in the whole world and who has written so excellently on philosophy, medicine, astronomy, and of Common Good and Right that I cannot believe otherwise. He has it either through innate influence or through an indescribable grace of God, if not through the intercession of evil spirits. For nobody knows what Theophrastus may think mean and improper.' I cannot remember having read anything more scholarly. Therefore, dear reader, receive these writings with good grace and do not fear to hold them higher than the ancients although they are thought to be novel. Vale.[18]

HOW TO BE A GOOD DOCTOR

The essence of what Paracelsus taught at Basle is probably contained largely in the *Archidoxa* and the medical books he wrote a few years later, *Paragranum* and *Opus paramirum* – to which we shall come shortly. But he was also concerned to impress on the students the qualities that were expected of a good doctor, which are laid out in the volume published by Toxites in 1571 as *An Excellent Treatise, by Philip Theophrastus Paracelsus, the famous and experienced German philosopher and doctor*. Paracelsus insisted that knowledge was not enough; the physician should also possess certain qualities of character (naturally, he himself flouted these guidelines flagrantly):

- He shall not consider himself competent to cure in all cases.
- He shall study daily and learn experience from others.
- He shall treat each case with assured knowledge and shall not desert nor give it up.
- He shall at all times be temperate, serious, chaste, living rightly, and not a boaster.
- He shall consider the necessity of the sick rather than his own: his art rather than his fee.

- He shall take all the precautions which experience and knowledge suggest not to be attacked by illness.
- He shall not keep a house of ill fame, nor be an executioner, nor be an apostate, nor belong to the priestcraft in any form.[19]

Elsewhere Paracelsus expanded on these tenets – sometimes with apparent humility and humanity, sometimes with delightful fastidiousness. A doctor should have 'a gentle heart and a cheerful spirit', and should 'despise no one'. He should have 'greater interest in being useful to his patient than to himself'. He should, of course, know 'all the properties of the flesh, the bones of the body, the blood vessels, the veins and arteries of the whole body, what injury can befall each organ, how injury affects each organ' and so forth. And as for remedies, he should know all the 'vulnerary herbs', all the 'tissue-forming remedies', all the essences, what to forbid the patient and what to permit him, how to plaster wounds and apply lotions. Yet beyond all this, the good doctor 'must not be married to a bigot, should not be a runaway monk, should not practise self-abuse', and 'must not have a red beard'.[20] Stringent qualifications indeed. And the doctor must be ever aware of the gravity of his position. 'Actors and the race of poets should not enter medicine,' Paracelsus claimed, 'they are too witty, and it is not good for them to be serious.'[21]

Indeed, he believed that 'a good physician must be a born physician'. For this reason, 'no one should be surprised that the medical faculty is full of students who contribute nothing to its good reputation, but only harm it and make it an object of contempt. A tree that has once borne fruit cannot be changed. And no more than an apple can be changed into a pear, will such people ever become good physicians. A life-long calling must be innate.'[22]

But Paracelsus is rather vague on what exactly constitutes a 'born physician'. There are three types, he says. Some are 'born of nature', which is to say, owing to a favourable horoscope – 'as also musicians, orators, and artists are born'. Others are 'given by God, and directly taught by God' – that is, their knowledge derives from a kind of divine revelation. The third kind, however, sounds rather like the average medical student: he 'is taught by men introduced to medicine and trained in it, to the extent that man in general and they in particular are capable of learning'.[23] What kind of physician did Paracelsus consider himself? After all, he insisted that his knowledge of medicine was derived from studying the natural world – not from books, nor from the stars, nor directly from God. 'The art of healing comes from nature, not from the physician,' he said. 'Therefore the physician must start from nature, with an open mind.'[24]

There is something of a disjunction here between Paracelsus' almost prosaic approach to the everyday business of learning medicine and treating patients and the cosmic principles on which all of his thinking is based. Perhaps it was hard for him to imagine that anyone could find their way to the heart of medicine by a path other than the one he had taken. After all, he despised the bookish physicians at the universities who had never stepped out of academia and barely deigned to touch a patient; yet here he was, at face value one of them, lecturing in the halls of Basle. And so he insisted that 'the art of medicine cannot be inherited, nor can it be copied from books; it must be digested many times and many times spat out . . . one must not doze like peasants turning over pears in the sun.'[25]

And more than this; for the true doctor must know the wider world:

> Therefore travel and explore everything, and whatever comes your way, take it without scorn, and do not be ashamed to do so on the ground that you are a doctor, a master.
>
> The physician does not learn everything he must know and master at high colleges alone; from time to time he must consult old women, gypsies, magicians, wayfarers, and all manner of peasant folk and random people, and learn from them; for these have more knowledge about such things than all the high colleges.
>
> The arts are not all confined within one man's country; they are distributed over the whole world.[26]

When Paracelsus writes like this, the filthy, vain, foul-mouthed drunkard vanishes and we can begin to see why so many fell under his spell.

11
The Battle of Basle

How Paracelsus Left Town

> 'But of you, base and vile rabble, I take no account,' he cried. 'Throw stones! Come on, attack! Assail me as hard as you can, and you will see what penalty you will have to pay for your insolent folly!'
>
> Miguel Cervantes, *Don Quixote* (1605)

Some say that Paracelsus wasted his opportunity at Basle; others that he was the victim of a vicious plot. Both miss the point. You might as well lament that an alcoholic squanders his money and his health at the bar. There is some wayward part of his character that compels him to do as he does, even though he protests at the injustice of it all. We watch Paracelsus in Basle as though seeing a man run headlong towards a precipice. Like an indestructible lunatic, he will do so again and again throughout his life.

Jung recognized this self-sabotaging aspect of Paracelsus' character, and found within it the key to his impossible temperament and his peculiar antics:

> Paracelsus was a little too sure that he had his enemy in front of him, and did not notice that it was lodged in his own bosom . . . He was so unconscious of the conflict within him that he never noticed there was a second ruler in his own house who worked against him and everything he wanted . . . And when one unconsciously works against oneself, the result is impatience, irritability, and an impotent longing to get one's opponent down whatever the means.[1]

Who, for example, can fail to see mischief in his purposes when, during his lecturing days at Basle, he announced to the doctors that he would reveal to them the greatest of medical secrets? They gathered in the lecture hall in their fine gowns and robes, eager to mock this impostor, who had already prepared a contemptuous gesture of his own. As he held up a dish and revealed its contents, the assembled crowd was confronted with steaming human excrement. Predictably, they stormed from the hall in outrage, pursued by Paracelsus' accusations: 'If you will not hear the mysteries of putrefactive fermentation, you are unworthy of the name of physicians!'[2] But it was not all calculated insult; Paracelsus genuinely believed that 'decay is the beginning of all birth'[3] – and of all health, for 'that which prevents putrefaction also will prevent health'.[4] When the body is working properly, he said, digested matter is broken down in the bowels. If this process is disrupted and undigested matter cannot be voided, sickness results. It sounds all very Victorian, but the notion is, as we have seen, ultimately alchemical: many chemical transformations were held to involve a process of decay of the raw ingredients in order that they should give rise to more virtuous substances. 'Decay is the midwife of very great things,' he said. 'It brings about the birth and rebirth of forms a thousand times improved. This is the highest mystery of God.'[5] Yet, however much he believed that, it is hard not to suspect some element of deliberate provocation in his decision to confront the conservative doctors with a bowl of shit.

As if he had not already enough enemies in the medical faculty, Paracelsus also incurred the wrath of the apothecaries. As town physician he had the authority – indeed, the duty – to oversee their activities. It is unlikely that the apothecaries of Basle were any worse than those elsewhere: a mixture of honest folk and swindlers, some competent enough, others whose shelves were stacked with useless or harmful remedies. But whether principled or deceitful, they had little medical training and were essentially merchants out to make a profit. Worse, in Paracelsus' view, was the fact that they conformed to the medical orthodoxy and sold mostly Galenical medicines.

And so he set out to reform the apothecaries. There is little likelihood that he approached this with any tact. 'I do not take my medicines from the apothecaries,' he once insisted, 'their shops are just foul sculleries which produce nothing but foul broths.'[6] The best and truest apothecary was not to be found in any town, but in nature herself:

> Herbs are gathered together in an apothecary's shop and can be bought there, and in one shop more numerous and varied herbs can be found than in another; similarly there is in the world a natural order of apothecary's shops, for all the fields and meadows, all the mountains and hills are such shops. Nature has given us all of them, from which to fill our own shops. All nature is like one single apothecary's shop, covered only with the roof of heaven; and only One Being works the pestle as far as the world extends.[7]

Anyone who thought like this was unlikely to have much respect for the peddlers of herbs, potions and powders. Paracelsus insisted that the apothecaries rid their shops of all false medicines and improve their standards of hygiene. He deplored the employment of uneducated people and children in these shops, and demanded that the apothecaries be required to take an examination before they could practise – an unthinkable imposition, since rather few of them would be likely to make the grade.

Moreover, Paracelsus observed that the apothecaries were in league with the doctors. 'The doctors', he said, 'take more trouble to screen their movements than to maintain what concerns the sick, and the apothecaries cheat the people with their exorbitant prices and demand a guilder for messes not worth a penny.'[8] He wanted to extend the supervisory powers of his position so that he might stamp out such abuses. The medical faculty responded by calling him a liar, a fool, a necromancer, the 'forest-ass of Einsiedeln'. But Paracelsus could play that game, and retaliated with the accusation that the physicians were 'a misbegotten crew of approved asses'.[9]

INFERNAL ACCUSATIONS

Paracelsus had some supporters, such as Boniface Amerbach, a wealthy merchant and respected humanist. Amerbach attended Paracelsus' lectures and took notes that have since provided the only record of what he said. He was a close friend of Erasmus, who left him his estate in his will. And as a professor of law at the university he was a valuable ally to Paracelsus from within the academic system.

But as for Oecolampadius, whatever feelings of obligation he

Boniface Amerbach, painted by Hans Holbein in 1519.

may have had towards the man he'd been responsible for appointing must have eroded fast as it became clear that the unpredictable doctor was scarcely serving the intended goal of strengthening the anti-Rome faction within the university. And so Paracelsus' position became increasingly precarious. Blithely unaware, he continued to alienate himself. The town aldermen, who had approved his appointment, could not have been much impressed by his entreaty (more of a demand, really) that those who were persecuting and slandering him be brought to heel:

> Honourable, austere, pious, provident, generous, wise, gracious, and favourable gentlemen who have appointed me as a physician and ordinarius:
>
> It has come to my knowledge that doctors and other physicians who reside here have commented unbecomingly in the streets and cloisters on the status which I receive through your kindness. This does great damage to my practice and my patients. They boast they are the faculty and dean and that your appointment of me, a foreigner, is without right and merit; so that I would rather not have

> followed the honourable call of your austere gracious wisdom.
>
> With the help of God, I have cured invalids whom the ignorance of other doctors almost maimed and I think that I should deserve honour instead of infamy. Your austere, honourable wisdom has appointed me physician and professor; you are my superiors, masters, faculty and dean, not they; and I should be entitled to graduate my disciples to be doctors as behoves a full professor.
>
> Should the other doctors have power to prevent that, I should not have abandoned [good positions with] princes and cities. If such is not the case, I humbly entreat your austere, honourable wisdom to make public the liberties of my status, privileges and rights.[10]

In other words, I wouldn't have bothered coming if I'd known things would be this bad; kindly put them right. This high-handed appeal indicated to the town council that their appointee was starting to make a nuisance of himself.

Perhaps his opponents sensed how this man, suitably goaded, could bring about his own downfall. And so they produced a suitable provocation. One Sunday morning, a notice appeared nailed to the door of the town cathedral and in other prominent places. This was no apocalyptic thesis in the manner of Luther, but a crude lampoon written in mock-Latin, allegedly by Galen himself speaking from hell. It was a direct attack on the teachings of Dr Theophrastus – or 'Cacophrastus', derived from the Greek *kakos* (bad) but with a decidedly more scatological connotation:

> *The Shade of Galen Against Theophrastus, or rather Cacophrastus*
> Hear, you who soils the renown of my glorious name:
> A talker, you say, an idiot am I?
> You say that of chemistry I have not the feeblest experience,
> Or knowing it, was not skilled enough to use it.
> Unbearable! Have I not known the commonest simples?
> Onion, garlic and hellebore, I know them well enough.
> Hellebore I send to you to cure your addled brains.
> True, I don't know your mad alchemical vapourings,
> I know not what *Ares* may be, nor *Yliadus*,

Nor *Archeus*, the spirit which you say preserves all living things.
All Africa bears not such fabulous fantasies.
And yet, you nonsensical fool, you think to speak with me!
Are you itching to measure your weapons against mine,
You who could give no answer to Wendolin's well-reasoned word?
I don't think you are worthy of carrying Hippocrates's piss-pot,
Or even to feed my swine . . .
What will you do, madman, when you are found out?
You'd do well to hang yourself . . .
The Stygian law here forbids me to speak further with you today.
Enough for you now to digest! Reader and friend, fare thee well!

Out of Hell.[11]

It was a childish gesture and cowardly in its anonymity. A wiser man would have laughed it off. But Paracelsus was vain enough to feel wounded by the slightest insult, no matter what its source. The infantile name-calling of 'Cacophrastus' seems to have rankled deeply, for he even brings it up himself in his *Paragranum* two years later:

> They abide, painted doctors, and if they were not painted with this title, who would recognize them? Their works would certainly not reveal them. Outwardly they are beautiful, inwardly they are squalid dunces . . . What, then, is the origin of that medicine which no instructed man desires, from which no Philosophy issues, in which no Astronomy can be noted, in which no Alchemy is practised, and in which there is no vestige of Virtue? And because I must point out these things essential in a physician, I must needs have my name changed by them to Cacophrastus, when I am really called Theophrastus, both for my art's sake and by my christening.[12]

One might think that the man who paraded dung before the academics of Basle might resign himself to finding it smeared into his name; but his response was typical self-important bluster. 'Knaves did it,' he fumed. 'Shall I be a lamb? Rather do they turn me into a wolf.'[13] A wolf, however, who wanted others to hunt and dispatch his prey. He insisted that the university authorities discover which of the students (or could it have come from higher still in the faculty?) were responsible for the libellous doggerel. And he sent another strident letter of complaint to the town council, although why this university matter should have been their responsibility was clear to no one else:

> In unbearable anger and distress it is fitting that the sufferer should call upon the magistrates to protect, counsel, and help him. If he has been silent concerning the many slanderous letters sent to him, it is now impossible for him patiently to suffer such an injurious libel and outrage as this which has now been openly posted up. From the tenor of the lampoon it is evident that the author is one of his daily listeners. He has already suspected that there were some who instigated and suborned other doctors of medicine to write against him. But now, he demanded that the whole body of his hearers should be summoned and examined so as to discover who wrote the lampoon that the libeller might be dealt with as he deserved.[14]

As if this were not obstreperous enough, Paracelsus added the threat that if he were not heeded he could not be accountable for his actions: 'He could not himself vouch that his temperament might not urge him to say or do something injudicious were he to receive no support in this matter, or were he to be further incensed. In no circumstances would he suffer more insolence.'[15] The aldermen did nothing, of course, for even if they had been inclined to, they could hardly have intervened so directly in the university's affairs. Paracelsus was incensed.

So he took the initiative. He decided that if the reformists on the town council would not support him, he might find more satisfaction from a higher authority. In the autumn of 1527 he headed south to Zurich, the centre of Zwinglian politics.

He seems to have become somewhat diverted from his mission by the temptation to drink with friends and students, fuelling the later charges of drunkenness. But he did manage to make arrangements to meet with Heinrich Bullinger, Zwingli's right-hand man and an able and sober humanist scholar, who succeeded Zwingli after his death. Unfortunately, Paracelsus was seldom in a fit state for such an encounter, and the impression he made on Bullinger could scarcely have been worse. 'I had several religious and theological discussions with Hohenheim,' Bullinger said later.

> If there was a trace of orthodoxy I failed to notice it. Instead, he talked a lot of magic of his own invention. Had you seen him, you would not have suspected a doctor in his appearance. He rather looked like a [drover]; and with such people, indeed, he liked best to associate. When he lived at the Stork Inn here, he always watched out for teamsters, and then he drank and ate with them, dirty as he was. When he had drunk enough, he would go to sleep on the first bench and sleep himself sober. In short, he was an extremely dirty, unclean man. He did not attend mass, nor did he seem to care much for any other divine things.[16]

Clearly, Paracelsus would get no help from that quarter. But it hardly mattered anyway, for events suddenly took another turn for the worse. Froben, his erstwhile advocate in Basle, evidently never quite recovered from his illness. Before Paracelsus left for Zurich, his host announced his intention to travel to the Frankfurt fair, where all Germany's publishers gathered annually (as they still do today). The doctor knew Froben was unwell, and advised against it. But Froben went anyway, and on the trip he collapsed and died. The circumstances are unclear; one account says that he plunged from a window, apparently in the grip of a fit or a stroke. Whatever the particulars, it was certainly regarded as no mere accident but the result of ill health.

When Paracelsus returned to Basle in November, he found his reputation in tatters. So much for his miraculous cure! Here was the perfect ammunition for his enemies in the medical faculty, who insinuated that the publisher's death was due to Paracelsus' incompetence and noxious 'medicines'. For a time, it looked as if he might even be indicted of murder by negligence.

In fact, he did somewhat blame himself – not for treating his patient poorly, but for failing to have anticipated the dangers he faced. In a letter to his *combibones optimi* in Zurich he seems to have been genuinely distraught by the loss of a friend, although he does not neglect also to establish his innocence:

> Alas, how wretched is the state of mortals, because there is scarcely any joy that is not presently followed by sorrows – a fine company of helpers! Hitherto I have not perceived my blindness, for I did not consider in the present that the wise man must most diligently observe, not only those things that are at his feet, but those that are behind him, like a two-headed Janus, and those also that are in all directions around him.
>
> The reason is that your most delightful assembly, that I lately enjoyed, and do still recall with gratitude, had so enchanted my heart that I forgot all about the future. My mind presaged no disaster: I thought the whole matter was well managed and deemed that joy would exist without the company of grief. Now, when I see these things that I ought to have foreseen, how, I say, shall I restrain myself from grief and mourning, since the dearest friend I had in Basle, whom I left in health and strength, has died through the accident of a sudden fall from an upper storey, where he was accustomed to sleep.
>
> He had been freed by me from the heaviest chains, into which he had been thrown by the petty doctors of Italy: by me he was restored to health, of which fact Erasmus of Rotterdam is a witness, with all his family, as the letter written by his own hand sets forth.
>
> Now, when I was feasting with you, and taking life easily, he died whom I had left in good condition: he, I say, whom I loved as my own eyes: being snatched away by the mishap I have mentioned, namely, Johannes Frobenius, the parent and tutor of all learned and good men, being himself also wise and good, the most diligent promoter of all kinds of learning. Wherefore also have I need to fear the same suddenness in death that has overtaken him. What shall I say to myself? Death is common to all. Wherefore, be warned. Watch, most excellent fellow-learners, and if to any extent we fail in our office, attribute it to that severe grief wherewith I am now tortured, and can find no relief.[17]

The wolves of Basle smelled victory.

THE FINAL STRAW

Predictably, they did not need to do very much to secure it, but only to give their victim enough rope and wait for him to follow the advice of the 'Shade of Galen': to hang himself by the neck.

Cornelius von Liechtenfels, a canon of Basle cathedral and loyal to the Roman Church, was one of the richest and most powerful men in the city. When in early 1528 he fell ill and was thought to be dying, he offered a reward of a hundred guilders to any doctor who could cure him. The tale is almost too fantastical, for we can see just what is going to happen. Paracelsus is summoned, he prescribes purgatives, a strict diet and a regular dose of his laudanum pills, and in a matter of days the canon is restored to health.

Browning speculated in his romantic poem that it was all a plot, a set-up by the Catholics to force Paracelsus to rash action. But perhaps the canon simply did not feel that so straightforward and brief a treatment warranted a hundred guilders, which was an immense sum. In any case, he did not pay the reward, but sent his saviour just six guilders, an amount more commensurate to the services offered.

It was neither the first nor the last time Paracelsus was cheated of his fee. But in his embittered frame of mind, he had no inclination to let the matter pass. He took his case to the city court; but the judges, whether through lack of evidence or dislike of the irksome plaintiff, ruled in the canon's favour. Paracelsus was now convinced that everyone was against him, and he threw to the winds what little caution he possessed. He openly reviled the city magistrates as corrupt and worthless; and he backed up his criticisms by having them printed in a pamphlet which he distributed freely around the town. 'How should *they* understand the value of my medicines?' he demanded. 'Their method was to vilify the physician. Should a sick man be healed, they must needs tell him not to pay for his cure, so that the sick and the law judged of healing as if it were shoemaking.'[18]

Amerbach tried to dissuade him from this impetuous course, but there was no stopping him. Once the pamphlet was out, Amerbach implored Paracelsus to apologize, but again to no avail. This time he had gone too far. Blatant defamation of the court of law was impermissible, and a warrant was drawn up for the arrest of the troublesome doctor. The intention, Paracelsus suggests, was to exile him on an island in Lake Lucerne; but it is not impossible that some more severe punishment was in store, perhaps even execution.

A friend – probably Amerbach – hastened to Paracelsus' lodgings on the eve of his arrest and encouraged him to flee the city. There was no time to pack up his manuscripts or his alchemical apparatus; these he hastily entrusted to Oporinus before saddling his horse and riding out from Basle in the early hours of morning.

ON THE RUN

His first destination was Neuenburg, where he found refuge at a country residence of the Amerbach family. But Boniface's brother Basil was there on his honeymoon – hardly the most auspicious time for a visit from a desperate fugitive. So Paracelsus did not stay long, but moved on to the house of another friend, the humanist and doctor Valentine Boltz, at Rufach in Alsace, about forty miles from Basle.

It says much for his insatiable curiosity that, outcast and virtually penniless, Paracelsus nevertheless found time to take a detour on his way to Rufach so that he might pass through Ensisheim, where a magnetic iron meteorite had fallen near the town gate in 1492. When a piece of heaven fell to earth, it became a sacred object: the Ensisheim meteorite, weighing about a hundred pounds, had been carried to the church amidst the chanting of psalms. Paracelsus deduced that it was made of stone and iron, and he described the object in a book called *Concerning Meteors*, which was published in 1569.

Boltz greeted his guest warmly, and indeed they remained friends throughout Paracelsus' life; he dedicated a later tract on the Anabaptists to Boltz. But he did not settle, passing on to the capital of Alsace at Colmar. Here he was welcomed by Lorentz Fries, a like-minded alchemist and physician, who secured permission from the town leaders for his guest to stay. Finding here 'what he had sought after the storm, safety and bearable quiet days',[19] Paracelsus sent for Oporinus to join him with his belongings, and prepared once again to set up his laboratory in the cellar of his lodgings. With his furnace, bellows, tongs, crucibles and retorts, he was back in business.

12

Against the Grain

Quicksilver and Wood

> In recent times I have seen scourges, horrible sicknesses and many infirmities afflict mankind from all corners of the earth. Amongst them has crept in, from the western shores of Gaul, a disease which is so cruel, so distressing, so appalling that until now nothing so horrifying, nothing more terrible or disgusting, has ever been known on this earth.
>
> Joseph Grunpeck, *Tractatus de pestilentiali scorra sive mala de Francoz* (1496)

All the bile and bitterness that Paracelsus accumulated from his defeat in Basle pours out in the opening pages of his second great work, *Paragranum*, which he began writing in 1529. The title may be aptly translated as *Against the Grain*, and the book begins as a diatribe against the injustices and abuses to which he feels he has been subjected. It is the outcry of a man who seems to have lost all sense of proportion: paranoid, repetitive, vain and self-aggrandizing. There was, without doubt, some foundation to Paracelsus' charges against the academic physicians, for some of them surely were complacent, pompous, snobbish and avaricious. But the *Paragranum* makes it clear that, as in Luther's dispute with Rome, dialogue was no longer possible. This is a battle that can be decided only by total, unqualified victory. The enemy cannot be persuaded, but must instead be obliterated.

Paracelsus begins with an accusation and a justification:

> They reproach me that my writings are not like theirs; that is the fault of their understanding, not my fault, for my writings are well rooted in experiment and evidence and will send forth their young shoots when the right May-time comes.

All the criticism he has suffered is just a sign of the weakness of the opposition:

> They cry out because their art is fragile and mortal; what is not mortal does not cry out . . . The art of medicine does not cry out against me, for it is immortal and set upon such an eternal foundation that heaven and earth shall be shattered before medicine perishes . . . They cry out because I wound them; it is a sign that they themselves are sick in medicine; this disease is their struggle against me, because they are not pleased to be discovered and exposed . . . And because I write from the true source of medicine, I must be rejected, and you who are born neither of the true origin nor of the true heredity must adhere to the spurious art which raises itself beside the true.[1]

It is time now for plain words:

> Who is there amongst the instructed who would not prefer what is grounded on a rock to what is grounded on sand? Only the abandoned academic drunkards who bear the name of doctor must suffer no deposition! They abide, painted doctors, and if they were not painted with this title, who would recognize them? Their works would certainly not reveal them. Outwardly they are beautiful, inwardly they are squalid dunces.[2]

These false doctors, Paracelsus declares, will be eventually cast aside by heaven itself, in favour of the alchemical physicians:

> Heaven will make different physicians who will know the four elements, and in addition magic and Cabbala which are a cataract before your [that is, the Galenists'] eyes. They will be dowsers, they will be adepts, they will be Archei, they will be spagyri, they will have quintessences, they will have arcana, they will have mysteria, they will have tinctures. What will become of your soup-kitchens in this revolution? Who will dye the thin lips of your women-folk and clean their pinched faces? The devil, with the cloth of hunger.[3]

We see him working his way by degrees into a vitriolic, bombastic frenzy:

> Follow after me, Avicenna, Galen, Rhazes, Montagnana, Mesuë and others. Follow after me, and not I after you, you from Paris, you from Montpellier, you from Swabia, you from Meissen, you from Cologne, you from Vienna and from the Danube, the Rhine, and the islands in the sea. Italy, Dalmatia, Sarmatia, Athens, Greek, Arab, Israelite, follow me and not I you. Not one of you will survive, even in the most distant corner where even the dogs will not piss. I shall be monarch and mine will be the monarchy . . . the stubble on my chin knows more than you and all your scribes, my shoebuckles are more learned than your Galen and Avicenna, and my beard has more experience than all your high colleges![4]

Well, he is not exactly Erasmus. But amidst the bluster there is a weighty claim which shows that Paracelsus now identifies with those who have suffered persecution for their spiritual beliefs:

> With what scorn have you proclaimed that I am the Luther of physicians, with the interpretation that I am a heretic. I am Theophrastus, and more so than him to whom you compare me. I am that and I am a monarch of physicians, and I can prove what you cannot prove. I will let Luther defend his cause, and I will defend my cause, and I will rise above the charges which you level against me.

This is not simply a facile analogy. Paracelsus, whose medicine is rooted in a cosmic theology, sees his argument with the doctors as much more than an academic matter. His supposed medical heresy is as fundamental as Luther's religious heresy, and perhaps he even believes that he faces comparable dangers: 'Who are Luther's foes? The very rabble that hates me. And what you wish him you wish me: to the fire with us both.'[5]

The great battle in which he is engaged has, in fact, an even more exalted parallel:

> Christ was the source of blessedness, for which He was scorned, but the true scorn overtook the scorners when

> neither they nor Jerusalem remained. And I may well compare the doctors of the Schools and the barbers and bathmen to the hypocrites who loved the highest seats in the assembly of scorners.[6]

This rhetoric is on the one hand a reflection of its time, when every conflict was apt to take on apocalyptic dimensions as the Final Days approached. But Paracelsus' identification with the martyrdom of Christ is also characteristic of the obsessive, in much the same way that scientific cranks today invoke the name of Galileo in self-justification. Paracelsus is no crank – or rather, he is not *merely* that – but he shares the trying conceits and quirks of the type.

FELLOW SPIRITS

In late February of 1528 Paracelsus arrived at the house of Lorentz Fries in Colmar, and at first they got on splendidly. On the 28th Paracelsus wrote to Boniface Amerbach in Basle saying that Fries was in good health. Amerbach tells his friend regretfully that there is little chance of salvaging anything of his position in Basle – a possibility that Paracelsus seems to have discounted anyway – and invites him to stay at Neuenburg. In his reply, Paracelsus is as unrepentant as ever: 'Perhaps I spoke somewhat too freely against the magistrates and others, but what does it matter since I am able to answer the accusations made against me, as I have always maintained? . . . Truth draws hatred, first hatred from my professional fellows, then hatred, anger, envy from magistrates and judges.'[7]

Lorentz Fries was a man out of the same mould as his friends Cornelius Agrippa and Paracelsus himself: another of Lynn Thorndike's intellectual vagabonds, an unconventional physician and a scholar of the occult, part folk doctor and part charlatan. He encouraged Luther and Melanchthon to heed astrology, and he published several astrological almanacs, prompting Rabelais to parody him in the *Pantagrueline Prognostication* (1533), in which the master Alcofribas predicts things that always happen every year anyway. Conrad Gesner observed of Fries's medical compendium *Synonyma* that it was 'stuffed with a thousand mistakes and unworthy of learned ears'.[8]

Fries studied at Vienna, Pavia and Montpellier, and after establishing a medical practice in Colmar he later moved to Strasbourg and Metz. His reputation was built on his *Spiegel der Artznei* (1518), possibly the first medical book to be published in German. 'It occurs to me', Fries said, 'that the German tongue is no less worthy, that everything might be written in it, than Greek, Hebrew, Latin, Italian, Spanish, or French, into which all sorts of things are translated. Is our own language supposed to be inferior?'[9]

Although this, like the book's dedication to 'the poor sick of the common people', has a Paracelsian tenor, Fries warns the reader against swindlers and empirics and advises them to trust only those doctors possessing academic qualifications and knowledge of the liberal arts. Indeed, part of his motivation for writing in German was that ordinary folk should be made aware of this precaution. In many ways he followed the orthodox medical canon, although he rated Avicenna above Hippocrates and Galen (his sense of history can be gauged by his accusation that the Greeks borrowed their ideas from the Arabs). One of Fries's more exotic distinctions is that he is said to have been the first person to label the New World 'America' on a map of the globe.

The *Spiegel* is a bizarre mixture of conventional dogma and wild fantasy. It dismisses Brunschwick's sound and careful books on distillation as the work of a 'coarse peasant who lived near the fish market in Strasbourg',[10] before descending into pure superstition. This is Fries's cure for rheumatism: pluck and clean a fat old goose and stuff it with chopped cat, lard, incense, wax, mutton fat, salt, and flour of rye and beans. Roast it well, and rub the drippings on the limbs of the patient. To cure stones, cook a young hare to a fine powder and administer it in half-ounce doses morning and evening. Leprosy should be treated with an oil made from green lizards. And so on.

What Fries might have lacked in judgement, he made up for in salesmanship. Memory, he says in the *Spiegel*, is like a sieve, and merely reading the book is like using it to bail water. Therefore the reader had better buy a copy and keep it with them. Follow the instructions therein, and 'your memory will be that of a god rather than a man'.[11] The book went through many editions.

BACK TO THE BENCH

At Colmar, Oporinus left his master, exhausted by his demands, and returned to Basle. Despite some attempts by biographers to cast this separation as acrimonious, there is no real evidence that the parting was other than friendly, and Paracelsus even gave to his former secretary a portion of his famous laudanum remedy. Oporinus became a man of fluctuating fortunes: he married four times, and his reputation as a scholar and printer apparently did not spare him from the persistent spectre of debt. When Paracelsus' posthumous fame began to grow, Oporinus deeply regretted having lent out the books of his former master and never getting them back.

For a time Paracelsus thrived in Colmar. He told Amerbach that he had quickly become busy with patients there, and Oporinus confirms how much Paracelsus was admired in the town. He was welcomed into the humanist circle of the town officials, and befriended the town provost, Hieronymus Boner, and the magistrate, Konrad Wickram. His commitments to clients and his chemical experiments did not prevent him from writing: he dedicated a book on syphilis, paralysis, boils and agues to Boner in June, and another entitled *On Open and Visible Diseases* to Wickram in July.

The relationship with Fries was less agreeable. Given the Colmar doctor's Galenist inclinations and veneration of Avicenna, it was inevitable that the two would quarrel sooner or later. Fries's astrological prediction for 1531 contains a jibe against physicians who reject the ancients and claim to teach a new kind of medicine, which was probably a barb meant for Paracelsus. In time he was to accuse his former guest of being a necromancer who did the devil's work.

Paracelsus' influential connections did not help him gain permanent residence in Colmar, for the authorities were aware of his chequered career and controversial ideas, and would not extend the temporary permit required for residing in the town. Nor would they allow him to publish his book on syphilis. And so he had no option but to move on again, which he did some time in the first half of 1529. He travelled to the Duchy of Württemberg, east of Alsace, and made his way to the small town of Esslingen, close to the ancient seat of his family at Hohenheim.

The alchemist's laboratory, as depicted by Pieter Bruegel in the mid sixteenth century.

The Bombasts still kept a house there, facing onto the meadow of St Blaise. It stood empty, and would suffice as a temporary lodging.

At least the place had a cellar where Paracelsus could install his furnace and alchemical equipment. Yet these were not happy days that he spent in the dark basement, slowly poisoning himself with the fumes from his experiments. In so small a town there were few patients who might pay for his services, and he always insisted that he treat the poor for free. He secured assistants, but they were not learned men – more probably vagrants and rogues with little interest in his work and teachings. Bruegel drew a vivid image of the kind of conditions under which Paracelsus must have laboured, and the seventeenth-century alchemist Johann Becher confirms this picture:

> If someone wants to take up working itself, there is nothing in the world that can bring him sooner into confusion than [inept assistants will], both because of the operations as well as the instruments and the materials, which are abundant and diverse. If he has the misfortune to meet with a Laborant or assistant who is disorderly, messy, and lazy, inside a month

> the Laboratorium will look like the Confusion in Babylon. The used glass vessels have been smashed into bits so that they cannot be rinsed out. For each operation they take new crucibles and glasses just so that they do not need to clean the old ones. Whole things, broken bits, clean, dirty, new, used, prepared materials, raw materials, wooden, clay, and glass utensils are standing all mixed together. The windows, tables and floor are full. And nothing is labelled as to what it is. Tongs, ladles, hammers and other instruments lie strewn everywhere in such a way that when you want anything, you must search for half an hour. The stink from the furnaces, the soot and dust from the coals, the sand, water, and lime do not help the work so much as aid the confusion . . . In this way the days and years disappear, as do the costs for materials, instruments, coals, and salary. In contrast, not only nothing remains in the crucible, but also the corporeal gold and silver become dirty and adulterated. When a year is over, you know nothing about what has been done, in fact you know less at the end than at the beginning, for all processes look very well on paper . . . Which all makes alchemy as harmful as it is vexing.[12]

Astonishingly, the remains of Paracelsus' experiments are said to have still been visible in the house in 1882: a ceiling blackened with soot and covered in astrological and cabbalistic symbols, a small hammer, and a mortar and iron pestle. It must have seemed to the nineteenth-century renovators as though they had stumbled across the threshold of the Renaissance, and the occult atmosphere was clearly unnerving: the proprietor filled in the cellars but had a portrait of Paracelsus painted on the gable wall, based on Scorel's painting in the Louvre, to propitiate his spirit.

Paracelsus hints that his days in Esslingen were bleak and lonely, and before long he was on the road again, probably even more ragged than before, heading for Switzerland. He made his way to St Gallen, where he stayed at Castle Horn, home of his rich friends Bartholomew and Hieronymus Schobinger. He renewed his acquaintance with his former mentor Vadianus, and with the support of the Schobingers he constructed another chemical laboratory. Here Bartholomew arranged for Paracelsus to have his portrait painted, which shows that he had grown a beard (just as the *Paragranum* suggests) – in all other portraits he is clean-shaven. He is richly attired and wears the great broadsword that became his most constant companion.

But before the year was over, Paracelsus was heading back through Württemberg and on to the north-east, through Nördlingen, to the city of Nuremberg, where he arrived in November. 'My calamities,' he later said, 'which began at Esslingen, were but confirmed and added to at Nuremberg.'[13] His arrival is recorded in the Chronicle of the German humanist writer Sebastian Franck:

> Dr Theophrastus von Hohenheim, a physician and astronomer. In the year 1529 the said doctor came to Nuremberg. A peculiar and wondrous man, who laughs at the doctors and scribes of the medical faculty. They say he burned the writings of Avicenna at the University of Basle and that he stands alone against nearly all the medical guild. He uses his own judicial physic, and has contrarieties with many. His practice is against all, and he is another Lucian,* so to speak.[14]

Franck was a kindred spirit, and the two men met in Nuremberg and later in Augsburg. A wanderer who argued for non-violent ecclesiastical and social reform, Franck believed that scholastic learning and adherence to dry logic made men foolish and blinkered: 'the more learned, the more perverted',[15] he was wont to proclaim. He shared Paracelsus' conviction that true wisdom is all around us, written in the great Book of Nature. A committed Neoplatonist, he claimed that Plato and Plotinus 'had spoken to him more clearly than Moses did'.[16]

THE FRENCH DISEASE

Since his time at Colmar, Paracelsus had become intensely interested in one of the most feared medical conditions of his time. In Nuremberg he found the opportunity to study and to write on it in detail. But unfortunately for him, this disease and its

* Lucian of Antioch (d. AD 312) was regarded as the architect of the Arian heresy, which denied that Christ was truly divine.

The effects of syphilis, depicted by Dürer in 1496. The date indicated in the globe alludes to the claim by Theodoricus Ulsenius, city physician of Nuremberg, that the disease was caused by a conjunction of Jupiter and Saturn in 1484.

treatment had by then become not just medically controversial but also clouded by the politics of commerce.

Girolamo Savonarola, preaching hellfire at the height of his power in Florence in 1495, had warned that God was about to punish the sinful princes and clerics of Italy with 'a great scourge'. 'You evil servants, filthy as you are,' he thundered, 'let your loins be rotted with lust.'[17] And so it came to pass. In 1494, the debauched French soldiers garrisoned in Naples began to fall prey to a terrible disease that no one had seen before. Marcellus Cumano, an army doctor to the Venetian forces, described the chilling symptoms that he observed among the enemy:

> several men-at-arms or footsoldiers who, owing to the ferment of the humours, had pustules on their faces and all over their bodies. These looked rather like grains of millet, and usually appeared on the outer surface of the foreskin, or on the glans, accompanied by a mild pruritis. Sometimes the first sign would be a single pustule looking like a painless cyst, but the scratching provoked by the pruritis subsequently produced a gnawing ulceration.[18]

The pains were so awful that the victims lay screaming all day and all night. As the disease progressed, the suppurating flesh rotted away to the bone. Ulrich von Hutten, the humanist knight who protected Luther after Worms, contracted the disease in Italy in 1509 or 1510 and described its dreadful effects:

> It was of such filthiness that a man would scarcely think this sickness, that now reigns, to be of that kind. There were boils, sharp and standing out, having the similitude and size of acorns, from which came such foul humours and so great a stench that whosoever once smelled it thought himself to be infected . . . If anything may cause a man to long for death, truly it is the torment of this sickness.[19]

According to a French doctor in 1495, 'So repulsive is the appearance of the whole body, so great is the agony, above all at night, that this malady surpasses in horror both leprosy and elephantiasis and threatens a man's life.'[20] From Naples the new disease spread to Bologna in 1495, and in the following year Swiss troops returning home from the Italian Wars carried it to Geneva. It moved quickly throughout France and Germany, and was known in England by 1497. In that year Maximilian I declared that it was God's punishment on impious humanity.

In Germany its origins among the soldiers of France led to its being called the French disease; in France, needless to say, the blame was laid elsewhere (generally Spain). To the Turks it was simply the 'Christian disease'. It wasn't until 1530 that it acquired the now familiar name, when the Italian physician Girolamo Fracastoro invented a legendary origin for the *morbus gallicus* in which a shepherd complained to the sun god about the pitiless heat of the sun, and in protest began a cult of a mortal king.

The god's retribution was to afflict the shepherd with 'buboes dreadful to the sight'.[21] The shepherd's name was Syphilis.*

Not all doctors were ready to accept the divine origin of syphilis. Leoniceno, the famed physician who may have lectured to the young Paracelsus at Ferrara, could not understand how an entirely new disease could appear so abruptly. 'Surely', he said, 'men have been always afflicted by the same diseases; neither can I comprehend how this disease has suddenly destroyed our age as none before.' He proposed that it was due to putrescence of the air, caused by the stagnant waters left behind after heavy rains had created 'a great overflow of waters through all Italy'.[22] Others felt that the cause must be astrological: Hutten noted that 'The astrologers fetch the cause of this infirmity from the stars, saying that it proceeds from the conjunctions of Saturn and Mars, which happened not long ago, and of two eclipses of the sun.'[23]

In 1526 the Spanish physician Gonzalo Fernandez de Oviedo y Valdés pointed out how syphilis had first emerged in a port filled with Spanish sailors very shortly after Columbus returned from the New World, and he suggested that Columbus' crew might have brought the disease back with them. The idea that syphilis came to Europe from the West Indies has gained wide acceptance, although the truth remains unclear – it was possibly an old and dormant European disease that developed into a particularly virulent form at this time. Curiously, Paracelsus himself suggested as much, saying that 'new diseases' like syphilis are transformed versions of older diseases, creating a need for fresh theories and remedies.

It tells us much about the prevailing social mores that the connection with sexual activity took some time to be recognized: promiscuity was not an obvious suspect when it was so widespread. Even once this idea had been suggested, it was sometimes denied. The physician Scanaroli of Modena was perhaps being too credulous of his patients when he claimed to have observed cases of syphilis in virgins and sexually inactive men.

Sexual licence – and its new hazards – reached to all levels of society. Popes Alexander VI and Julius II were both rumoured to

* The origin of the name is not clear, but Fracastoro hints that it has Greek roots. In Alexandrian Greek, *syphlos* means shameful, hideous or deformed – certainly a plausible derivation.

suffer from syphilis, as were Erasmus and Charles V.* Francis I, an energetic philanderer (his enormous nose was believed to advertise his virility), was said to have been infected via a vengeful lawyer whom the French king had cuckolded: the man deliberately contracted the disease from a prostitute in the knowledge that his wife would pass it on to Francis.

It is equally telling that once syphilis was revealed as a sexually transmitted disease, blame was soon attached to women. One belief was that a woman became a carrier of syphilis if, having 'sores' in her womb, she had intercourse with a leprous man. 'This thing as touching women', said Hutten, 'resides in their secret places, having in those places little pretty sores full of venomous poison, being very dangerous for those that unknowingly meddle with them.'[24] Women, it was thought, did not actually suffer from the disease, but only harboured it and transmitted it to men – all the more readily because a syphilitic woman was believed to be more lusty. Meanwhile, a man could cure himself of the disease, so many thought, by having intercourse with a virgin (as many Africans believe of AIDS today).

Martin Luther, observing syphilis run rife in monasteries, called for the female culprits to be viciously punished:

> I must speak plainly. If I were a judge, I would have such venomous, syphilitic whores broken on the wheel and flayed because one cannot estimate the harm such filthy whores do to young men who are so wretchedly ruined and whose blood is contaminated before they have achieved full manhood.[25]

But the rapid spread of syphilis was due mostly to the mobility of armies criss-crossing Europe. Army camps were full of prostitutes, and even fear of the pox evidently could not quell the lust of the soldiers.

Syphilis was not in fact a major killer; compared to the death toll of the plague, it was insignificant. But its unpleasant and all too visible effects made it deeply feared and loathed in the early sixteenth century. What could the doctors do to cure it?

* It would not be surprising if such rumours were true, but that does not make them more than rumours. Renaissance society could be no less gossipy than ours.

As this was not a disease known (or, at any rate, described) in antiquity, Avicenna and Galen were of no help. That, indeed, was probably a part of the attraction for Paracelsus, who (as an apparent celibate) had no reason to fear it. The Arab physicians had recommended mercury remedies for some skin complaints, and since syphilis presented itself as an 'external' disease, these medicines were applied to treat it. One form of treatment involved rubbing the skin with mercury-based lotions; another advised inhaling the fumes of a bowl of warm mercury. 'A night with Venus', ran a popular saying, 'leads to a life with Mercury.'

Mercury inhalations had unpleasant side-effects: after a week or two the patient's teeth fell out and he began to shake uncontrollably. According to Hutten, patients treated in this way found that 'all their throats, their tongues, the roofs of their mouths, were full of sores, their jaws did swell, their teeth were loosed, and continually they voided the most stinking scum and matter that could be, and whatsoever it ran upon was polluted and infected by it.'[26] Folk remedies, while perhaps less harmful, were hardly more appealing. Gaspar Torella of Valencia recommended a particularly lurid cure that involved applying to the ulcerated penis a cock or a pigeon plucked and flayed alive, or a live frog cut in two.

Around 1517 a new remedy was found. Sailors returned from Hispaniola (Haiti) with a dense wood known as guaiac, from which an anti-syphilitic medicine could be prepared. Its origin in the West Indies was consistent with the belief (which Paracelsus advanced in other contexts) that local diseases have a local cure. Fracastoro celebrated this in his mythical account by explaining how a nymph called Ammerice urged people to look for a cure to Syphilis's disease among the sacred plants of the forest. Guaiac may have been discovered during a mission to the New World sent by Cardinal Matthäus Lang of Salzburg, who (like many cardinals and bishops) was himself syphilitic.

A decoction was boiled from shavings of the wood. The resulting frothy scum was applied to the syphilitic sores, and the patient drank the liquid, ideally while confined for thirty days in an enclosed, stuffy chamber. Hutten himself underwent the guaiac treatment and wrote a pamphlet in praise of its benefits. The syphilis nevertheless killed him eventually, but demand soared for the marvellous wood, this *lignum sanctum* or *lignum vitae*.

It was not a cheap cure, however, for it had to be imported

from the West Indies. (The poor had to continue to make do with the noxious mercury remedies.) The high price that could be demanded for guaiac wood opened up a profitable market. Seeing this opportunity, the enterprising Fuggers of Augsburg extracted from Charles V a monopoly on the sale of guaiac, in return for the loan with which Charles bribed his way to the Imperial throne. (Guaiac wood failed to cure the Emperor's own dose of the pox, and he is said to have resorted to an even more exotic medicine based on a Chinese root.)

WOOD OR METAL

When Paracelsus began to challenge the efficacy of the guaiac cure in his treatises on syphilis, he was therefore threatening not only what had become the conventional practice of the medical profession but also a lucrative business run by the most powerful merchants in Europe. Even the Church endorsed the guaiac cure, largely because the Fuggers' business was conducted under their auspices, but no doubt also because so many clerics were taking the treatment themselves. (Fracastoro's eulogy to the virtues of the guaiac wood, which drew heavily on Hutten's account, was added to his poem at the prompting of his patron, Cardinal Bembo.)

Paracelsus made a thorough study of the disease, and his description of its symptoms shows that he was quite capable of casting light as well as heat. He was convinced that the guaiac cure was worthless; but neither did he place any trust in the administering of great doses of mercury, which he saw did more harm than good. 'You anoint patients with quicksilver,' he said in 1538, 'you fumigate with its cinnabar, you wash with its sublimate and do not wish people to say it is poison; yet it is poison and you introduce such poison into man.' In mercury, he claimed, 'there is a concealed winter, coldness and snow'.[27]

Mercury *is* effective against syphilis – indeed, it was virtually the only known remedy until the arsenic-based drug Salvarsan, discovered in 1909 – but Paracelsus had the crucial insight that its efficacy does not simply increase with dosage. Rather, it is a matter of finding the correct dose. What is a poison if taken in excess, he argued, may be a medicine if taken in moderation, and in a form carefully prepared by the physician so that its toxicity is minimized:

> Is not a mystery of nature concealed in every poison? What has God created that He did not bless with some great gift for the benefit of man? Why then should poison be rejected and despised, if we consider not the poison but its curative virtue? . . . He who despises poison does not know what is hidden in it; for the arcanum that is contained in the poison is so blessed that the poison can neither detract from it nor harm it . . . In all things there is a poison, and there is nothing without a poison. It depends only upon the dose whether a poison is poison or not.[28]

This conclusion pleased no one. The Fuggers would not stand to see their profitable West Indian wood disparaged, and although they had interests in the sale of mercury too, at that time the market for it was not favourable and they had not (yet) obtained a monopoly. So when Paracelsus published a pamphlet in Nuremberg criticizing the use of guaiac wood, he was once again courting trouble.

He was already out of favour with the town doctors. They had been suspicious of him since his arrival, and soon challenged him to a public disputation. Mindful of his past experiences, Paracelsus made an alternative proposal: if they would let him treat a syphilitic patient, he would show them what he could do.

Outside the city gates was a quarantine hospital for 'lepers' – at that time leprosy was something of a catch-all term for wasting diseases of the skin and flesh, including syphilis itself and the affliction known as elephantiasis or 'Greek leprosy'. Apparently Paracelsus cured nine of the fifteen inmates of the hospital – something attested in the Nuremberg city archives, although we must assume that all he really did was alleviate their symptoms.

This miraculous feat did nothing to silence the hostile doctors, who continued to malign him. He returned the compliment, calling them cheats who were married to 'fat buxom wives', and accusing them of affecting cures with sorcery which they concealed by invoking the names of saints. We need only know that he was at the same time aligning himself with political radicals and launching abusive attacks on the Lutheran party among the town authorities to recognize the familiar pattern repeating itself.

Paracelsus followed up his pamphlet on syphilis with a more extensive work, *Essay on the French Disease*, of which the subtitle

A wood cut from Paracelsus' *Essay on the French Disease*, showing the preparation of the guaiac remedy.

– *About Impostors* – reveals its general tone. It included a damning appraisal of all concerned with the guaiac trade, including the credulous Nuremberg doctors who dispensed the wood:

> The spiritual and worldly traders (odd as it is to make wood an article of trade) have brought you doctors a wood and you have taken your medical theory and practice from them . . . The red hat [the Cardinal] and the Fuggers' wagons have brought the wood but not its virtue . . . Didn't you learn anything at school, that you must now learn your art from the Fuggers and let the Cardinal play the schoolmaster to you?[29]

A woodcut from the book's title page reveals that its target is not just the spurious guaiac cure, but the conduct of doctors in general. They bleed their patients in more ways than one, while confining them to a bleak and joyless life. On the left of the picture a ragged, barefoot syphilitic is being offered an unappealing chicken leg by a richly dressed physician while his guaiac remedy boils in the pot. On the right, a sumptuous meal and wine is laid out by an inn-keeper. I will not starve you like the Galenists and piss-pot doctors, Paracelsus implies, adopting the compelling visual language of the propagandizing pamphleteer.

Frontispiece of Paracelsus' *Prognostications Concerning Europe 1530–1534.*

Practica D. Theo phraſti Paracelſi/ gemacht auff Europen / anzufahen in dem nechſtkunffti gen Dreyſſigſten Jar/ Biß auff das Vier vnd Dreyſſigſt nachuolgend.

The first chapter of the *Essay* was printed in 1529; the other two were subsequently suppressed by the town censor. Nothing could be published in the city without the censor's approval, owing to a by-law passed in 1523 to stamp out distribution of the slanderous pamphlets of both the Lutheran and the Catholic parties. Not entirely lacking in guile, Paracelsus had dedicated his book to the censor and city chancellor, 'The Honourable and Estimable Master Lazarus Spengler', after the first chapter had received Spengler's approval. It was not enough, however, to secure his good opinion in the face of the objections from the outraged medical faculty at the university.

Rather than crying foul as he did in Basle, Paracelsus sought a more effective response. The prospective publisher of his book was the Nuremberg printer Friedrich Peypus, who had already brought out his *Prognostications Concerning Europe 1530–1534*, written in Esslingen. This short book, his first proper publication and the first known occasion on which he refers to himself by the name of Paracelsus, brought him to public attention as an astrological forecaster rather than a physician.

Paracelsus somehow persuaded Peypus to print the work on syphilis as an 'unauthorized' edition. It appeared in early 1530 under the title *Three Chapters on the French Disease*, still with its dedication to Spengler. It was not a good job: printed in haste, it was poorly proof-read and littered with errors. All the same, it provoked outrage among the doctors and the town council.

Paracelsus was already unpopular with the authorities, for he was now also speaking out against Lutheran orthodoxy. Around this time he began to resume the apocalyptic religious preaching that he had practised at Salzburg. In defying the town censors and releasing his banned book, Paracelsus finally overstepped the mark again and had to flee the city.

He did not let it rest there. Taking up lodgings in the small Protestant town of Beratzhausen, south of Nuremberg on a tributary of the Danube, he vowed in a letter to a friend in Nuremberg that he would re-enter the fray: 'When the ulcer [that is, the furore] subsides, I shall return to Nuremberg, in order to face these men as a man and to enjoy your friendship.'[30] And from Beratzhausen he sent for publication in Nuremberg an even more extensive treatise, *Eight Books on the Origin and Causes of the French Disease*.

The town authorities predictably forbade its publication too. But Spengler and a few of Paracelsus' supporters requested that it be sent to an independent medical expert for an assessment of its merits. The town aldermen acquiesced. This 'expert' was, however, anything but independent. He was Heinrich Stromer, the dean of the medical faculty at the University of Leipzig. Stromer's interest in the guaiac cure was not purely professional; he owned shares in the Fuggers' import business. He advised the Nuremberg town council that neither this nor any other book by the impudent doctor should appear in their city.

On hearing this judgement, Paracelsus responded in characteristic style, doing more to vent his spleen than to further his cause. 'It is not your business', he wrote at first to the Nuremberg aldermen:

> to judge or forbid without careful consideration and discussion. As a matter of fact, you are not able to judge of my work, you have not enough intelligence. If the University has any reason to complain of me, let it appoint a Disputation, not forbid public publication. Until I am vanquished in a Disputation such a prohibition is repression of the truth. Printing is for the bringing of truth to light. My writing concerns neither government, princes, lords nor magistrates, but occupies itself with the deceptions of medicine so that all men, rich and poor, may be set free from abomination . . . I do not speak blasphemy or calumny against anybody. There are others who revile all authority, secular and clerical, noble and common, in print and otherwise, and it is being tolerated. I do not do that. I only denounce the abuses of medicine in order to protect the common man from robbery.[31]

He was quite right, of course. But somehow he managed to check his impulses, and did not send the letter. On 1 March 1530 he wrote another, somewhat more courteous, one. All he had, he said, was a great desire:

> to write what would really benefit the sick, who were so grievously maltreated and allowed to perish. I trusted that a city like Nuremberg, which was celebrated for its action in protecting the truth, would also protect the men who made the truth known, and would grant them room and refuge. Let those who doubt the truth of my statements meet me in an open Disputation, which, as formerly, so now, I would willingly attend.[32]

He received no reply.

13
The Alchemist Inside

A Hermetic Biology

> We may consider the whole animal Body as an instrument, which, from the nourishment it receives, collects materials for continual chemical processes, and of which the chief object is its own support.
>
> Jöns Jacob Berzelius, *Progress and Present State of Animal Chemistry* (1818)

Although the *Paragranum*, the book that Paracelsus completed in high dudgeon in Beratzhausen, begins as a naked outpouring of rage, it is much more than that. After its opening rant against the foolishness and ignorance of the medical profession, it settles down to provide the first clear systematization of Paracelsus' theory of medicine.

How remarkable, and yet how characteristic, to find amid the bile and fury of this period of Paracelsus' writings a declaration in his *Hospital Book* (1529) that 'the highest ground of medicine is love'.[1] The doctor can do nothing if he does it without love. But if it is hard to imagine the belligerent and volatile Dr Paracelsus administering his potions and cures with love, we should remember that in the sixteenth century the Greek word *agape* signified neither individual infatuation nor a kind of vague, generalized benevolence. Rather, it conveyed a profound spiritual concern for the well-being of others. Unconditional and unlimited, *agape* has a religious quality, the 'fruit of the Spirit'[2] as the Bible expresses it: it is the love of Christ for humankind. For Paracelsus, the love of a physician for his patients is the love of which St Paul speaks in the first book of Corinthians: 'it always protects, always trusts, always hopes, always perseveres.'[3]

Paracelsus brings this idea of Christian love directly into medicine, insisting that the doctor's task should go beyond the Hippocratic oath, which constrains the physician only to avoid intentional harm. Yet Paracelsus hardly lived his life according to saintly notions of a love which 'does not envy, does not boast, is not proud; is not rude, is not self-seeking, is not easily angered'.[4] He seemed to see no contradiction in writing the kind of vituperative polemic with which the *Paragranum* commences (a second draft was more moderate) and yet entreating the doctor to act always out of love. For the latter is a Christian duty, and in practising medicine the physician is doing God's work. In dealing with the medical traditionalists, on the other hand, he clearly felt better advised to follow Luther's dictum and use 'a tough axe on coarse wood'.[5]

In the *Paragranum*, Paracelsus presents the 'four pillars' of medicine: philosophy, astronomy, alchemy and virtue. He sounds unexpectedly Aristotelian in the way that he links this schema to the four elements:

> The first pillar, Philosophy, is the knowledge of earth and water; the second pillar, Astronomy together with Astrology, has a complete knowledge of the two elements air and fire; the third pillar, Alchemy, is knowledge of the experiment and preparation of the four elements mentioned; and the fourth pillar, Virtue, should remain with the physician until death, for this completes and preserves the other three pillars.[6]

But these four pillars are not simply a list; they spell out a kind of hierarchy or a sequence of steps by means of which the doctor turns knowledge into medicine. 'Philosophy' here implies an over-arching theory about how the world ('earth and water') works: it provides the 'scientific' fundamentals of the healing arts. We have seen that for Paracelsus this philosophy was grounded in Neoplatonism and a belief in correspondence between the microcosm and the macrocosm: it was, in short, the set of ideas underlying natural magic.

I shall deal with the second pillar of medicine – astronomy – in due course. Paracelsus does not mean astronomy in the modern sense, for in the Renaissance it was barely distinguished from astrology: one might say that astronomy was knowledge of the celestial bodies and their movements, while astrology pertained to the interpretations of these things as they apply to human affairs. Like his contemporaries, Paracelsus thought of astrology as a central aspect of medicine – although, as we shall see, he stopped well short (as some others did not) of the belief that a person's fate was preordained in the stars. Without a knowledge of astrology, one could not hope to discover the healing virtues of herbs and minerals, nor to understand how they should be applied.

So far, his contemporary physicians would have found little to dispute in this list of medicine's foundations. But alchemy? That was no part of the regular university training. It was viewed as a lowly and futile pursuit, a business for apothecaries and charlatans. But as the *Archidoxa* shows, Paracelsus considered that making medicines was not simply a mechanical process of boiling herbs or grinding minerals, as Dioscorides and Rhazes had advised. Rather, alchemy was a process of separation to remove the impure from the pure, using techniques such as distillation to concentrate the potency of nature's materials. Nor was this early chemotherapy a matter of simply selecting the right medicine and telling the patient to take it thrice daily. For by 'virtue', the fourth pillar, Paracelsus means the skill of the doctor. This involves knowledge and experience, but also something akin to an innate spiritual quality: the love with which treatment must be given. Just as there are 'virtues' in plants which the alchemical physician must extract, concentrate and mobilize to bring about healing, so the virtue of the doctor was like an invisible force that supplied him with intuition and special powers.

Virtue was the mark of the magus. It was not something that could be learnt from a book, although it could be developed by studying the occult arts. Just as in the Gnostic tradition knowledge of God is delivered by revelation, similarly an understanding of the physical world and its mechanisms arrives by divine providence and reveals that the individual is favoured by God. It makes the doctor special not because he has credentials to proclaim his learning and is entitled to wear an expensive hat, but because he has a privileged insight into the way the universe is governed. 'The doctor', said Paracelsus, 'is not subject to human law, but only to God through the medium of nature.' And he does God's work: 'Medical science too is full of mysteries, and must be studied like the words of Christ. These two callings – the promulgation of the word of God and the healing of the sick – must not be separated from each other. Since the body is the dwelling place of the soul, the two are connected and the one must open access to the other.'[7] The healer is truly a holy man, as the Tartars believed. Paracelsus' sacred view of medicine perhaps makes his headstrong fanaticism more comprehensible. Like Luther, his reforms were all the more urgent because he was on a mission from God to save the damned from suffering.

It has been suggested that the privileged status of the magus in Neoplatonic tradition is linked to the Renaissance notion of genius, which held that great men such as Titian or Leonardo were blessed with uncommon abilities. But this specialness is really a medieval concept whose roots lie in the miraculous and ecstatic visions of saints. It is concerned with the individual not as a remarkable personality gifted with great creative powers but as one of God's elect. 'I have been chosen by God to extinguish and blot out all the fantasies of elaborate and false works, of delusive and presumptuous words,' said Paracelsus, 'be they the words of Aristotle, Galen, Avicenna, Mesuë, or the dogmas of any among their followers.'[8] He even goes so far as to say that, when he is abused by the other doctors, 'they would not punish me, they would punish Christ'.[9] By 1530, when Paracelsus was completing the *Paragranum*, Martin Luther had come to much the same conclusion and was openly defying the authority of the Church in the belief that God had selected him to be its 'reformator'.

A SYSTEM OF HEALING

The beginnings of Paracelsus' synoptic view of medicine appear in his very early works, such as the *Volumen medicinae paramirum* (*c.*1520) (not to be confused with the grand *Opus paramirum* that followed the *Paragranum* in 1530). Here he states that there are five 'active principles' – the five Enses or *entia* – that influence our bodies and give rise to disease:

- *ens astrorum*, the 'virtue of the stars'
- *ens veneni*, poisoning or disturbance of the metabolism
- *ens naturale*, a person's natural constitution
- *ens spirituale*, which causes 'diseases of the spirit' such as mental disorders
- *ens Dei*, diseases sent by God.

The first of these – the importance of the stars and planets – was no more than general knowledge. 'It hath been many times experimented and proved', wrote Jean d'Indagine in his textbook on chiromancy and physiognomy in 1575, 'that that which many physicians could not cure or remedy with their greatest and strongest medicines, the astronomer hath brought to pass with one simple herb, by observing the moving of the stars.'[10] Astrologers cast horoscopes for diagnoses (see Chapter 15), forecast the most favourable times for treatment, and predicted the course of a pregnant woman's labour and the sex of the child.

'The physician who does not understand astronomy cannot be a complete physician,' Paracelsus asserted in his major work on the subject, *Astronomia magna* (1537–8), 'because more than half of all diseases are governed by the heavens.'[11] The power of the stars, he says, is transmitted mechanically through the emission of 'vapours' that seep into the world and influence its nature. Some of these 'exhalations' cause illnesses. There is nothing remarkable about such astral poisons, which are akin to those already familiar to the doctor – they are like the red arsenic that poisons the blood, the mercurial poisons that hurt the head, the orpiment (arsenic sulphide) that produces dropsy and tumours. And so the physician can combat them with chemical remedies and negate the baleful prospects of the stars. To this extent, Paracelsus takes a scientific view of the astrological character of aetiology: the principles that operate in the heavens are no different from those on earth.

Poisons have many mundane origins too. They are hidden in food, for example, where they are identified as the *ens veneni*. In this class of diseases we can begin to discern the outlines of a rudimentary biochemistry: for Paracelsus suggests, in a brilliant and original insight, that in the process of ingestion and digestion there is a kind of alchemy going on.

We have seen how alchemy has its origins in metallurgy, which at face value suggests that it has nothing to do with living organisms. But in prescientific times there was no clear division between the inorganic and the vital worlds, for even metals were thought to have a kind of sluggish, unresponsive 'life', and were considered to ripen and decay in the earth just as a fruit does on the branch. It was surely this vitalistic picture of the universe that allowed Paracelsus to construct what we might call a bio-alchemy: to intuit that man and the universe are chemically related.

A vague natural animism would have contributed nothing new.

Yet Paracelsus developed the idea further, mapping out detailed correspondences between the chemical processes of the alchemist's retorts and alembics and those that happened in the body. Human bodies, he proposed, were themselves alchemical laboratories.

It has often been suggested that the Scientific Revolution replaced man the spiritual being with man the clockwork mechanism, a system of pumps and levers. According to Fabricius ab Aquapendante, William Harvey's teacher at the University of Padua, 'the mechanism which nature has devised is strangely like that which artificial means has produced in the machinery of mills.'[12] Harvey himself conceived of the heart as a pump or 'water bellows'. But Paracelsus' 'alchemical man' stands between the two extremes of spirit and matter – and is more than either. It might appear that his vision of the body as a series of chemical vessels heating, distilling, subliming and refining their contents is every bit as mechanical and lifeless as the 'machinery' of Fabricius. Yet we must remember that the vessels of the alchemical laboratory were themselves regarded as hosting a process that was as much spiritual as it was material. There was no essential distinction (or rather, the distinction was subtle) between the spirit of wine (alcohol) condensed in a distilling apparatus and the spirit of man. Both were purified essences of something material and mundane.

For Paracelsus, purification by separation was the essence of alchemy. Alchemy, he says, 'is nothing but the art which makes the impure into the pure through fire . . . it can separate the useful from the useless, and transmute it into its final substance and its ultimate essence.'[13] And just as the alchemical adept labours to do these things in his workshop and thereby to create gold or medicines, so there is an alchemist inside man himself, whose job it is to separate those elements in food which nourish from those which poison. Paracelsus called this inner alchemist the *archeus*.

'There is no poison in the body,' Paracelsus explains in the *Volumen medicinae paramirum*, 'but there is poison in what we take as nourishment. The body is perfectly created, but food is not. Other animals and fruits are our food, but they are also a poison to us'[14] – as they were all too often in that time when few could afford to throw away a piece of meat just because it had gone a little rank.

> Because everything is perfect in itself, but both a poison and a benefit to another, God employed an alchemist, who is

> such a great artist at dividing the two from each other, the poison into his sack, the goodness into the body . . . Just as a prince knows how to employ the best qualities of his servants and to leave the others alone, so the alchemist uses the good qualities of our food for our nourishment and expels those things that would harm us.[15]

It is the archeus that performs these most basic processes of life. It resides in the stomach, and knows how to make our flesh from what is good in food while rejecting what is bad:

> The alchemist takes the good and changes it into a tincture which he sends through the body to become blood and flesh. The alchemist dwells in the stomach, where he cooks and works. The man eats a piece of meat, in which is both bad and good. When the meat reaches the stomach, there it is the alchemist who divides it. What does not belong to health he casts away to a special place, and sends the good wherever it is needed . . . That is the virtue and power of the alchemist in man.[16]

This alchemical power of transformation within the body explains why one substance, when eaten, appears in another form – for 'we do not have to eat hair to grow a beard'.[17] Therefore, Paracelsus concludes:

> all our nourishment becomes ourselves; we eat ourselves into being . . . For every bite we take contains in itself all our organs, all that is included in the whole man, all of which he is constituted . . . We do not eat bone, blood vessels, ligaments, and seldom brain, heart, and entrails, nor fat, therefore bone does not make bone, nor brain make brain, but every bite contains all these. Bread is blood, but who sees it? It is fat, who sees it? . . . for the master craftsman in the stomach is good. He can make iron out of brimstone: he is there daily and shapes the man according to his form.[18]

What an inspiration this was to the alchemist, to see that such wonders are possible! The central tenet of Paracelsus' alchemy is that man can learn to perform the transformations that nature effects, of which the maturation of metals to gold is only one. If alchemy has the power to turn barley and turnips into flesh and blood, what might not be possible in the alchemical kitchen? Might we even aspire to create life itself?

The poisons separated by the archeus are not necessarily (as might be supposed) expelled from the body as excrement, but rather they leak away each through its own particular exit: 'sulphur through the nose, arsenic through the ears'[19] and so on. Which neatly explained the foul breath of so many of his contemporaries, one supposes.

This force of life operates in animals too:

> The ox eats grass, man eats the ox. The peacock eats snakes and lizards, animals complete in themselves but not good food except for the peacock . . . Every creature has its own food, and an appropriate alchemist with the task of dividing it. The ostrich has an alchemist to separate its excrement from its nourishment. The salamander eats fire and he needs his own alchemist. The pig will eat human excrement because the alchemist of the pig is more subtle than the alchemist of man and can still separate nourishment from the excrement. Pig excrement is eaten by no animal because there is no cleverer alchemist to extract the nourishment than the alchemist of the pig.[20]

Paracelsus' writing is never more clear or delightful than when he has a really good idea.

The archeus, Paracelsus tells us later in *On the Nature of Things* (1537), leaves 'signs' on the creatures it fabricates so that we may make deductions about what they are and what they have experienced. It 'signs the horns of the stag with branches by which its age is known . . . So, too, the signator marks the horns of the cow with circles from which it is known how many calves she has borne.'[21] God, then, is not the only power capable of leaving signatures in nature that can be understood by the magus. Indeed, man too is a signator in the prosaic sense that he creates signs such as the coats of arms or the coloured cloaks of knights, which denote their rank and allegiance. Nature leaves signs of a person's character in the face and hands, making possible the arts of physiognomy and chiromancy. It is not only the human hand that can be 'read', but also 'all herbs, woods, flints, earths, and rivers – in a word, whatever has lines, veins, and wrinkles'.[22] The archeus, here promoted from a digestive alchemist to an organizer of all nature, also 'marks the clouds with different colours, whereby the tempest of the sky can be prognosticated. So also he signs the circle of the moon with distinct colours, each one of which has its own special interpretation.'[23] All of nature *signifies*, all is purposive, once our eyes are opened to it.

But if the archeus in man exists to remove the poisons from what we eat, why then is there an *ens veneni*, the source of ailments caused by these toxic residues? Well, no mechanism is foolproof:

> Sometimes the alchemist does his work imperfectly and does not divide the bad from the good thoroughly, and so decay arises in the mixed good and bad and there is indigestion. All maladies from the *ens veneni* arise from defective digestion.[24]

It is possible to see all of this as a prescient forebear of today's materialistic view of the organic world, where all life arises from nothing but the interactions between atoms and molecules. But Paracelsus' theory of the archeus and the alchemy of life inevitably led him into deeper waters. It would scarcely have been feasible for a philosopher of the Reformation to have overlooked the theological implications of these ideas, which appear to allude not only to the 'daily bread' of the Lord's Prayer but to the mystery of transubstantiation, the incendiary issue that divided Luther from Zwingli. What does the archeus find in the sacramental wafer? Paracelsus does not say (though it was surely not diplomacy that held him back). But in the interchangeability of bread and body he saw a sign of God's mercy and love:

> He receives his first nourishment from his mother through mother-love, and then he receives it by the mercy of God, to whom his daily petition rises: 'Give us this day our daily bread', which also means 'Give us this day our daily body' . . . It is for this that Christ taught us to pray, just as if he had said: 'The body received from your mother is not sufficient: it might have died today, yesterday, or long ago.' Bread is now and henceforth your body: you live no longer by the body of justice, but by the body of mercy: therefore pray

> your heavenly Father for your daily bread, that is, for your daily body which is the body of mercy: we eat ourselves daily not in justice but in mercy and prayer.[25]

To Paracelsus, the material was spiritual, and the spiritual material.

As for the *ens naturale*, what else is it but an admission of the sad and inevitable frailty of corporeal mankind, of the fact that one day our bodies will falter and die? Except that, in Paracelsus' view of the world, things do not just *happen*; he regarded the *ens naturale* as a power connected to astrological forces. We should not imagine this means our fate is fixed by our planets, however. Even though each vital organ corresponds to a planet – the liver to Jupiter, the brain to the Moon, the heart the Sun, the spleen Saturn, the lungs Mercury, the gall bladder Mars and the kidney Venus – yet the one is not governed by the other: 'Saturn has nothing to do with the spleen, nor the spleen anything to do with Saturn.'[26] Rather, these correspondences are simply a manifestation of the cosmic mirror that makes man a microcosm of the universal macrocosm. The two are analogues, but are not causally related. From a scale model of a building you can read the proportions and relationships of the building itself; but crushing the former does not raze the latter.

Paracelsus says that the seven organs, like the seven known celestial bodies, are autonomous and form a kind of solar system characterized by 'circuits' like the planetary orbits. Exactly what this means is somewhat vague – he did not intend to imply that the organs actually move around the body. Yet each, like the heart and lungs, has a characteristic rhythm appropriate to our human scale: for 'astronomical time is long, human rhythms are short'.[27] These 'circuits' of the organs produce various conjunctions, oppositions and so forth, each of which denotes a particular state of good or ill health just as planetary conjunctions bring about war and pestilence (see Chapter 15). The trajectories are set at birth, and they encode the life history of the person – including the day when all courses are run, the whole mechanism comes to a halt, and life ends:

> When a child is born, his firmament and the seven autonomous organs like the planets are simultaneously brought into being . . . At the time of birth this firmament of the child acquires predestination – that is, how long the *ens naturale*

> will run for. Take the example of an hourglass which one sets up and allows to run. As soon as it begins, you know the time it will end. Nature also knows how long the *ens naturale* will run.[28]

Thus some people live to a ripe age simply because that is how their internal 'clock' is set; for the same reason, others die when they are but infants. Those whose rhythms run more slowly last longer:

> If a child is predestined to live but ten hours, its bodily planets will complete all their circuits, just as they would if it had lived for a hundred years. The bodily planets of a centenarian, on the other hand, perform exactly the same number of circuits as those of a child which survives an hour, only at a slower rate.[29]

Here is a poignant echo of the modern realization that every creature's life is counted in more or less the same number of heartbeats, more or less rapidly executed.

By constraining man to run according to the same material principles as the rest of the cosmos, was Paracelsus ultimately demoting humankind and helping to usher in the Enlightenment idea of man as machine? Or was he, by making of man an entire autonomous microcosmos, reflecting the humanist desire to ennoble humanity and free it from medieval subservience to higher fates and forces? We can read him either way, but both are retrospective interpretations. For Paracelsus, this was simply how things were.

In the *Volumen medicinae paramirum*, after explaining the *ens naturale* Paracelsus unexpectedly invokes the four Galenic humours. But he instantly dismisses 'the traditional teaching that these are derived from the stars or the elements'.[30] Instead, he says, they are manifested in the body in the form of the four flavours: sour (melancholy, identified with the Aristotelian principles of cold and dry), sweet (phlegm: cold and moist), bitter (choler: hot and dry) and salt (blood: hot and moist). How strangely apt these assignations seem, reminding us that the classical partitioning of material experience was far from arbitrary. With this sleight of hand, Paracelsus can give a conventional account of the four-humour theory, with its notions of imbalances, by speaking of predominances of salt and so forth and thus withholding any support for the despised Galen.

It is hard to decide whether the *ens spirituale* was conceived more out of superstition or insight. The spirit, says Paracelsus, may be invisible but 'it can suffer from all diseases just as the body can . . . But do not forget that when the spirit suffers, so does the body.'[31] Who will argue with that?

Yet we should not be too ready to identify such spiritual diseases with mental illnesses (discussed in Chapter 16), although surely the two were sometimes linked. Rather, in Paracelsus' scheme the spirits of humankind take on the pale mantle of ghosts: each person possesses one, and 'they converse with each other, not through our speech but through their own'.[32] Spirits may fall out and injure one another, he says, and then the sickness manifested in the body calls for a 'spiritual medicine'. The harm caused by attacking a person's spirit was precisely what necromancers and sorcerers wrought:

> You know that if a wax image of another person is buried and weighted with stones, then that person will suffer pain in the places where the stones lie and will not recover until the image is unburdened. If a leg is broken in the image, then that person will also break his leg; the same applies to stabs, wounds, and other things.[33]

Anthropologists acknowledge that this kind of witchcraft is 'real' enough in societies that believe in it, insofar as people who are convinced they are the victim of such sorcery can suffer immensely and may indeed be 'cured' by a magical treatment that they recognize. That Paracelsus credited such things is not at all surprising: it was common knowledge in the sixteenth century and supported by frequent individual experience.

The fifth and final *ens* – the *ens Dei* – illustrates one of the most profound paradoxes of the late Middle Ages. Paracelsus was not at all unusual in attributing plague, syphilis, dropsy, epilepsy and so forth to specific *natural* causes that were susceptible to medical intervention. Yet it was obvious to others, some physicians among them, that great pestilences like the plague were sent by God to punish sinful mankind. How then could material remedies prevail against divine wrath? And if they did, was that proper? Shouldn't men in all their wickedness merely accept the fate God allotted to them, rather than try to frustrate Him with medicine?

A minister, said Luther, should not flee from a city wracked with plague, but should stay and tend his flock, trusting that God would spare him if he were worthy. It was not so much a matter of Christian duty to the sick (although depriving a community of spiritual guidance left it vulnerable to the influence of the devil) as of an obligation not to try to evade God's will. 'This is God's decree and punishment', Luther said, 'to which we must patiently submit and serve our neighbour.'[34] According to the *Ancrene Riwle*, a thirteenth-century treatise on the rules and duties of monastic life:

> Sickness helps man to understand what he really is, and to know himself . . . Sickness is your goldsmith, who, in the bliss of heaven adds gilding to your crown. The more intense your suffering, so the more elaborate the goldsmith's work becomes: and the longer it lasts, the brighter shines the gold in the dusk when the martyrs gather in heaven, because of the pain you have endured on earth with a good will.[35]

Yet Luther was pragmatic enough not to advocate total laissez-faire. Indeed, this might be seen as arrogantly tempting God to strike one down. For after all, God has provided medicines for us, so it was well to use them. 'You ought to think this way,' he advised. '"Very well, by God's decree the enemy has sent us poison. Therefore I shall ask God mercifully to protect us. Then I shall fumigate, help purify the air, administer medicine, and take it. I shall avoid places and persons where my presence is not needed in order not to become contaminated . . . If God should wish to take me, He surely will find me."'[36]

Part of the confusion about God's role in disease was scriptural. The Bible states clearly that plagues were sometimes sent by God to punish the wicked, but also that He sometimes saved people from the plague as if rescuing them from some malevolent threat. Why would He need to do both?

There was also uncertainty about the *operative* causes of pestilence and disease – not where they came from, but how they were manifested and transmitted in the world. Were they sent directly from the hand of God? Or, if an ill wind from a star was to blame, was that God's doing? This wider question continued to impose itself between theology and science for hundreds of years: did God work within the mechanisms he created in the beginning, or outside them? Andreas Osiander, the leading Lutheran in Nuremberg in the late 1530s, decided in 1537 that although God was the first cause, He availed himself of secondary motive forces:

> I will not enter against them, that speak naturally thereof, and say: Such plague cometh out of the influence of the stars, out of the working of the Comets, out of the unseasonable weather and altering of the air, out of the South winds, out of stinking waters, or out of foul mists of the ground: for such wisdom of theirs we will leave unto them undespised, and not fight there against: But (as Christian men) we will hold unto us the word of God . . . Namely, that this horrible plague of the pestilence cometh out of God's wrath, because of the despising and transgressing of His godly commandments.[37]

Paracelsus seems to take a similar view. 'All health and disease comes from God and not from man,'[38] he says – only then to make a careful distinction between 'natural' and 'purgatorial' afflictions. The former are due to the first four *entia*; the latter are to be regarded as 'chastisement', and the physician has no power to combat this *ens Dei*. Yet at the same time, Paracelsus is careful to spell out that *whenever* a physician attempts a cure, it will not work unless God wills it, for:

> every disease is a chastisement: therefore no physician can heal it, unless this chastisement is ended by God. The physician should be one who works in the consciousness of predestined chastisement. Considering that every disease is a chastisement, a physician should reflect that he cannot determine the hour of recovery nor the hour of his medicine's effectiveness . . . God created medicine to combat disease, and also the physician, but He denies them both to the patient, until the hour has come when Nature and art can take their course. Not until the time has come, and not before.[39]

One way of reading this is as a convenient excuse for medical failure. But although it is unlikely that all doctors overlooked that convenience, it would be fairer to say that these are the words of a physician struggling to comprehend what science, as an inquiry into nature, can mean in an age when all of nature was created and presided over by an attentive God. In that respect, what is remarkable is that the model of nature that Paracelsus devised did not dovetail even more awkwardly with theology.

SAINTS AND SINNERS

Paracelsus continued to wrestle with the tensions between medicine and Christian theology all his life, and it is futile to look for consistency in his conclusions. Despite his notion of *ens Dei*, for instance, he insists elsewhere that even 'incurable' diseases were natural and not heaven-sent. His most persistent and enduring message, however, was that there was no room in medicine for religious superstition. The powers of heaven might be remarkable, but they were not inexplicable – indeed, they were not qualitatively different from the forces that the physician summoned in his treatments. When Christ performed his miracles, Paracelsus insisted, he drew on powers that are in principle available to all – for there were miracles in the Old Testament too. Thus, the miraculous cures that were constantly being reported in Paracelsus' time were nothing more than good physic. After all, why should God, who had in His wisdom filled the world with medicines so that the magus might discover and use them, rely on the haphazard dispensation of prayer-invoked marvels to redeem sick people?

The hope of miraculous cures drew great crowds to the shrines and relics of saints, where they prayed fervently and made offerings. Paracelsus denounced their belief as superstitious – indeed, he regarded it as impious and dangerous, since it was a form of idolatry. The beautiful images of saints, he argued, tempted people towards sorcery, a temptation perceived and encouraged by the devil. The cult of saints had a strong hold on the public imagination in the sixteenth century, and was actively cultivated by priests intent on drawing worshippers (and thus donations) to their own shrines. Some of these places would be piled with wax models of afflicted limbs and body parts left there by people hoping for a cure. Zwingli condemned this as idol worship, and

Erasmus remarked sardonically that there seemed to be a handy saint for every misfortune: one cured toothache, another gave protection to women during childbirth, others watched over sailors or domesticated animals. Certain diseases became associated with saints, such as the nervous condition called St Vitus's dance, which Paracelsus referred to instead by its classical name, chorea lasciva.

Paracelsus' response was pithy: 'The saints are in heaven, not in wood.'[40] In *On the Causes of Invisible Diseases* (1531–2) he dissected the relationship between saints and illness, arguing that there was always a naturalistic explanation both for a disease and its cure. Even in those cases (and there was then, as now, no end of them) where a miraculous cure seemed to result from a visit to a shrine, Paracelsus preferred a naturalistic explanation. The bodies of saints, he argued, were particularly rich in those same curative forces that gave herbs and minerals their medicinal powers. These corpses also possessed a kind of magnetic attraction that drew people to them: a natural phenomenon that the Church exploited shamelessly by surrounding it with ritual and mysticism. Visiting a saint's shrine was in fact no different from visiting a spa.

It was a brave man indeed who, in the face of the seemingly random efficacy of Renaissance medicine, tried to produce a rational explanation. Faced with a redundant Galenism on the one hand and a world dominated by God, saints and demons on the other, Paracelsus' attempt to steer a course that rejected useless tradition while acknowledging Christian spirituality must command admiration. It is hardly surprising that he rambled, that he invented wildly, that he contradicted himself, became incoherent, and that, most of all, he occasionally despaired. For the sad truth was that no matter how diligently he applied himself, no matter how great his skill and his knowledge, yet still the chances were that his efforts would come to nothing and his patient would die.

14
Beyond Wonders
The Matrix of the World

I did not know what to say, my mouth
had no way with names,
my eyes were blind.
Something knocked in my soul,
fever or forgotten wings,
and I made my own way,
deciphering that fire,
and I wrote the first, faint line,
faint, without substance, pure nonsense,
pure wisdom
of someone who knows nothing,
and suddenly I saw the heavens
unfastened and open,
planets, palpitating plantations,
the darkness perforated,
riddled with arrows, fire and flowers,
the overpowering night, the universe.

Pablo Neruda, 'Poetry'

While Paracelsus wrote in Beratzhausen, the spread of Lutheranism throughout the German states provoked Charles V at the Diet of Speyer in 1526 to threaten religious dissenters with severe punishment. At another Diet three years later, the Edict of Worms outlawing Luther was implemented. The reformers protested to the Diet, after which they were popularly known as Protestants. When Melanchthon's moderate creed, the *Confessio Augustana*, failed to make reconciliation with the Roman Church at the Diet of Augsburg in 1530, the Lutheran princes concluded that further negotiation was pointless. The following year the princes of Saxony and Hesse joined with other Lutherans in the League of Schmalkalden, vowing to support one another in defending their religious rights. (Luther himself, conservative as ever, did not approve.) Thus the Reformation descended into a bloody and Machiavellian squabble between the Church, the empire and the princes.

Some time in the early part of 1530, Paracelsus left Beratzhausen and rode along the Danube to the city of Regensburg. He may have been summoned to that region by a wealthy man called Bastian Kastner, who lived in the nearby town of Amberg. Paracelsus went to treat Kastner's sick leg, which no other doctor had been able to heal. It was not a happy encounter, as Paracelsus himself relates:

> [Kastner] had no rest, and below his knees nothing was left but stench and suppuration. You know how such cancerous ulcers develop. This man asked me to see him, promising a good fee as sick people always do when they need us. However, the closer to the cure, the less they think of their promises. So it was here, too. Now it would have been decent to reimburse me for the eight-mile ride on horseback. But he wouldn't hear of it, as is customary with the rich. So I decided to have nothing to do with such a man. For what should I expect if in the beginning the patient was so filthy. However, Bernhard, who had first persuaded me to go there, undertook to pay a sizeable sum in case the man should be healed. So I took the patient on, and put up at his house. Then the following happened, in view of which I advise all physicians to beware of patients who offer room and board. I healed first his arm . . . however, the art is not in closing the wound. The doctor has to prevent it from opening again. Now Dr Burzli, a brother of the patient, did not know this, but was trying to hasten the cure. He thought he had learned enough from me, broke into my room like a burglar, stole my medicine, and also used other deceptions.[1]

Paracelsus gives two separate (and not entirely consistent) accounts of the episode. In one he goes on to explain how Burzli, satisfied that he had deduced Paracelsus' secrets, deliberately provoked an argument that led to Paracelsus' dismissal, and then 'finished the job himself'. But Paracelsus gloats that he knows Burzli misunderstands the medicine, so that the hapless Kastner, deprived of the full treatment, 'thinks his skin is whole, but inside he is all rotten'.[2] None of this gives much support to Paracelsus' claim 'to consider beneath my dignity a person who cheats me out of my fee'.[3] On the other hand, it accounts for another of his resolutions: 'not to treat any prince or gentleman, unless I have my fee in my purse'.[4]

Paracelsus seems to have gone back and forth between Beratzhausen and Regensburg throughout that summer and autumn. Towards the end of the year he headed south via Munich to St Gallen, home to his old teacher Vadianus, who referred Paracelsus to the burgomaster, Christian Studer. Studer was in ill health, and while Paracelsus treated him, he lodged in Studer's house (which still stands opposite the abbey) until the summer of 1531. Studer was related by marriage to Paracelsus' friend Bartholomew Schobinger, and the two of them resumed their chemical studies in a laboratory set up in Schobinger's castle. Bartholomew was apparently a somewhat timid and studious chemist, preferring the pragmatic art of distillation over chrysopoeian alchemy, which he believed to have 'brought evil to many'.

This was a productive time for Paracelsus: he continued to write furiously, working on his great, all-encompassing book *Opus paramirum*, which might be loosely translated as the 'work beyond wonders'. Johann Rütiner, a diarist in St Gallen at the time, observed that Paracelsus was 'very diligent, sleeps but little; with boots and spurs and fully dressed, he throws himself into bed and rests merely for three hours or so, then writes on again'.[5]

Rütiner, who claims credulously that Paracelsus had been a gypsy for five years, recalls one of the doctor's unorthodox cures. He was asked to heal the injured hand of a boy named Caspar Tischmacher, whereupon he operated on it and removed a small bone. But the boy's hand swelled up, and Caspar's father angrily summoned Paracelsus to explain himself before the city magistrates and surgeons. Typically, Paracelsus ignored the summons, provoking the High Senate of St Gallen to give him an ultimatum: the boy's hand must be healed within fourteen days. When this had still not happened, Caspar's father went again to the courts, who seemed to have lost interest in the case. But Paracelsus advised that the hand be bound with living earthworms for a night. This 'cure' was finally successful. Anne Stoddart suggests that Paracelsus knew the wound was nearly healed anyway, and was merely humouring the superstition of the senior Tischmacher; but one can imagine how the worms (maggots?) might have proved efficacious by eating away diseased flesh. In any event, the episode seems to have persuaded Rütiner that Paracelsus knew everything.

THE STUFF OF NATURE

The *Opus paramirum* is a rather fluid affair, for it is actually a collection of five books, some completed at St Gallen and some later. None was published in Paracelsus' lifetime, and their subsequent printing took place in a piecemeal manner. In 1562 part of the *Paramirum* was edited and published by Adam von Bodenstein; the first full edition was made by Michael Toxites in 1575.

The core of the *Opus* appears in the first two books, which begin with a preface in which Paracelsus implores his friend Vadianus to abandon faith in the old medicine and embrace the new, so that 'I may not spend my time in St Gallen in vain'.[6] (The two men later fell out; perhaps this is why.) Here Paracelsus collects together his thoughts on the chemical constitution of man and the world. It is, you might say, the first textbook of biochemistry, and it establishes the material basis on which Paracelsus aims to unite alchemy with medicine.

He asks the age-old question: of what are things made? According to Aristotle, there are four elements, each a manifestation of the still more fundamental 'first matter', the *prima materia*. As we have seen, the Islamic alchemists added another layer of theory by proposing that all metals are comprised of the two 'principles' sulphur and mercury. It was implied without ever being too precisely explained that these principles subsumed the Aristotelian elements much as they in turn incorporated the *prima materia*.

Paracelsus unified metallurgical alchemy with organic medicine by adding a new alchemical principle: salt, which engenders solidity. 'From Salt the diamond receives its hard texture, iron its hardness, lead its soft texture, alabaster its softness, and so on. All stiffening or coagulation comes from Salt'[7] – including the corporeal substance of the human body. 'Without Salt', he said, 'no part of the body could be grasped.'[8]

'The first thing the physician should know', Paracelsus explains:

> is that man is composed of three substances. These three form man and are man and he is them, receiving from them and in them all that is good and all that is evil for the physical body . . . The names of these three things are Sulphur, Mercury, and Salt. These three are combined to make a body and nothing else is added save life and that which pertains to it. If you take an object in your hand, you have these three substances concealed within one body. A peasant can tell you that you are holding a piece of wood, but you also know that you have a compound of Sulphur, Mercury and Salt.[9]

We should not interpret these three substances as *elements*; rather, he says, they are all three *within* the four classical elements. They should properly be regarded as propensities or qualities – spiritual entities that, as it were, give form to the classical elements. Salt, for instance, is the principle that directs materials towards a solid form. With that understood, we can dispense with Aristotle's scheme and explain everything – including medicine – according to the system of the three principles, the *tria prima*. At least, that is what some later Paracelsians implied. It is not exactly how Paracelsus saw it. The fact is that his theory of matter is at best confusing, and at worst frustratingly contradictory.

Perhaps this is inevitable, since he was not simply supplanting one elemental scheme with another. Aristotle's elements, like those of modern chemistry, are bluntly materialistic: they are passive ingredients, mixed in various proportions in all things. But Paracelsian matter is vitalized: it is an active substance, a living entity, filled with creative potential. Moreover, in Paracelsus' theory mundane substances are not *composed* by the admixture of simpler materials but are *disclosed* by chemical processes, in particular by separation.

This idea that matter comes into being through a kind of alchemical revelation is, fundamentally, not scientific but theological. When God created the world, he separated the water under the sky from the water above. Then he parted the seas and exposed the dry land. To Paracelsus, Genesis is an alchemical tale: it is the original Great Work, reflected in humble microcosm by the distillation and coagulation of substances in the retorts and crucibles of his laboratories.

That is why, for Paracelsus, as for the Ionian philosopher Thales two millennia before him, water is the first substance. 'When the world was still nothing but water, and the Spirit of the Lord moved upon the face of the waters, the world emerged from the water; water was the matrix of the world and of all its creatures,' he wrote. 'This water was the matrix; for it is in the water that heaven and earth were created, as in no other matrix.'[10] But even here, 'water' is not exactly the Aristotelian elemental fluid. Rather, it is something more akin to the *prima materia*, the formless chaos from which all creation was shaped. It is, in fact, the yolk of the universe, pregnant with life, which Paracelsus called the Mysterium Magnum and which 'contains all creatures in heaven and on earth . . . all elements live from it and in it'.[11] There are many such mysteria: milk is a mysterium of cheese and butter, and cheese in turn a mysterium of maggots, which were thought to form spontaneously in rotting food. In other words, every mysterium possesses some kind of life-giving agency. And they are all related to the Mysterium Magnum, the 'one mother of all things'[12], the 'wonderful beginning'[13] out of which came all of creation in a marvellous alchemical separation. 'It is the greatest wonder of the philosophies', Paracelsus wrote, '[that] when the Mysterium Magnum in its essence and divinity was full of the highest eternity, *separatio* started at the beginning of all creation.'[14]

But there remained a vexing question for philosophers and theologians of the Renaissance: did God make this 'first matter', or did it exist already? That is to say, was it *created*, or *uncreated*? The Bible did not supply an unambiguous answer. In the Gnostic tradition that informed Neoplatonic philosophy, the primeval 'water' was deemed to be uncreated. That seemed to be Paracelsus' view too, although he leaves room for debate. The Mysterium Magnum, his *prima materia*, he says, existed in the beginning 'with God' – that is to say, it was alongside Him from the start. Yet he implies elsewhere that this prime matter is the Fiat itself: God's Word, as though the Lord spoke it into existence. It isn't clear, however, that this is the same stuff as the 'dark abyss' of uncreated, primeval water mentioned in Gnostic texts.

Paracelsus cannot resist adding further complications. For example, he takes pains to distinguish *prima materia* from *ultima materia*, ultimate matter – but seems unable to decide to which he should give primacy. In *Labyrinth of the Physicians* (1537–8) he says that *prima materia* is converted by the great alchemist of nature, the Vulcan, into *ultima materia*; in his *Book on Minerals*, God is described as performing the opposite transformation.

There is a similar confusion about the *tria prima*: Paracelsus asserts that they are themselves the three constituents of the four ancient elements, whereas later Paracelsian writings say that Aristotle's quartet *comprises* the three principles. Given that the latter are not simply tangible substances, however, the question is not so unambiguously stated in any case – separating these prescientific 'elements' was never envisaged to be like sifting gold from sand in the sediments of a river.

It appears that, as far as the genesis of matter is concerned, Paracelsus' response to these deep and difficult questions was to invent ever more strangely named concepts, piling them up as if to avoid loopholes in his theories through sheer redundancy. And so, to the sum total of all species that can be derived from the *tria prima* in any matrix (mysterium), Paracelsus gave the name *Iliadus*. Earth, air, fire and water are an Iliadus in the matrix of the world – they are the Iliadus of the Mysterium Magnum, brought forth by processes of separation and condensation. Man too is an Iliadus, his body composed of an array of organs and limbs. An Iliadus, Paracelsus tells us, is liable to grow 'fruit': in the matrices of elemental earth and water, the fruits of the Iliadus include minerals. In man, these fruits may be bitter: they include diseases such as ulcers, regarded by Paracelsus as the products of a chemical process comparable to that of mineral formation. Just as rust appears in water, and dust in air, so skin disorders such as alopecia grow in man.

There are many more of these lexical inventions – Aquaster, Scaiolae, Adech, Aniadus, Enochinum, Ilech, Gamonymum, to name a few – and attempts to decode them are not terribly illuminating. Why this profusion of new and peculiar words, which exasperated Paracelsus' contemporaries and led later commentators to assume he was mad? His supporters, both historical and modern, have tended to suggest that there is nothing at all elusive in these schemes and classifications, if only one takes the trouble to study them carefully. Such assertions are rather undermined by the fact that Paracelsus on occasion fails to define a neologism even if he uses it only once in all his writings. Is this, as some critics have implied, a smokescreen behind which he conceals his ignorance? Jung has a more probable explanation. The unconscious and unresolved conflicts between the Christian and the pagan, the humble and the proud man, the creator and the destroyer, left Paracelsus with a pathology that Jung recognized from his clinical experience:

> Generally certain symptoms appear, among them a peculiar use of language: one wants to speak forcefully in order to impress one's opponent, so one employs a special, 'bombastic' style full of neologisms which might be described as 'power-words'. This symptom is observable not only in the psychiatric clinic but also among certain modern philosophers [even then!], and, above all, whenever anything unworthy of belief has to be insisted on in the teeth of inner resistance: the language swells up, overreaches itself, sprouts grotesque words distinguished only by their needless complexity. The word is charged with the task of achieving what cannot be done by honest means.[15]

Jung comments on the irony 'that Paracelsus, who prided himself on teaching and writing in German, should have been the very one to concoct the most intricate neologisms out of Latin, Greek, Italian, Hebrew, and possibly even Arabic'.[16] But I doubt that there is anything *wilfully* dishonest or dissembling in Paracelsus' arcane terminology, and if it does not amount to a self-consistent scheme, it does at least provide some kind of framework on which to hang a bold new vision of the world. That, however, was probably small consolation to Paracelsian doctors of the seventeenth century as they laboured to understand what all this meant they should do to their patients.

Out of the Mysterium Magnum emerged the four ancient elements, and these provided the matrices, the secondary mysteria, for all things in the world. Ever keen to break with tradition, Paracelsus rejected the Aristotelian idea that these elements each have two characters or complexions – fire being hot and dry, water cold and moist, and so forth. Rather, he says, the elements have only one nature: earth is cold, fire hot, water moist and air dry.

Yet these elements do not exist uniquely, he said, but in countless different forms – as many as there are objects in the world. There are, for example, 'many thousand kinds of water in the element aqua'.[17] Each has the potential to bring forth different objects: some waters nurture stones, others corals, others form the flesh of aquatic animals. Plants are 'growing water': a striking premonition of the famous experiment performed in the seventeenth century by the Paracelsian physician Jan Baptista van Helmont (page 392). Of course, stones can arise from some of the many types of the element earth too. Indeed, so can water: the waters that flow underground belong not to the element water but to earth.

Each object in the world, Paracelsus says, belongs to a specific element, and returns to that mysterium. The matrix of humankind is the earth, for that is what the Bible says:

> Man was not born out of a nothingness, but was made from a substance . . . The Scriptures state that God took the *limus terrae*, the primordial stuff of the earth, and formed man out of this mass. Furthermore they state that man is ashes and powder, dust and earth; and this proves sufficiently that he is made of this primordial substance.[18]

From such considerations he seems to extract his concept of *limbus*, another nebulous and fertile matrix. Indeed, elsewhere Paracelsus writes 'The *limbus* is the primordial stuff of man . . . What the *limbus* is the man is too.'[19] But the *limbus* is also described in the next breath as 'heaven and earth, the upper and lower sphere of the cosmos, the four elements, and everything they comprise'.[20] This apparent vagueness makes a kind of sense once we recall Paracelsus' belief that the macrocosm of the universe is reflected in the microcosm that is man. Or we might rather say that this extreme degree of holism in Paracelsian cosmology makes it very hard for him to tell anything apart or to distinguish one thing from another in terms of its materialistic composition.*

In consequence, it is hard to find here the basis for any truly scientific thinking about the nature of matter. And that is not

* Elsewhere Paracelsus seems to use *limbus* in a wholly different context to denote a 'margin', alluding to its Latin meaning as the border or hem of a garment.

surprising, for this theory is best described not as proto-science but as chemical theology. It is no mere coincidence that the *tria prima* form a trinity, nor indeed that one of these is the corporeal and mundane salt, one the potent and fiery sulphur and one the volatile, ghostly mercury – the Son on earth, the Father in heaven, and the Holy Spirit. Paracelsus is explicit about this aspect of his philosophy: in the *Book of Meteors* he writes, 'God made everything from three . . . for the origin of this number is from God . . . also the Word was threefold, for Trinity has spoken it and the Word is the beginning of heaven, earth and all creatures . . . each creature can thus be divided into these three parts.'[21] Nor is this a uniquely Paracelsian notion, for medieval alchemists also ascribed a body, soul and spirit to metals.

Moreover, in the Alchemical Creation the process of separation is allied to the idea of the 'Fall'. This breaking away, this emergence of the individual and the differentiated from the whole and homogeneous, is a sort of corruption. It is inevitable that such degeneration happens – mankind could not have come into existence without it, and it is a common phenomenon in the chemical laboratory, where substances were often regarded as 'dying away' when they separated into volatile spirit and solid residue. But it was a 'fall' nonetheless, as Paracelsus emphasized by referring to the sum total of the disruptive and degenerative processes that lead to the differentiation of matter as the Cagastrum (literally, 'bad star', although *astrum* may refer more generally in his works to a power or influence). In the human body, this kind of degeneration creates disease, for example as a poison is separated out from food in the stomach or as a carbuncle blooms like a mineral formation in the deadly signature of the plague. The opposite process of homogenization is, in medical terms, curative; in theological terms it restores purity of soul and redeems humankind. This perspective stimulated the alchemical quest for a 'universal solvent' that would dissolve and homogenize all substances – a solvent that Paracelsus named the alkahest. His (discerning) follower van Helmont sought in vain for this spirit that dissolved all bodies 'as warm water dissolves ice',[22] and they were still searching for it in the eighteenth century – even after the German chemist Johann Kunckel pertinently pointed out around 1702 that if anyone were to find it, there would be nothing to keep it in.

In contrast to the scientists of the seventeenth century, Paracelsus was looking to explain the world not through quantitative theories but on the basis of analogy, which he intended as much more than a pedagogical aid. Because of the relationship of the microcosm to the macrocosm, analogy seals discrete phenomena into concrete relations. 'Paracelsus', says the Renaissance historian Brian Vickers, 'pushes the correspondence between macrocosm and microcosm to the point where they become interchangeable, identical, as if fused with each other.'[23] In consequence, according to the medical historian Owsei Temkin:

> With Paracelsus the analogy almost takes the place of the parable in the New Testament. To make his reader see the truth of his interpretation, Paracelsus has no other means but to lead him as near as possible through examples. Hence the style of Paracelsus is marked by a series of statements connected by analogies or by open or hidden biblical references. It fits none of the great scientific methods, be they scholastic, mathematical, classificatory, purely descriptive, or even experimental in the modern sense. And these pictures, these visions are offered as interpretations of what is otherwise hidden and obscure in its causes.[24]

Which is a way of saying, perhaps, that you had to start somewhere. Copernicus began with mathematics, Leonardo and Vesalius with careful observation. But Paracelsus' over-brimming creative imagination, allied with his impatience and visionary ambition, stood him in very poor stead for instigating any kind of scientific method. Here he is, struggling to do something like science with a miner's coarse lexicon and the mind of a poet.

THE ELEMENTS OF MEDICINE

By adding salt to sulphur and mercury, Paracelsus produces the first real system of chemistry that could accommodate both organic and inorganic matter. Yet at the same time, this apparent innovation was not doing much more than taking the old sulphur–mercury system back towards the spirit of the Aristotelian quartet. For sulphur was considered to be the principle of flammability, the component that makes things burn; mercury stood for the fluid component of things, their volatile 'spirit'; and salt was the body, the principle of solidity. So what we have here is nothing other than fire, 'fluid' (water) and earth. In his theory of *tria prima*, Paracelsus was simply restating what the Greeks knew already: that substances can be transformed between solid and fluid forms.

The theory proved persuasive to many precisely because it was rather hazy, and thus easy to generalize. The processes of sublimation, distillation and evaporation in the alchemical laboratory, for example, typically produced a volatile substance and a solid or coagulated residue: the 'salt'. Paracelsus illustrates the three-principle scheme with reference to the burning of wood: 'The flammable part is the Sulphur, the smoke is the Mercury, and the ash is the Salt.'[25]

The *tria prima* formed the centrestone of Paracelsian chemistry in the seventeenth century, and they persisted (in an ingenious form – see Chapter 20) right up to the advent of the modern age of chemistry. But in the *Opus paramirum* Paracelsus is concerned to elaborate on what the theory implies for medicine:

> The physician must know these three [principles] and must understand their combinations, their maintenance, and their analysis. For in these three lie all health and all sickness, whether whole or partial. In them therefore will be discovered the measure of health and the measure of disease . . . Death is also due to these three, because if life be withdrawn from the primary substances in whose union life and man exist, man must die. From these primary substances therefore proceed all causes, origins, and knowledge of disease, their symptoms, development, and specific properties, and all that is essential for a physician to know.[26]

Each of the three principles tends to produce certain kinds of diseases when present in excess:

> Salt is subject to four processes: resolution, calcination, reverberation, and alkalification, which all occur in the body as in the outside world. Thus too much Salt goes into solution in people who indulge in overeating or lechery. In these, Salt is converted into fat. Obese bodies are like land

> which has been over-manured, it brings fruit too quickly on, or like land which has suffered an excess of rain, causing the fruit to decay.[27]

Salt that is calcined (heated) in the body produces perspiration, itching, scabs and sores: 'All skin wounds are occasioned by salt: as it rusts iron, so it rusts the tissues of the body.'[28] Salt that is 'reverberated' causes mucus and pustular wounds. It can lead to ulcers, cancers and leprosy. In the same way, mercury may cause gout, arthritis, brain fever, gall trouble, syphilis, paralysis and depression. Sulphur is responsible for all manner of ailments: when cold and congealed, it gives rise to 'diseases having a similarity to snow, frost, and gravel';[29] but if it turns hot, moist or dry, other dire consequences follow, such as fever and the plague.

The distinction between imbalances of the *tria prima* and imbalances of the four Galenic humours is of course rather slight. The key to the chemical physiology of the *Opus paramirum*, however, is that it relates directly to the chemistry of the world: the microcosm reflecting the macrocosm. This means that the observations made in an alchemical kitchen can help to rationalize the vicissitudes of human health. What is more, it helps us to identify cures.

Medicines, says Paracelsus, must observe the principle that 'like seeks its like': within the chemistry of the disease lies the cure:*

> It would be a wild disorder if we were to seek our cure in contraries. A child asks his father for bread and he does not give him a snake. God has created us and He gives us what we ask, not snakes. So it would be bad medicine to give bitters where sugar is required. The gall-bladder must have what it asks, and the heart too, and the liver. It is a fundamental pillar upon which the physician should rest to give to each part of the anatomy the special thing that accords with it . . . For know that you are the father rather than the physician of your patients: therefore feed them as a father does his child.[30]

* We've seen earlier, however, that this was not an invariant principle of Paracelsian medicine.

Yet physic may not be so simple a matter as merely identifying the material cause, the 'principle', that lies behind a disease and then administering some substance containing that same fundamental essence. For 'Blood is one Sulphur, flesh is another, the major organs another, the marrow another, and so on . . . There is [also] one Salt in the bones, another in the blood, another in the flesh, another in the brain, and so on . . . Mercury has as many forms as Sulphur and Salt.'[31] What this means is that Paracelsus' *tria prima*, insofar as they can be given a material interpretation at all, are explicitly not substances but *categories*. Indeed, chemists today speak routinely of 'salts', recognizing more varieties than the 'rock salt' we use to season and preserve food. In this respect Paracelsus sounds surprisingly modern when he says:

> There are many kinds of salt . . . Some salts tear and heal again, as do the salts of alum . . . Vitriolic salts are also visible, and in many forms. Salts of lead are of many kinds and of different natures: also salts of lime and salts which can be separated from other bodies.[32]

These salts are tangible, they are the compounds of today's chemistry.* But the Salt of the *tria prima* is, like Sulphur and Mercury, more akin to a property. These three 'components' are admixed into every substance: 'they do not exist for themselves, but are mixed as a third body in everything to complete it.'[33]

No wonder, then, that it is such a complex and subtle thing to concoct chemical remedies from the materials of nature, since it involves cooking with intangible ingredients which no balance can weigh. 'Nature is so careful and exact in her creations that they cannot be used without great skill,' Paracelsus warns. 'For it is not God's design that the remedies should exist for us ready-made, boiled, and salted, but that we should boil them ourselves, and it pleases Him that we boil them and learn in the process.'[34]

Yet the physician must remember that his medicines do not actually cure in themselves; rather, they create the conditions that

*Salts, to the chemist, are traditionally compounds of a metal and non-metal elements, such as sodium chloride (sodium and chlorine), copper sulphate (copper, sulphur and oxygen) and calcium carbonate (calcium, carbon and oxygen).

allow the body to heal itself. And how could it be otherwise? – for 'there is nothing in heaven or in earth that is not also in man'.[35] The cure is, in this sense, within us already. The business of healing is like that of gardening: one must till and nourish the ground, tend and feed the seedling. The gardener does not make the seed grow; he simply creates the conditions under which that might happen. That which shapes the tree is the 'craftsman in the seed'. And this craftsman works on the elements in the environment, within which the plant's essence is already inherent: 'Rain has the tree in itself, and so has the earth-sap: rain is its drink, *liquor terrae* its food by which the tree grows.'[36] In the same way, food has within it the whole man, and it is our internal craftsman, the archeus, that works on these ingredients to make the living body. And likewise, the healing power of a medicine is released not by the doctor but by the physician inside each of us:

> Therefore man is his own doctor; for as he helps nature she gives him what he needs, and gives him his herbal garden according to the requirements of his anatomy. If we consider and observe all things fundamentally we discover that in ourselves is our physician and in our own nature are all things that we need. Take our wounds: what is needed for the healing of wounds? Nothing except that the flesh should grow from within outwards, not from the outside inwards. Therefore the treatment of wounds is a defensive treatment, that no contingency from without may hinder our nature in her working. In this way our nature heals itself and levels and fills up itself, as surgery teaches the experienced surgeon.[37]

It is understandable that Paracelsus was apparently so successful as a doctor. His was an age when the cure was often worse than the condition: cutting open of veins, trepanning the skull, cautery, amputating the limb. There was a common belief among surgeons that wounds and incisions should be stuffed with wadding, a procedure almost guaranteed to produce infection. In contrast, Paracelsus argued for minimal intervention, for keeping wounds clean, feeding the patient well, and letting nature take its course. Does the doctor knit the tissues to close a wound? Of course not; only the body can do that. When even minor ailments might elicit life-threatening medical treatment, the doctor who did nothing could seem like a wonder-worker.

Paracelsus called this healing power of nature the *mumia*. As is so often the case with his coinages, it has potentially confusing associations: for it was not uncommon for a piece of mummified body, or of the body's embalming wrappings, to be added to a medicine, presumably to confer preservative qualities. Paracelsus' abstract *mumia* was something else, but that did not stop him from listing 'mumia' in his recipes (including a preparation for haemorrhoids), where it does indeed refer to mummy powder.

This substance was in great demand as a medicine, and by the late Middle Ages a lucrative trade had developed in mummified remains from the Middle East. It was inevitably an invitation to fraudsters – who, however, might be considerably more sophisticated than to pass off any old crumbly brown stuff as 'mummy'. In his *History of Druggs*, the botanist Pierre Pomet reveals that:

> The mummies that were brought from Alexandria, Egypt, Venice, and Lyons are nothing else but the bodies of people that die several ways. Those from Africa called white mummies, are nothing else but bodies that have been drowned at sea, which, being cast upon the African coast, are buried and dried in the sands, which are very hot.[38]

The apparent concern to use the bodies of people who had met an abrupt, premature end (as opposed to those who died naturally or through illness) implies that the mummy counterfeiters were not so much making 'false mummy' as 'low-grade' mummy, less effective simply because it had been artificially 'aged' by heat. For it was widely believed that mummy acquires its medicinal virtue from the life force that remains in the body of one who has died suddenly and unnaturally, through accident or violence. Indeed, the Paracelsian physician Oswald Croll, in his *Basilica chymica* (1609), relates openly how to make an artificial mummy:

> Take the carcass of a young man (some say red-haired), not dying of a disease but killed, let it lie twenty-four hours in clear water in the air, cut the flesh in pieces, to which add powder of myrrh and a little aloes, imbibe it twenty-four

> hours in spirit of wine and turpentine, take it out again and hang it up for twelve hours, then imbibe it again, twenty-four hours in fresh spirit, then hang up the pieces in a dry air and a shady place.[39]

Given the stipulations on the nature of the body, what shady practices might have sometimes been involved in procurement of the brown powder on the apothecary's shelf – next, perhaps, to the tubs of human fat (good for rheumatism) and the skulls coated with fungus or 'moss', ideally taken from the gibbet?

But the true *mumia*, said Paracelsus, was not embalmed flesh but an 'inner balsam', the force of life that resides in the body and restores it to wholeness and health:

> The surgeon should not interfere with the *mumia*'s working; he must protect it. Flesh possesses an inner balsam which heals, and every limb has its own cure in it . . . For the *mumia* is the man himself, the *mumia* is the balsam which heals the wound: mastic, gums, glaze will not give a morsel of flesh; but they can protect the working of Nature so as to assist it.[40]

Nevertheless, the metaphysical *mumia* was in fact linked to powdered mummy in Paracelsus' mind, since it was precisely this inner balsam that made mummy potent. That is to say, there was *mumia* in mumia – for, so far as the corpse of a slain man was concerned, 'The whole of the body is useful and good, and can be fashioned into the most valuable mumia. Although the spirit of life has gone forth from such a body, still the balsam remains, in which life is latent.'[41] One who has died 'a natural and predestined death', on the other hand, has expended his *mumia*, and is of no value to the physician: 'Let him be cast to the worms.'[42]

As Jung points out, we would be unwise to imagine that Paracelsus had these slippery associations and distinctions laid out clearly in his own mind but merely forgot or chose not to explain them to us. If the terminology is confusing, that is because he is hazy about it himself. 'We have no reason to suppose', Jung says:

> that behind his neologisms there was a clear, consciously disguised concept. It is on the contrary probable that he was trying to grasp the ungraspable with his countless

> esotericisms, and snatched at any symbolic hint that the unconscious offered.[43]

The blithe lexicons that some Paracelsian scholars have provided in the past are therefore to be taken with a pinch of Paracelsian salt.

Paracelsus' *mumia* is nonetheless evidently a potent natural source of health and vitality. Yet his theory does not lead him, like some New Age healer, to leave every illness untouched and trust to this mystical agency to make it better. That, indeed, is the remarkable thing. His ideas supported both a commendable hands-off approach and also the use of potent (even virulent) medicines like mercury in cases (such as syphilis) where the condition demanded it. There was no contradiction in this, for if man is a microcosm then there is nothing 'alien' to him in the world, and a chemical remedy is, if properly prepared, also a natural remedy. 'We must understand therefore', he said, 'that when we administer medicine, we administer the whole world: that is, all the virtue of heaven, earth, air and water.'[44]

VAPOURS AND PRECIPITATES

Just as Paracelsus regarded the Creation not as the assembly of the world from its components but as a process of separation to differentiate its elements, so too his medicine was predicated on the idea of separating good from bad. Medicines were not just assembled from raw ingredients but were prepared by alchemical means: by distillation, sublimation, coagulation and so on, by *extracting* rather than composing:

> [Medicine] lies in the knowledge of what is inside and not in composing and patching up pieces to make it. What are the best trousers? Those which are whole; those patched up and pieced together are the worst ones. Who is so stupid as to believe that nature has distributed so much of a virtue to one and so much to another herb, and then commissioned you doctors to put them together? . . . Nature is the physician, not you; from her you take your orders, not from yourself; she composes, not you.[45]

Paracelsus passed all fields of medicine through this alchemical filter. Some emerged the better for it, clarified, rid of their murky detritus. For others, the old theories were merely replaced by equally spurious new ones.

Such was the case with uroscopy. Traditional uroscopy – the visual analysis of a patient's urine – was not just a form of naïve crystal-gazing; it claimed a theoretical basis. Urine was regarded as a kind of overflow of liquid from the blood, and so it was assumed to reflect the blood's physical characteristics. The practice of urine inspection was allied to the ancient 'catarrh' theory of disease, which attributed all illnesses to 'vapours' that ascend through the body from the stomach to the brain, where they condense as mucus. This noxious fluid then flows down through the nose and throat and into the lungs, joints and other organs, causing symptoms ranging from coughing to rheumatoid pains.

Paracelsus gave this catarrh theory an alchemical inflection. The vapours ascending to the brain and the catarrh descending from it became Mercury, which could precipitate to cause gout and arthritis or sublimate to induce brain fevers. Other vapours, like 'sulphur vitrioli', might overwhelm the brain and cause madness and epilepsy.

The presence of catarrh was supposed to be revealed by shaking a urine sample. If small particles were seen to sink slowly, the illness had taken hold. Paracelsus, for all that he abused the piss-prophets, did not dismiss urine inspection entirely: he demanded rather that it be conducted more 'chemically', using techniques like distillation. This foreshadows the use of quantitative analysis in medicine, but it was in itself no great advance for science. Paracelsus' followers, such as the late-sixteenth-century physician Leonhard Thurneisser zum Thurn, elaborated fantastically on his schemes for urine analysis, for example advising that it be distilled in a tall cylindrical vessel scaled to the proportions of the human body, whereupon changes in colour or appearance would reveal the seat of disease.

Jan van Helmont was dismissive of the arcane philosophy that was meant to underpin chemical medicine, observing that not only did the Paracelsians want to make man a microcosm but they wanted his urine 'to enjoy the privilege'.[46] Even so, van Helmont persisted in believing that urine had a diagnostic value (see page 393). He recommended that the doctors measure its specific gravity (its density relative to water), a procedure first proposed by Nicholas de Cusa in the fifteenth century.

The third book of the *Opus paramirum* introduces a source of chemical imbalance in disease to which Paracelsus was to award great significance in his subsequent teachings. When old wine sits in a barrel, a solid residue may collect on the wood: a hard, mineral-like substance formed from salts of some of the organic acids present in the grape (that is, compounds of these acids combined with metals, primarily potassium), one of which now goes by the name of tartaric acid.

Paracelsus noted that similar hard deposits form in the body (the substance that accumulates on teeth is still called tartar today). 'Stones' can form in the bladder and gall bladder, causing excruciating pain (which, in the sixteenth century, was nothing compared with the pain of having them surgically removed). And the common condition of gout was the result of tiny crystals precipitating in the joints. None of these solids is in fact composed of tartaric acid salts: gallstones are made mostly of crystallized cholesterol, while gout is due to precipitated salts of uric acid, and kidney stones are primarily calcium oxalate. To Paracelsus, however, all these processes seemed akin not just to each other but to the formation of salt deposits in wine barrels and of minerals from salt-rich natural waters 'by the action of water on the earth'.

The stuff that causes gout 'burns like hell', Paracelsus said – so he named it 'tartar', after Tartarus, the domain of the Greek Underworld where awful punishment was meted out to those who had offended the gods. Paracelsus' 'tartaric diseases' included not only gout and stones but arthritis, rheumatism, lumbago and sciatica. The accumulation of tartar pointed to a deficiency in the archeus, which cannot break down this recalcitrant component of food and drink:

> There are refuse materials in the body which are neither faeces in the ordinary sense nor subject to incorporation in man. They cannot be broken down, yet they are not man and they remain in man. Now there are such diseases which are caused by this refuse and vary according to the degree of separation and the location of the refuse. These diseases are stone and sand, glue and mud . . . These four are four types of refuse from food. They are all known as tartar.[47]

Paracelsus identified the foods and drinks that lead to 'tartar' deposits:

> It is in the vegetables like barley and peas which you can deduce from the mucus they produce and their thick texture. All kinds of food with this mucus lead to stone in the body. Milk products, meat, and fish have sand in them and this becomes tartar.[48]

It also comes, he said, from pear and apple juice, wine and beer. Some foods from particular locations are rich in it: 'Thus it may happen that a Swiss suffers from a Nuremberg or Westerburg tartar owing to the consumption of cereals and vegetables imported from those places.'[49] He goes on to explain how tartar is deposited in the body, attributing particular significance to the kidneys and the urine they generate: 'Thus the kidneys also have their particular excrement which is contained in the urine and is excreted with it and is the hypostatis [deposit].'[50]

Paracelsus came to lay great stock in tartar, even going so far as to imply that it is a universal cause of disease. In his *Book of Tartaric Diseases* (1537–8) he planted an important biochemical signpost when he suggested that acid may be involved in digestion in the stomach. But Paracelsus did not consider this to be a feature of healthy digestion; rather, he said, stomach acid can produce tartar by 'overcooking' the food. To rid the body of tartar, Paracelsus used potassium tartrate (an example of treating like with like) or oils and resins that made 'tartaric' stones 'as soft as honey'. 'Tartar emetic' (the mildly toxic potassium antimonyl tartrate), which induces sweating and vomiting, was still in use in the 1960s.

Perplexingly, Paracelsus also recommends an intake of acid as a *protection* against tartar deposition, saying that it boosts the stomach's digestive power so that it might dissolve tartar 'as an ostrich digests iron, as a blackbird a spider'.[51] This, he claims, explains the medicinal virtue of sulphuric acid, as well as the efficacy of some mildly acidic mineral waters. It was left to van Helmont a century later to identify the crucial role of acids in normal digestion.

FEMALE MYSTERIES

The fourth book of the *Opus paramirum* concerns a topic that most doctors preferred to avoid. 'I am not embarrassed', Paracelsus says defiantly, 'to be the first one who dares to write on the diseases of women.'[52]

One of the most profound mysteries of the universe was the creation of a new living being. But because that process was hidden away in a woman's lower body, it was considered unseemly to investigate it, or even to think about it too closely. As a result, not only conception and childbirth but also many ailments specific to women received little attention from the medical community; only midwives bothered with such things. (It was a woman, Cleopatra, who allegedly put aside such scruples to investigate the messy business of gestation by having female slaves killed and opened up at various stages of pregnancy.) Luther expressed an opinion that typified the view of physicians: 'If women die in childbed, that does no harm. It is what they were made for.'[53]

When male doctors did sometimes write about female medicine, they sought to entice their (male) readers by promising to reveal the *secreta mulerium*, 'women's secrets'. Paracelsus, in contrast, wrote plainly about the value of self-knowledge:

> Thus, o women, take notice of yourselves and the nature of your illnesses . . . This is why I describe them . . ., so that you may recognize that you carry a double microcosm within yourselves, since your body is the same as that of men and [yet] you are the world of human birth.[54]

Paracelsus displays a deeply ambivalent attitude towards women. On the one hand, Oporinus tells us that he never looked at a woman; certainly he never married. But his celibacy does not stem from the misogynistic prudishness that characterized, for example, many of the leading lights of the Royal Society in the seventeenth century. There is pride in his avowal of chastity as the calling of a man of science – but not, apparently, distrust of women:

> Chastity endows a man with a pure heart and power to study divine things. God himself, who bids us do this, gave man chastity. But he who is unable to be his own master does better not to live alone.[55]

Women in the Renaissance were both denigrated and feared, and misogyny was woven into the fabric of society. The

Florentine writer Vespasiano da Bisticci proposed that women should observe the following rules:

> The first is that they bring up their children in the fear of God, and the second that they keep quiet in church, and I would add that they stop talking in other places as well, for they cause much mischief thereby.[57]

Women were thought to be constitutionally weak and feeble-minded, for which reason courts of law did not always hold them accountable for their actions, treating them as though they were children. Meanwhile, the figure of the nagging and domineering wife was a common caricature in art, literature and theatre (Oporinus was reputedly shackled to such a harridan). As for female sexuality, the stereotype was quite the reverse of later forms: women were supposed to be voracious, while men could control their ardour. Syphilis, as we have seen, was considered to be spread by wanton women, who were made all the more insatiable by the disease. In order to disown their own sexual dreams and fantasies, men (especially clerics) imagined nocturnal temptations by wicked female spirits such as night hags and succubae. The famous fourteenth-century account of the travels of the English knight John Mandeville provides a characteristic fable, speaking of an imaginary land in which men let others take the virginity of their young brides for fear that they will be fatally bitten by serpents in the women's bodies. The key archetypes of womankind were the virtuous and immaculate Virgin, the pox-ridden prostitute, and the witch.

While regarding women as different from men, Paracelsus does not appear automatically to dismiss them as inferior. Each sex has its strengths and weaknesses:

> A woman is like a tree bearing fruit. And man is like the fruit that the tree bears . . . The tree must be well nourished until it has everything by which to give that for the sake of which it exists. But consider how much injury the tree can bear, and how much less the pears! By that much is woman also superior to man. Man is to her what the pear is to the tree. The pear falls, but the tree remains standing.[58]

Cornelius Agrippa argued that women would be found quite as intelligent and capable as men, if it were not for male oppression and lack of education. Agrippa, who married at least three times, wrote a remarkable book whose title speaks for itself as an attack on the prejudices of the age: *Female Pre-eminence, or the Dignity and Excellency of That Sex above the Male.* To the Renaissance man, his plea must have seemed bizarre:

> Custom spreading like some epidemic contagion hath made it common to undervalue this sex, and bespatter their reputation with all kinds of opprobrious language, and slanderous epithets . . . Let us no longer dis-esteem this noble sex, or abuse its goodness . . . Let us re-enthrone them in their seats of honour and pre-eminence . . . and treat them with all that respect and veneration which belongs to such terrestrial angels.[59]

Such even-handedness is equally apparent in Paracelsus' Paramiran writings on pregnancy and women's diseases. He adopts the traditional view that God created woman 'out of man'; but this does not render her inferior – just different:

> Thus her body cannot be compared with that of man, although she was taken from him. It is true that she resembles him, for she has received his image, but in all other things, in essence, properties, nature, and peculiarities, she is quite different from him. For man suffers as man, woman suffers as woman, both suffer as two creatures beloved of God. He proves this with the twofold medicine which he gives us: a masculine medicine for men, a feminine medicine for women.[60]

As an aspect of the unique nature of femininity, Paracelsus invokes the very ancient association of women with water and the sea:

> What causes the sea to rise? Just as the sea expels things and falls, one may understand woman as a mother of children. The sea is the mother of water. Because woman is a mother, she produces such rivers in herself, which rise up and flow out every four weeks . . . Thus the menstrual blood is an

> excrement of things flowing into the matrix, which die there and are then expelled. Some doctors have mistakenly called the menstrual blood a blossom of women. Women blossom at the moment of conception, and the fruit – that is, the child – follows as in the case of all blossoms . . . As long as pregnancy lasts, there is no excrement, for all things are quiet and bide their time . . . for this is the nature of woman, that she is transformed as soon as she conceives; and then all things in her are like summer, there is no snow, no frost, and no winter, but only pleasure and delight.[61]

Where does it come from, this extraordinary tenderness in a man who has never had a lover, let alone a child, and barely knew his mother?

Paracelsus' view of conception contrasts with that of Aristotle, which provided the orthodox theory until it was challenged by William Harvey in the seventeenth century. To Aristotle it was obvious that the menstrual blood, withheld during pregnancy, coagulated into the fetus under the agency of the semen. The woman was passive in all this: the blood was merely the soil in which the male seed grew, providing justification for the tradition of male inheritance.

For Paracelsus, on the other hand, man and woman play an equal part in conception:

> When the [man's] seed is received in the womb, nature combines the seed of the man and the seed of the woman. Of the two seeds the better and stronger will form the other according to its nature . . . The seed from the man's brain and that from the woman's brain make together only one brain; but the child's brain is formed according to the one which is the stronger of the two, and it becomes like this seed, but never completely like it.[62]

So the child is derived not from one seed but from many, each specific to a limb or organ. These seeds are arranged within the matrix of the womb, which puts 'each in its proper place, and thus each single member is placed where it belongs, just as a carpenter builds a house from pieces of wood'.[63] Paracelsus is less concerned than Aristotle with maintaining any balance of materials as the fetus grows – for where is there any such thing in the growth of a tree from a shoot? It is 'the action of material nature' that makes the child grow, ensuring that 'what belongs to the flesh develops into flesh'.[64]

Paracelsus views fetal development as fundamentally akin to the growth that takes place everywhere in nature: for not just plants but rocks and all other things have an organic existence in his world. The concept of the *seed* is central to all such procreativity: seeds lie everywhere in nature, which is abundantly fecund. 'In the case of insensible creatures like foliage and grass, stone and ore, and other things,' he says in an early tract called *Das Buch von der Gebärung der empfindlichen Dinge in der Vernunft* (*c.*1520), 'the seeds for their procreation are found mixed and embedded in the essence of their nature, and are inseparable from it.'[65] But humans are not constantly carrying their seeds, 'in order that lust might not lead them out of the light of Nature'. The seed is formed only when the desire arises: 'if a man wants to procreate, this speculation creates the desire, and the desire creates his seed.'[66] That desire must itself be 'kindled by an object' – and since God wishes us to multiply, he has provided man with the 'object', in the form of woman. But this is not to depict women simply as vessels for men's desires, for Paracelsus stresses that the relationship is symmetrical: 'it is the same with woman'. Thus women are neither the passionless beings of the nineteenth century, nor the shameful inciters of lust that they appeared in the eyes of medieval monks.

Despite (or perhaps because of) his interest in gynaecology, Paracelsus never trusted midwives. That might have been a consequence of the superstition that surrounded their task, which all too easily spilled over into accusations of witchcraft. Or maybe it reflected the high incidence of death in childbirth, both of infants and of mothers – a situation that surely reflected more on the primitive surgical techniques of the time than on the competence of midwives.

Yet it would be both anachronistic and simplistic to portray Paracelsus as an early champion of women's rights. He shared some of the prejudice of his times, for example, in his assertion that 'Man is the Little World, but woman . . . is the Littlest World . . . Thus the cosmos is the greatest world, the world of man is the next greatest, and that of woman the smallest and the least.'[67]

The fifth book of the *Opus paramirum*, which speaks of the

'invisible diseases' of mental illnesses, was apparently written later than the others, and is sometimes treated as a separate work. We will return to Paracelsus' views on these matters in Chapter 16.

SCIENCE OR MYSTICISM?

The *Opus paramirum* lays out a theory of a universe pervaded by inherent creativity. Just as the body has its *mumia*, a life force that closes wounds and restores order and health, so the world too is permeated by an invisible force that drives it towards a state of perfection. This is the *Iliaster*, a word apparently derived from the Greek *hyle* (matter) and the Latin *astrum*, alluding to the powers of the stars and destiny. Walter Pagel's definition of the Iliaster is close to incomprehensible – 'matter which essentially is and expresses the sum total of specific actions possible and realizable in nature'[68] – but it seems to be another term intended to denote a kind of primeval matter and to convey the inherent productivity of the natural world. Matter itself, says Paracelsus, is creative: it seeks form. This form – that is to say, the attributes of real, individual things – is imprinted on the Iliaster by the alchemical agency of nature, personified as Vulcan. If this sounds close to pantheism (not to say animism), well, so it was with Paracelsus, who revered nature as though it were synonymous with God Himself.

But it is not mere mysticism. These invisible forces of nature – the *mumia*, the arcana, the Iliaster – may have been teleological (and how could they not be, in a world created by God?), but they were also universal and they had an explanatory, mechanistic function. They enabled man to understand his world: 'everything occult shall become apparent,'[69] Paracelsus believed. His explanations liberated the world from being regarded as an endless succession of unique events explicable only by divine or diabolical intervention. As he wrote in the Paramiran book on invisible diseases:

> You have seen how natural bodies, through their own natural forces, cause many things [deemed] miraculous among the vulgar. Many have interpreted these effects as the work of Saints; others have ascribed them to the Devil; one has called them sorcery, others witchcraft, and all have entertained superstitious beliefs and paganism. I have shown what to think of all that.[70]

The *Opus paramirum* is one of the first attempts to envisage a *biochemistry*, a chemistry of life. It goes beyond the alchemical *Archidoxa* by explaining how chemical medicines interact with the body. Although the explanations rely on the misconceived notion of correspondences and sympathies between processes in nature and those in the human body, yet every so often they approach a kind of truth, because the basic assumption of the *Paramirum* echoes the modern belief that there is nothing *special* about biochemistry; it follows the same laws and principles as the rest of chemistry. Although Paracelsus remained convinced that a kind of life force infused everything in nature, from minerals to stars, his medical alchemy can be regarded as the beginning of the end for the concept of vitalism.

The *Paramirum* is not, however, a work of science. It is a wondrous vision, an ecstatic revelation. 'Strange, new, amazing, unheard of, they say are my physics, my meteorics, my theory, my practice,' he admits in the conclusion of the first two books, addressed again to Vadianus. 'Unto whom the gift is given he receives it: who is not called I need not call.'[71] If indeed Paracelsus worked on the *Opus paramirum* without taking his boots off to sleep, that is no wonder; for this kind of writing does not spring from careful, sober labour. It conveys, as few other works of that era do, the strangeness and the beauty of a magical universe, of a world created by a loving God, where marvels and mysteries lurk under every rock and behind every hill:

> For God, who is in heaven, is in man. Where else can heaven be, if not in man? As we need it, it must be within us. Therefore it knows our prayer even before we have uttered it, for it is closer to our hearts than to our words. God made his heaven in man great and beautiful, noble and good.[72]

15
Star and Ascendant

A Science of Prophesy

> But he who supposes . . . that all the force of the universe pertains to those spheres, is as foolish as the one who, entering a man's residence, thinks it is the ceilings and the floors that govern the household, and not the thoughtful and provident good-man of the house.
>
> William Gilbert, *De magnete* (1600)

In August 1531, while Paracelsus was in St Gallen, a new star blazed across the sky. This was the same comet that Edmund Halley saw a century and a half later, and which we now know by his name. Just as the comet of 1665 was interpreted by Halley's contemporaries as an augury of catastrophe (and indeed the Great Plague and the Fire of London followed soon after), so did its appearance in 1531 excite a spate of prognostications. Among them was Paracelsus' *Interpretation of the Comet*, which he dedicated to Zwingli at a pivotal time for the Swiss Reformation. Paracelsus sent the manuscript to Zwingli's associate Leo Judae in Zurich, who had it printed. But the tract contained little to console the reformers: it predicted that bloodshed would follow the comet's arrival. Two months later, Zwingli was killed at the battle of Kappel.

Genuine astronomy – the study of the movements of heavenly bodies, a science of measuring devices, numerical tables and geometrical calculations – was not to Paracelsus' taste. The primary value of astronomy, he felt, was for prognostication and prophecy. He considered this the highest form of magic, and many of his writings are devoted to it. Of the twenty-three works published during Paracelsus' lifetime, sixteen were examples of the yearly astrological forecasts known as *practica*.

Paracelsus' astrological prophecies spoke of impending apocalypse. Like many of his contemporaries, he believed that the world

Vßlegung des Com=
meten erſchynen im hochbirg/ zů
mitlem Augſten/ Anno 1531. Durch
den hochgelertenn Herren
Paracelſum.rc.
17

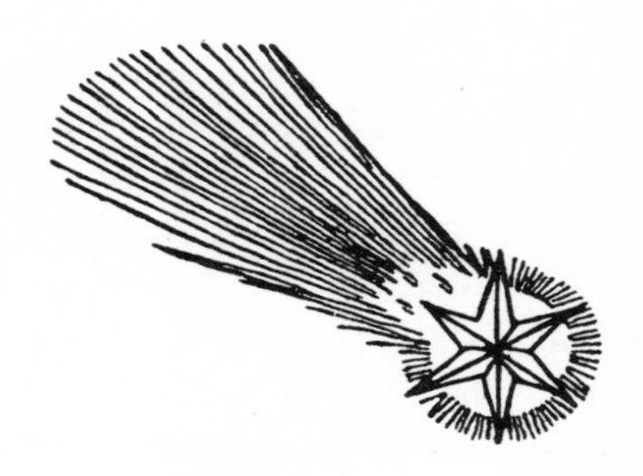

The title page of Paracelsus' *Interpretation of the Comet* (1531).

was entering the Last Days and that the sages could anticipate when and how this would happen. If Lutheranism was somewhat more receptive to astrology than Catholicism was, this was partly because suggestions of the imminent Second Coming suited the Protestant temper. Although Luther's prophecies were biblical rather than astrological – he was always somewhat sceptical of astrology – Melanchthon was an advocate of star-gazing, as were many other Lutherans. Johann Arndt, a Protestant minister and an enthusiastic Paracelsian, made much use of astrology in his chiliastic forecasts in the late sixteenth and early seventeenth centuries.

Luther himself was widely regarded by his followers as a prophet – a second Elijah, the mythical figure sometimes called the 'Elijah of the last times'. Paracelsus contributed to this perception. An old prophecy foretold that a 'third Emperor Frederick' would emulate the messianic Frederick II Barbarossa (1194–1250) by challenging the papacy and renewing Christianity. Many thought that this must refer to Frederick the Wise of Saxony, the man who sheltered Luther after the Diet of Worms. In Paracelsus' *Chronicle and History of Carinthia* (1538) he claimed that Frederick Barbarossa had discovered the image of a monk in a Carinthian monastery identified by the name of 'Lutherus'.

Although Paracelsus believed that the end was nigh, he was not sure *how* nigh. In one of his earliest works on astrological prophecy, a commentary on the widely read *Prognostication* of Johannes Lichtenberger (1488), he rejected the common suggestion that the Holy Roman Empire was about to collapse. He later concluded from a study of another millenarian text, the 'Nuremberg Figures', that Rome's power was doomed after all – but without saying much about when this downfall would happen. Elsewhere he implied that the Second Coming might still be several generations hence. But by the time he wrote *Prophecy for the Next Twenty-Four Years* in 1536, he was prepared to predict the dissolution of the existing authorities and the arrival of a 'time for the people' in the next three decades. There would first be the usual bout of calamities – an eclipse of the sun, floods and wars – after which the 'light of God' would emerge in 1555 and the Antichrist would be overthrown. By 1560 there would be universal peace and joy, and a new Golden Age would begin.

READING THE HEAVENS

Opinions on astrology in the fifteenth and sixteenth centuries were diverse, and consistency was rare. Those who doubted astrology could never be entirely immune to its allure, and its practitioners seldom agreed with one another. The attitude of Nicholas de Cusa typifies some of the contradictions. Nicholas, as we saw earlier, was one of the most original thinkers of his age: a pioneer of careful experimental technique and a thinker of a strikingly rational turn of mind. Arguably the most remarkable of his ideas concerned the heavenly bodies, for he anticipated Copernicus (who knew of his works) in making the earth a modest and rather ordinary place. It was a sphere rotating on its axis, he said in the ironically (or perhaps cautiously) titled *On Learned Ignorance* (1440). Our world is larger than the moon but smaller than the sun. Shining with reflected light, it is not unlike a star, and indeed the stars too may have inhabitants. And most remarkable of all, the earth might not be at the centre of the universe, for it does not stand still.

In the mid fourteenth century such observations could be made without too much fear of censure (it helped, perhaps, if like Nicholas you were a cardinal). Nicholas de Cusa's dismissal of the astrologers as 'fools with their vain imaginings'[1] adds to the 'modern' impression that he makes. But elsewhere we find him using the Book of Daniel, a favourite millenarian text, to calculate when the world will end (he reckoned on 1700 or 1734). At the Diet of Nuremberg he claimed that the council of Basle was operating under an evil constellation. And what are we to make of his mystical assertion that the centre of the universe is one and the same as its circumference, and that both are manifestations of God?

Negative assessments of astrology and its practitioners were not uncommon in the late Middle Ages. Most notable were the attacks made by Nicolas Oresme (*c.*1320–1382), a French cleric who became bishop of Lisieux. In the typical style of the medieval scholastic, Oresme carefully enumerated fifteen points in favour of astrology and fifty-five against, as well as eighteen 'considerations' and eleven conclusions. He accused the astrologers of being 'frivolous' and 'fraudulent', and argued that in any case we cannot know the movements of the celestial bodies accurately enough to make sound predictions or to draw up reliable horoscopes (a criticism of practice rather than principle). Oresme then enlisted the ancients in support of his case. Aristotle does not rely on astrology to explain the world and its elements in his *Meteorology*, and neither he nor Plato wrote any works specifically on that science. Oresme cites Virgil in support of his contention that princes should devote themselves to government rather than to reading the stars. Yet despite all of this, it is clear that Oresme retained a belief in the core principle of astrology: that the world of human affairs is influenced by, and prefigured in, the heavens. He allowed that one might use the stars or comets to forecast general catastrophes such as famines, floods and war (although not to the extent of dating them too accurately), and that the broad inclinations, if not the detailed characteristics or fates, of individuals can be deduced from horoscopes.

The humanists shared this ambivalence. When Pico della Mirandola, in a dramatic volte-face, launched a crippling assault in his *Disputations Against Astrology* (1492), Marsilio Ficino endorsed Pico's critical views while remaining thoroughly enchanted by astrology. To live successfully, he believed, one must learn one's horoscope and see what strengths and weaknesses the stars have invested you with, and then live accordingly. In his correspondence Ficino constantly attributes particular events to planetary influences, and he often planned his movements and actions around the stars, deferring travel if the heavens were not auspicious. Yet according to Pico, no one had encouraged him to uncover the fallacy of astrology more than Ficino, and he recorded that the two of them often laughed together about the idiocies of astrologers. Ficino himself wrote against astrology a decade or so before Pico, even while he remained an enthusiastic practitioner.

Such attacks on astrology are thus analogous to Robert Boyle's critique of alchemy: they lambast the over-indulgence and credulousness of the ignorant, who practise the art without judgement or understanding, while continuing to accept the essential principles relied upon by true adepts. Astrology was, even for its critics, the central ontological framework of Renaissance natural philosophy. Paracelsus, Luther, Erasmus and even Copernicus took it for granted that, to some degree or other, the stars and planets influenced human lives. They could not conceive of a universe that was anything but teleological – for after all, they could not imagine it without a God. Our purposeless, mechanical universe would not just have appalled them; it would have been incomprehensible.

This belief in the power of the stars is not surprising. The heavens are, after all, constantly wondrous and astonishing, especially in a world where fire cast the only light at night. And the classical identification and personification of constellations shows how ancient is the impulse to try to make sense of this scattering of bright pinpricks: the gods live up there, and their stories are written across the sky. Across this backdrop, which rotates slowly with the seasons, the planets thread their courses like stars that have come loose, making their way through the twelve signs of the zodiac. These trajectories are baffling: predictable, hinting at simplicity, yet in the end strangely complex, with loops and retrogressions. (It was Plato who named the bodies 'planets', meaning wanderers.) They are telling us something, surely. Their real message is that we live in a system of planets surrounding a star; but can we be surprised if people once interpreted this celestial dance as a kind of divine semaphore?

Whenever this regular pattern was broken, people inevitably expected that these cosmic disruptions were reflected by events in our world. Comets came and went – ephemeral new stars, challenging the idea that the heavens were eternal and unchanging. On rare nights, all the more auspicious for their rarity, the moon would darken to blood red or even seem to be swallowed up: a lunar eclipse. Most terrible of all were the days when the sun, the source of all light and warmth and fruitfulness, would disappear in a solar eclipse and a shadow would fall over the earth. The feeling of awe, of a disjunction in the natural order of things, is still great enough today to draw people across the world in order to witness a solar eclipse, even though we know it is just the effect of the moon passing between the earth and the sun (and how strangely perfect is the fit!). It is hard for us to imagine the impact of such occasions when magic still filled the world.

And our forebears were more attuned to changes in the sky. The working day was delimited by the rising and the setting of the sun, and the hours by its angle as the day progressed. The sailor and the fisherman knew how the tides followed the phases of the moon, and on the open seas they navigated by the stars.

But the real attraction of astrology was its 'explanatory' power. The explosion of interest in the occult arts in Europe in the fifteenth and sixteenth centuries was surely linked to the fears and hardships of the times – the greater scale and frequency of war, the threat of famine caused by expanding populations, the schism of Christianity, the overturning of the medieval social order, the threats of new diseases. The only antidote to despair was an illusion of control, and that is what astrology provided. On earth there was chaos, but in the heavens order reigned, and men looked to them to find consolation for their terrors and sorrows. Before the rise of science, astrology filled an intellectual vacuum. He who did not believe in astrology could make sense of nothing. In astrology's promise to provide control over and insight into one's fate lies much of its continuing appeal today.

Ordinary people might find solace in the Grand Design that astrology imagines, but for scholars its ultimate justification was both philosophical and political. Aristotle introduced a two-tiered cosmology in which a perfect, invariant celestial realm presides over an ever-changing and imperfect terrestrial sphere. In the hierarchical society of medieval Europe, 'presiding' was interpreted in a very literal way, so that the world was considered to be *governed* by the heavens: the nobler ruling the humbler. Thus the macrocosm of the sky dominated the microcosm of man. The seventeenth-century herbalist Nicholas Culpepper explained:

> If you do but consider the whole universe as one united body, and man an epitome of this body, it will seem strange to none but madmen and fools that the stars should have influence upon the body of man, considering he, be[ing] an epitome of the Creation, must needs have a celestial world within himself . . . Every inferior world is governed by its superior, and receives influence from it.[2]

This natural order seemed so legitimate to the medieval mind that human experience was selected, pruned, organized and shuffled to fit. A belief cannot be disproved if it is not first doubted.

So in an age without science, astrology seemed like a science. If astrologers with their charts and calculations could indeed predict where planets would be on a certain day, and when they would line up in conjunctions – if they could *see into the future* in this manner – it was easy to imagine that they could also know what lay in store for humankind. Wise astrologers emphasized that they could identify only dispositions, tendencies, likelihoods – not certainties. In this way, astrology, like magic, became infallible, since failed predictions could be explained away by contingent particulars.

Predictive or 'judicial' astrology had several aspects. Forecasts of large-scale calamities or social circumstances – disasters (literally 'bad stars') such as terrible weather, crop failures, epidemics, wars, changes of government – were made on the basis of eclipses and planetary conjunctions, which, thanks to detailed astronomical observations and calculations, could be predicted well in advance.

Great Conjunctions – the alignment of Saturn, Jupiter and Mars – were particularly auspicious, heralding widespread destruction and conflagration. The degree of alignment varies from one event to the next: 'regular' conjunctions happen every twenty years, but 'major' conjunctions only every 240 years; 'maximum' conjunctions occur only every 960 years. The rarer a conjunction was, the more potent it was deemed to be. Astrologers considered this lining up of the planets as a kind of resetting of cosmic time in which the planets were returning to the positions they occupied when the universe was created.

So conjunctions were anticipated with great foreboding. The Joachimites believed that these alignments were harbingers of the end of the world. In the late fifteenth century, movements like Joachimism shared a sense of approaching apocalypse. The passing of the half-millennium did not quell this unease; if anything, it was heightened by a major conjunction of the planets in Pisces in February 1524.

This event gave rise to numerous astrological prognostications and debates. Some, noting 'the watery sign' in which the conjunction would happen, foresaw a flood of biblical proportions. In contrast, Pico's nephew Agostino Nifo (who had defended astrology against his uncle's attacks) published a treatise entitled *On the False Prognostication of a Deluge* (1519), and other astrologers issued assurances that the beginning of 1524 would bring nothing worse than heavy rain and snow. But anxiety was widespread, and to some extent the astrological prophecies of disruption may have become self-fulfilling, fuelling the social unrest that led to the Peasants' War in Germany. Leonhard Reynmann's *Practica* for 1524 was particularly inflammatory, depicting on its cover an illustration of scythe-wielding peasants confronting the clergy, while above them all looms a gigantic fish denoting Pisces and marked with the figure of death. As it happens, February 1524 was mostly fair according to contemporary records. But there were thunderstorms in early summer, while July saw hailstones 'as large as hens' eggs', convincing some that the deluge was on its way.

The astrological prognostications of the *practica* gave rise to the almanac tradition. The almanac not only listed future eclipses, conjunctions and other astronomical events, along with forecasts of what these signified, but it also indicated the dates of religious festivals, markets and other social occasions, and supplied a miscellany of interesting information ranging from recipes and gardening hints to summaries of great historical events. By the end of the sixteenth century pocket almanacs were widely available and often outsold the Bible. *Poor Robin*, a mock almanac published in England in the mid seventeenth century, deliciously satirized the way in which, as Jonathan Swift later put it, the forecasts were worded 'so cunningly and equivocatingly that, be the event what it will, still the words shall be capable of imitating it'.[3] In February 1664, said *Poor Robin*, 'we may expect some showers or rain either this month or the next, or the next after

The provocative title page of Leonhard Reynmann's *Practica* for 1524, showing peasants confronting the authorities.

that, or else we shall have a very dry spring.'[4] As far as astrology goes, *plus ça change*.

But astrology did not always paint with so broad a brush: the stars were thought to govern even the microscopic details of the world. The fate of individuals depended on their 'nativity': the positions of the celestial bodies at the time of their birth. This personal horoscope influenced a person's character and constitution and could be used to furnish predictions of subsequent events in his life. By considering a person's nativity in the light of the movements of the heavens, astrologers could make 'elections': predictions of the best time to undertake a certain course of action, such as a journey. Many princes would consult astrologers before passing judgement or declaring war; some sought astrological guidance even for trivial decisions, such as when to take a bath. Physicians made elections to find the most auspicious occasion to administer medicine or conduct an operation. 'Being grounded in astronomy', says Chaucer of his physician:

He watched his patient's favourable star
And, by his Natural Magic, knew what are
The lucky hours and planetary degrees
For making charms and magic effigies.[5]

The university of Bologna stipulated in 1405 that all medical students should study astrology for four years; in Paris, Charles V of France established a joint college of astrology and medicine in 1371. A person's humoral type was supposed to be dictated by the time of year he or she was born: springtime produced sanguine personalities, summer choleric, autumn melancholic and winter phlegmatic.

And astrology was a means of divination, answering any question by taking account of the positions of the stars and planets at the moment it was asked. Such enquiries were known as horary questions. 'As the nativity is the time of the birth of the body,' wrote the renowned English astrologer John Gadbury in 1658, 'the horary question is the time of the birth of the mind.'[6] This aspect of astrology impinged on medicine too, since an horary question could be a request for diagnosis, in which case the doctor might answer it by inspecting not just the arrangement of the heavens but also a sample of the patient's urine, bearing in mind when it was passed or when it was brought to him.

The Church was traditionally suspicious of astrology, since it seemed to trespass on God's territory. For one thing, if the future was already written in the inevitable patterns of the heavens, what room was there for free will and personal accountability? What room, indeed, for God to exert His omnipotence? But the astrologers were on the whole only too happy to accept that God could, if He willed it, direct human affairs in contradiction to the stars; it was an effective explanation for failed prognostications. Moreover, astrologers were quick to defend their art by pointing out that the biblical magi who paid homage to Christ shared their profession.

Nonetheless, there seemed to be something impious about the way that astrology promised almost boundless knowledge. 'Heaven', said John Chamber in his *Treatise against Iudicial Astrologie* (1601), 'is God's book, which we must leave to him. To what end has God placed us so far from the stars, if with astrolabes, staves and quadrants we can do all things as if we were nearer?'[7]

Notice that this diatribe could be equally well aimed at scientific astronomy as at astrology: it stems from the old objection that it is not only vain but also blasphemous to study our world too closely and imagine we can deduce how it works. Moreover, priests probably objected to astrology mostly for the same reason that they opposed natural magic in general: people went to the astrologers for advice, and no doubt for a fair degree of reassurance and comfort, whereas that was supposed to be the role of the clergy.

THE BREATH OF THE STARS

To most, astrology was just the way the world worked. But the philosopher needed a reason why – a mechanism. Even if the mechanism was spurious, the fact that the need was felt shows that astrology, like natural magic, was not simple superstition but shared of the same impulse that gave rise to science. It was in some ways more 'scientific' than astronomy as such, since astronomers could offer little more than Aristotle's tautologies to explain why the heavens moved as they did: circular motion was 'natural' to celestial matter, since the circle was a perfect figure, and so stars and planets moved in circles (or at least, in the combinations of circles known as epicycles) because it was in the nature of their matter to do so.

The influence of the stars on worldly matter, meanwhile, was the result of 'emanations'. One had only to consider the sun, which quite evidently did emanate heat and light that affected the climate on earth. Leonardo da Vinci perceived how the sun's heat causes moisture to rise through the air, having the 'power to stir up the dampness of the low places and draw this to a height in the same way as it draws the clouds and calls up their moisture from the expanses of the sea'.[8] This was not merely a matter of evaporation – it was thought that moisture can itself come from the sun and the stars. Aware of how warmth and dampness cause things to rot, astrologers asserted (and it was generally believed) that plagues were the result of an excessive amount of heat and moisture radiated by the stars, leading to decay. Other invisible emanations or forces were presumed to connect the moon's motions (and the sun's too) to the daily rise and fall of the seas.

Comets were considered to be 'sublunar': closer to us than the

Albrecht Dürer's *Melencolia I* (1514) is filled with occult and alchemical symbolism.

moon, and therefore more easily rationalized than an interruption of the distant, eternal heavens. They were thought to emit hot, dry emanations that scorched the earth and spoilt harvests, bringing famine. By drying men's blood, comets created an excess of the choleric humour, making people argumentative and more liable to go to war: thus even social calamities found a 'rational' explanation in astrology. Dürer suggests the cometary influence on the humours in his *Melencolia I* (1514), a woodcut laden with alchemical and occult imagery.

Paracelsus had his own theories: some conventional, others very much his own. As he says in the *Volumen medicinae paramirum*:

> The stars have their own nature and properties, just as men have upon the earth. They change within themselves: one sometimes better, sometimes worse, sometimes sweeter, sometimes sourer, and so on. When they are good in themselves no evil comes from them; but infection proceeds from them when they are evil.[9]

This evil comes from the poisons that leak from the stars into the air that bathes the world: for 'the stars surround the whole world just as a shell does an egg: the air comes through the shell and goes straight to the earth.'[10] Again, this is a thoroughly mechanical process, like the tainting of river water with sewage, and there is nothing supernatural about it: 'the stars themselves do not act: they only infect through their exhalations.'[11]

It is not, however, the air that transmits these astral influences, but the cosmic matrix of the Mysterium Magnum. 'The air is held in the firmament,' says Paracelsus cryptically, 'and if it were not in the firmament, the firmament would melt away, and this we call the Mysterium.'[12] By way of illustration, he adds:

> A fish pond which has its right Mysterium is full of fish; but if the cold becomes too great it freezes and the fish die, because the Mysterium is too cold opposed to the nature of the water. But this cold comes not from the Mysterium, but from the heavenly bodies whose property it is. The heat of the sun makes the water too warm and the fish die on this account also. Certain heavenly bodies affect these two things, and others make the Mysterium acid, bitter, sweet, sharp, arsenical, and so on, a hundred flavours. Every great change in the Mysterium changes the body, and so note how the stars contaminate the Mysterium, so that we fall ill and die of natural exhalations.[13]

This is all wrong, of course, but it is not unreasonable.

DOUBTING THE STARS

In his *Astronomia magna* (1537–8), Paracelsus lays out the case that astronomy is truly a Christian practice. The book provides a detailed and rather baffling taxonomy of the 'sciences of astronomy', which he first divides into four categories. Natural astronomy 'comes from heaven and was created by God the Father',[14] and that is what natural philosophers pursue with their instruments, observations and calculations. The second category, 'Supera', 'has its seat in heaven and is given to all those who shall rise again from the dead. It has its origin in Christ and it is practised, used, and administered by Him.'[15] Thus astronomy, like medicine, is vindicated by Christ's example.

The third, curiously named, category is even more nebulous, although also of religious origin. Olympi Novi ('new Olympus') 'derives from faith: whatever the natural heavens can do, this astronomy can achieve by faith, and it is used by and given to the faithful; it is practised and made manifest by them.'[16] The fourth type of astronomy is used by demonic spirits ('because they are natural astronomers'[17]), and it is called Inferorum. Clearly, this is in fact a categorization not of 'astronomy' as such but of the ways in which natural magic offers the capacity for directing and manipulating the influences of the stars.

Paracelsus goes on to explain that there are nine ways in which men might practise the astronomical arts – among them are astrology, medicine, necromancy and the art of recognizing signatures in nature. And there are a further ten 'members' of astronomy 'which are not human arts but ethereal arts'. That is, he explains, 'the heavens themselves are an astronomer and can practise without the aid of man'.[18] Notwithstanding the fact that he has already listed astrology as a 'human art' (one that seems to consist simply of 'understanding the motions of the stars'), it appears that for Paracelsus these ethereal astronomical arts are those practices we would normally associate with the prognosticating astrologer. *Divinatio*, for example, is the business of foretelling the future, a form of judicial astrology. So too is *augurium*, the practice of interpreting the auguries of fate such as comets or the movements of birds. *Inclinatio* is concerned with the stars' government over man; *impressio* is the virtue of the stars impressed on natural bodies, *meteorica* the causes of thunder and mist, but also of dragons, haloes, mirages and other amazing things.

There is no doubting Paracelsus' enthusiasm for 'astronomy'; but how closely did he consider humankind's fate to be bound to the heavens? Early in the *Astronomia magna* he seems to imply that the link is strongly deterministic:

> If there had been no Venus, music would never have been invented . . . And not only the arts, but also all wars, all governments, and everything which our brains produce, receive their guidance from the stars now and for ever.[19]

Yet it is not easy to reconcile these remarks with the scepticism Paracelsus shows elsewhere. On the whole, he is rightly regarded as a critic of astrology's excessive claims. While accepting that the stars exert an influence on the earth and its creatures, he is generally at pains to emphasize that this does not make human affairs preordained. 'The stars control nothing in us, suggest nothing, do not irritate us, incline to nothing,' he argued in one of the most extreme formulations of this point of view, 'they are free from us and we are free from them.'[20] It is not clear that this is meant as literally as it sounds; rather, he wants to imply that we can evade or override the astral influence and take control of our own fate: a thoroughly humanist attitude. As he says in the fourth book of the *Opus paramirum*:

> One man excels another in knowledge, in wealth or in power, and you ascribe it to the stars; but that we must banish from our minds: good fortune comes from ability, and ability comes from the spirit. Every man has a special spirit according to the character of which he has a special talent, and if he exercises that talent he has good fortune.[21]

Thus, he warns, we should be wary of those doctors who explain all disease in astrological terms rather than looking for the specific, proximal causes:

> Men speak of an *Inclinatio*: it is nonsense. They say that man receives an *Inclinatio* from Mars, Saturn, the moon, and so on: this is deception and error. It would be more reasonable to say 'Mars imitates man', for man is greater than Mars or the other planets . . . A man may become fat and it is not the fault of his food: or he may become thin and his food does not help him. And the doctors set it down not to the specific influences, but declare with the ignorant astrologers that it is *melancholia* due to Saturn in the ascendant: man owes nothing to the ascendant; he owes it to the *limbus* [see page 266] and he is made by the hand of God, not by the ascendant, nor by planets, nor by constellations and the like, as if these could compel him to be either lean or fat.[22]

THE EARTH MOVES

In 1543, as Oporinus published Vesalius' groundbreaking book on anatomy, an even more radical work rolled off the printing presses of Nuremberg, overseen by the Lutheran preacher Andreas Osiander. It was only just over a decade since Osiander and the other Nuremberg dignitaries refused permission for the publication of Paracelsus' works, judging them to be too inflammatory and subversive. But the book that Osiander committed to print was far more shocking than that.

So much so, in fact, that the preacher added an anonymous preface explaining that the reader should of course regard everything in this book as mere hypothesis, it being hubristic for man ever to assume that he can know God's work. Not that the author himself had much need to fear recriminations: he was already on his deathbed, having prevaricated over publication for thirteen years. (Legend has it that he saw the printed book only on the day he died.) Osiander, however, was uncomfortably aware that his own reputation, indeed his own skin, might be at stake. At face value, this work looked dangerously like heresy to Lutherans and Catholics alike.

The book was *De revolutionibus orbium coelestium* (*On the Revolution of the Heavenly Spheres*), and its author was the Polish astronomer Nicolaus Copernicus.* What it dared to say was that the earth is in motion, and so cannot be at the centre of Creation.

De revolutionibus was patiently extracted from its author by Georg Joachim Rheticus, a Protestant mathematician from Luther's Wittenberg who travelled in 1539 to Frauenburg on the Baltic coast of Prussia (now Frombork in Poland), where Copernicus hid away from the world as he laid the groundwork for a revolution every bit as explosive as Luther's. The Protestant leaders did not approve of Copernican astronomy (news of which circulated well before *De revolutionibus* was actually published), and in the year that Rheticus set out, Luther called its architect a fool who sought 'to reverse the entire science of astronomy'.[23] Even the mild and intellectual Melanchthon agreed: 'certain men,' he said, 'either from love of novelty, or to make a display of ingenuity, have concluded that the earth moves . . . Now, it is a want of honesty and decency to assert such notions publicly, and the example is pernicious.'[24]

* Although generally described as Polish, Copernicus was born to a German family in the German city of Thorn.

Copernicus' book rearranged the universe. Since ancient times, it was a virtually unchallenged belief that the earth sits motionless at the centre of everything while the heavens revolve around it. These revolutions of the heavenly bodies – the sun, moon, planets and stars – were obvious to everyone who ever gazed up at the sky. The only problem was that the heavenly orbs did not seem all to move in perfect circles. The planets, for instance, executed looping orbits around the earth, sometimes seeming to reverse direction temporarily. Medieval natural philosophers considered this to have been explained by the celestial system devised by the Egyptian astronomer Ptolemy of Alexandria (*c.* AD 90–170). Ptolemy's *Algamest* showed how these curious motions of the heavens could be understood in terms of a complex system of circular trajectories called epicycles. Ptolemy constructed a scheme of epicycles to account for how all the heavenly bodies moved around the stationary, central earth.

Ptolemy was to astronomy as Aristotle was to physics and Galen to medicine: his word, once reconciled with Church doctrine, became dogma. But astronomers had long been possessed of the uneasy feeling that the Ptolemaic system was (literally) more than a little convoluted, and not blessed with the elegance and economy that one might expect from the Divine Designer. Copernicus' world-changing idea began as an attempt to find a simpler solution to the motions of the heavens. In the end the astronomer was forced to conclude that 'it is more probable that the Earth moves than that it is at rest'.[25] The earth, like the planets, was a rotating orb.

But rotating around what? Copernicus found that the most parsimonious arrangement was one in which the sun, not the earth, was placed at the centre of the universe: his astronomical scheme was *heliocentric.* Now, if one arranges the planets as we do today, with the earth third from the sun, everything works harmoniously when the planets execute near-circular (ellipsoidal) orbits: one can then account for the way the heavens look from earth. But this was not exactly what Copernicus proposed. Rather, because he remained determined to use strictly circular motions, his heliocentric scheme still required epicycles – thirty-four of them – but they were simply small modifications to a system of heliocentric circular orbits. By eliminating Ptolemy's major epicycles, Copernicus was able to draw basically the plan of the solar system that we recognize today. He was observing the logical principle attributed to the fourteenth-century philosopher William of Ockham: the scheme that explains the observed facts in the simplest possible way is the most likely to be correct.

Historians of science have often cited *De revolutionibus* as the beginning of what is traditionally called the Scientific Revolution. Here is a break from the past that finally organizes nature in a way that is recognizably valid today. They point out how the new astronomy could not fail to find acceptance eventually – even in the face of the ban imposed by the Church on *De revolutionibus* in 1616, and Galileo's subsequent tribulations with Rome – because heavenly motions were too important in daily life for an incorrect theory to survive the emergence of a better one. An understanding of the heavens was of great practical significance: to tell the time, to plan the calendar, to navigate the oceans.

That is all true enough, but it neglects both what Copernicus actually said and how it was perceived at the time. For one thing, Copernicus stressed that because the heavens are essentially of infinite size, a slight displacement of the earth from their centre was equivalent to no displacement at all. And the primer on Copernican astronomy, *Narratio prima*, that Rheticus wrote in 1540 betrayed other themes entirely: no mere synopsis of heliocentric theory, it was an altogether more occult tract. As well as including a Cabbalistic celebration of the number six, it described how a moving earth could explain the changing fates of monarchies and how this new system might be used for astrological prediction. One of the main problems with the Ptolemaic system was not that it was cumbersome *per se*, but that it could not be trusted for making prognostications. It is possible that some readers of *De revolutionibus* were disappointed by it, rather than elated or outraged, because they found that Copernicus himself had not really pursued the astrological implications of his theory.

All the same, a Neoplatonist would not have felt entirely excluded by this founding text of modern cosmology. Copernicus reminds us how Hermes Trismegistus called the sun 'the visible god'; and this is not a historical aside, but a preparation for his placement of the sun at the centre of things ('as if resting on a kingly throne'[26]). While Copernicus' calculations suggest that this is the proper place for it (and remember that mathematics is at this time as much the language of the Cabbala as of science), a heliocentric arrangement also fits with the Neoplatonic notions of perfection and hierarchy. 'The state of immobility', says Copernicus, 'is regarded as more noble and godlike than that of change and instability',[27] and is therefore more suitable for the perfect heavens than for the earth, where all things are mutable.[27]

Rheticus himself was thoroughly familiar with Hermetic tradition, and was enthusiastic about the revival of Paracelsus' ideas in the 1560s and 1570s. He claimed to have composed seven books on alchemy, and to have met and been impressed by Paracelsus in person. In 1574 he announced his intention to publish Paracelsus' great alchemical work, the *Archidoxa*, being dissatisfied with the version produced by Gerard Dorn; but he died before this project could be accomplished.

THE END OF THE ASTRAL REIGN

That metaphysics and occult ideas feature in the schemes of Copernicus and his disciples is a reminder that Renaissance astronomy was fully interwoven with the art of astrology. Today, astronomy is the science of the stars: the study of how they are arrayed in the heavens, how they evolve and behave, what they are made from. Astrology, on the other hand, is a pseudo-science: an attempt to read human nature and destiny in the ephemeral and arbitrary juxtapositions of the stars and planets as they appear from the earth. That it persists at all is a testament to the experiential validity of this ancient art – something compels us to look to the stars for hope, guidance, explanations. In all probability, this is a displaced religious impulse: an expression of the fact that the cosmos dwarfs us, that the stars look down on us, aloof and indifferent and yet displaying patterns and rhythms that we cannot but regard as significant. Wrong ideas do not survive for so long unless they respond to some human need.

So long as mankind remained at the centre of the universe, the world literally revolved around him: everything was focused on man, and the cosmos was an enormous cryptogram arranged for his benefit. Once the earth was displaced from the centre, it became possible to imagine, and then impossible to deny, a grander scheme in which mankind was not the object but merely a bit-part player. We were diminished. Astronomers today speak of the Copernican Principle: the assumption that our place in the universe is nothing special, that it is no different from any other place. Once you accept that idea, astrology becomes untenable, because it is utterly contingent. The stars and planets look the way they do because we happen to be located where we are – but we could just as easily be somewhere else, in which case the heavens would look very different and any attempt to 'read their message' would tell us something else entirely.

Clearly, this fact did not bother Copernicus' supporters. Tycho Brahe, the great cartographer of the heavens whose measurements of planetary motion enabled Johann Kepler to formulate his planetary laws, was an enthusiastic astrological prophet. Indeed, his careful observations were made not so much to understand the celestial bodies in themselves as to obtain a sound basis for his forecasts. Kepler himself was entirely comfortable with the idea that comets were portents of dramatic events on earth. Astrological prediction was an important motivator for the flourishing of astronomical theory in the seventeenth century.

It is fair to say that Paracelsus was one of the Renaissance thinkers who helped to cut the puppet strings between man and the stars. We should not regard that as equivalent to a denial of astrology; but it was another step towards a world in which humankind, controlling its own destiny, could enquire into nature and become informed about what it contains – which is to say, towards a world of science.

Yet it was astronomy itself, so essential for astrology, that ultimately destroyed it. There was nothing in Copernicus' theory that could not be accommodated within the teleology of an astrological cosmology; and even Newton famously formulated his *Principia*, the blueprint for the clockwork cosmos, while working on an interpretation of the Biblical prophecies in the Book of Daniel. It was not theory but observation that made astrology untenable. Paracelsus believed that the heavens, once considered perfect and immutable, were in fact in a state of dynamic flux like the earth, growing and mutating and populated by monstrous beings called *penates*. Tycho Brahe was strongly influenced by these ideas, and his important observations of a comet in 1577 were made in collaboration with the Paracelsian scientist Petrus Severinus. He pronounced the comet a sign that the Last Days and the Golden Age were (once again) imminent.

These studies showed Tycho that comets were in fact superlunary – further away than the moon. This confirmed Paracelsus' notion of change in the outer heavens, dealing a fatal blow to Aristotelian cosmology and its attendant hierarchy. Once Edmund Halley proved that cometary apparations were generally periodic and predictable, one could no longer easily regard them as portents sent by God to herald specific events. The celestial bodies gradually became both less mysterious and less wonderful, and thus less powerful – a change accelerated by Galileo's telescopic observations of the pock-marked moon and his discovery of the four major moons of Jupiter.

Paracelsus' use of and views on astrological prognostication were representative of the age in which he lived: a mixture of credulity and scepticism, they ultimately became part of a movement to wrest mankind's fate from the control of higher powers and place it in our own hands.

16
Demons of the Mind

Invisible Diseases

The poet's eye in a fine phrensy rolling
Doth glance from heav'n to earth, from earth to heav'n
And, as imagination bodies forth,
The forms of things unknown, the poet's pen
Turns them to shape, and gives to aiery nothing
A local habitation and a name.
William Shakespeare, *A Midsummer Night's Dream* (*c*.1594)

In late 1531, when Zwingli was run through by a lance at Kappel, Paracelsus' patient Christian Studer died in St Gallen, and Paracelsus seems to have been plunged into despair. Leaving behind his manuscripts, he resumed his life of wandering. But it was no longer a combative and polemical man who followed the road west towards Lake Constance that October – here instead was a lost soul, adrift in the world.

He appears to have reached a peculiar stage in his life. If we cannot confidently call it a breakdown, it seems at the very least to have been marked by a deep fracture in his spirit. Paracelsus became a wretchedly depressed wanderer travelling in rags through the hills and forests of Switzerland – a mad prophet who had seen too much of the world.

These were his 'days in the desert', during which time he declared, 'I do not know where I shall have to wander now; I do not care either, so long as I have helped the sick.'[1] Yet for the first time he seemed to doubt whether even that much was in his gift. Medicine, he said, is 'but guesswork'. In the end, God alone has the power to give and to take away our health, and the doctor labours in vain against that stark reality. If his medicines work, that is because God has willed it, and it is futile for the doctor to look for any other reason. Here now is a genuine Luther of medicine, who insists that only through faith shall man be saved. Do not dare to question how God works through nature, for:

> one may find a medicine for this hot pain and another for that cold disease; that one for young people, that one for old; this one in this country, that one in another. Whosoever finds a medicine and its virtues, knows just that this medicine has such a power against such a disease. He should not think: This disease originates from that cause or this one. For the physical bodies are of such a nature that it is hard to reason about them. One may imagine he knows something that resembles a theory; but he will find contradictions, and he does not know the substance and origin. Therefore we shall know what medicines there are as though God had given them to us. God has given power to the stones, but He might have willed it differently. The doctor should take this for granted and not ask the reason why. Does the stone care why it cures the eye? Or the eye what colour the stone is? So why should the doctor ask questions? Why should he seek behind nature? It is destruction that makes the disease, so why should we ask for the cause?[2]

His earlier trust in chemical remedies seems to him now to have been a delusion: 'Where I had seen flowers in alchemy, there is but grass.'[3] This is a terrible capitulation, tragic and pitiful: not Christ in the wilderness, but Lear in the storm.

It is not so much as an itinerant healer that he drifts through the rural Swiss cantons and the German lands, but as something closer to a lay preacher, sermonizing to the common people 'in taverns, inns and public houses'. 'I gave up medicine', he says in the third (later) book of the *Opus paramirum*, 'to ply other trades.'[4] He spends some time in the poor communities of Appenzell, with its painted houses and its steeple-topped hills. He wanders in the alpine valleys around Urnäsch and Hundwil, and passes up into southern Germany: Württemberg, the Black Forest, Swabia. He travels north to Prussia, where in 1532 he is 'driven out'. By 1533 he is in Silesia, by way of which he comes back to Appenzell and St Gallen: a round trip with no apparent purpose and nothing to show for it except sore feet and an empty purse. And from this period there are no accounts of furious disputes, no diatribes, no medical miracles: one might even mistake this traveller for a humble man, disconsolate and yet teetering on the verge of ecstatic religious mysticism. 'I started out in the Light of Nature', he avers, 'and finished in the Light of Eternity.'[5]

When Paracelsus' detractors over the centuries have called him a madman, it is generally simple abuse. But perhaps he really did suffer from some kind of mental illness. When we consider the awkwardness of his relations with other people, the blind rages, the recurrent self-inflicted catastrophes, the inability to settle down, the periods of manic activity, the apocalyptic visions – all may make us suspect something like a manic-depressive disorder.

Isaac Newton's breakdown has been attributed to his alchemical experiments: too many hours spent in a stuffy chamber breathing toxic fumes. Perhaps the dark shadows stalking through Paracelsus' head were also chemically induced. And yet it is not difficult to imagine that his wild and unpredictable moods had some deeper psychic origin, and that in the early 1530s they finally overwhelmed him.

ROUGH ROADS

For most people in the Middle Ages, the longest journey they would ever make took them no more than fifteen miles from home. You did not travel unless you had no option, because it was generally a deeply unpleasant experience. Not that it was necessarily slow: by water one could make good time, and on horseback it was quite possible to cover fifty to seventy kilometres between dawn and dusk. There are accounts in the fifteenth century of pilgrims passing from Zurich to Strasbourg in a single day. Even at a fairly leisurely pace, a Parisian could travel to Calais in four days or so, to Toulouse in eight to ten, and to Turin in two weeks.

The drawbacks were the danger and the discomfort. Thomas More wrote admiringly of Erasmus' journeys that he 'defies stormy seas and savage skies and the scourges of land travel'; he goes 'through dense forest and wild woodland, over rugged hilltops and steep mountains, along roads beset with bandits . . . tattered by winds, spattered with mud, travel-weary'.[6] A German merchant in the late sixteenth century complained, 'I have had so little respite that my backside has been constantly a-fire from the saddle.'[7]

The rigours of travelling are revealed in a treatise called *Regimen of Health for Travellers*, written by the Italian physician Guglielmo Gratarolo and published in Basle in 1561. The book is a strange mixture: part practical manual, part guidebook (complete with place names and recommended travel routes), part medical text and part magical treatise. There are descriptions of charms said to protect the traveller: a wolf's eye, a vulture's heart, gems and herbs. Betony confers protection against witches and ambushes during night journeys; other charms ward off wild dogs. Eating lettuce will guard against storms at sea (a tip filched from Pliny), and Gratarolo gives a list of particularly ill-omened days in early spring for sea voyages.

If passing through uninhabited lands, says Gratarolo, take with you some compact rations to keep you going: marzipan, 'placenta of Genoa' (a kind of cheese cake), pills made of almonds, liver and oil of violets, dried bread soaked in wine, powdered meat, prunes for the bowels. A sprinkling of the herb pennyroyal will sweeten tainted or bitter water. The remedies that Gratarolo recommends are a reflection of the medical problems travellers could expect to encounter: sleeplessness, diarrhoea, constipation, headaches and sunstroke, foot sores, snow blindness (goggles were advised), frostbite, intoxication and vermin.

As for roadside inns and taverns, Gratarolo confesses that it would take someone even more widely travelled than he to do justice to the dangers one should anticipate. Thieves were everywhere, and the proprietors weren't to be trusted either. He recounts how his purse once went missing as he slept in an inn near Brescia, and when he drew his sword and stood at the front door refusing to let anyone leave until it was returned, the landlord 'found' the purse in his own bed. Gratarolo cites a process described by John of Rupescissa for turning gold and silver into an innocuous earth-like substance (alchemy in reverse!) so that it would not attract unwelcome attention on the road.

Lodgings could be foul and rowdy, the food execrable, the wine sour, the beds shared with lice and possibly with fellow travellers too. Gratarolo tells how, in a Milanese inn where he slept in a room with several others, he retired at night to find broken glass placed between the sheets: a 'practical joke' presumably played by one of his room-mates. Drunken fights were common. Paracelsus recalled one such from his wander-years in the 1520s: 'At Friaul I saw how (in a soldier's brawl) at a public

Tavern Scene, attributed to Jan Sanders van Hermessen (*c.*1540), shows the kind of inn brawl that would have been familiar to travellers in Paracelsus' time.

house a man's whole ear was chopped off; a barber came and stuck it on again with some mason's paste . . . but the ear soon fell off again, dripping with blood and matter.'[8]

The country inns were the worst, according to Gratarolo. Away from the towns it was a cut-throat world where the law of the Empire counted for nothing. Highway crime was rife, and to discourage it bandits were often hung from gibbets along the roads leading out of towns. And since the uprisings of the 1520s those roads were scattered with the dispossessed, such as exiled Anabaptists. 'They are driven from land to land', Paracelsus says of these persecuted outcasts, 'and the door is slammed to as soon as they approach.'[9] That, surely, was his experience too, and he confessed that he was often reduced to begging for a bowl of soup.

These were intolerant days, with each branch of the Christian faith hardening into mutually exclusive dogma. In England, Henry VIII became the self-appointed head of the Anglican Church in 1532, and Thomas More was beheaded for his refusal to recognize him as such. Two years later Ignatius Loyola founded the Jesuits, and in 1536 Jean Calvin established a hard-line, puritanical version of the Reformation in Geneva. The liberalism and tolerance of humanism had bloomed and withered; Erasmus died in 1536. To Paracelsus the increasing sectarianism was 'beastly and often fierce and mad',[10] and he wanted no part in it.

If Paracelsus was as argumentative in theology as he was in medicine, these desert-years see him transformed from firebrand into pious penitent. He is no longer the rebel, but fasts and moralizes like the 'abortive saints' that Nietzsche tells us filled the asylums of Jerusalem during the Crusades. 'Don't shrink from my rags, reader,' he implored. 'Let me carry my cross. I have sickness in me, my poverty and my piety.'[11] In his humility he is, dare one say, almost conservative: 'All things that we use on earth, let us use them for good and not for evil . . . Add nothing, take nothing away, spoil nothing, and likewise better nothing.'[12]

He preaches on the 'beatific life', the life of saintliness, which forswears riches, power and ostentation and looks only to serve and honour God. It is the voice of the puritan and redolent of the heresies of the religious communitarians. Now his tone is, however, no longer that of a revolutionary but of a martyr who has seen enough war and suffering in the name of God:

> If today's Christianity had prevailed at the time of Christ, one would have learnt that neither Christ nor St Peter would have allowed Catholic splendour but would have prevented it by atonement, prayer and preaching. It is easy to see that things today are not as they were at the time of Christ. If Christ were to see the present violence of the Empire and the justice of all the kings, princes, lords, towns and provinces, He would say this is not the law of God but the law of the devil . . . Blessed and more than blessed is the man to whom God gives the grace of poverty. But he who does not possess this grace thinks he is a rich man with much property, money and pleasure, and that he is mighty beside Emperor and Pope. But they are false Christians, they govern arrogantly, they have bad laws, they protect each other in their mischief, and you are abetting them and doing likewise following their laws and teachings. But it is the life of the devil . . . Become poor, as poor as a beggar, then the Pope will desert you, and the Emperor will desert you, and they will henceforth take you for a fool. But then you will have peace, and your folly will be great wisdom in the eyes of God.[13]

He suggests that people should never presume that they act of their own accord in the slightest degree, for God wills everything.

Gone is the liberating – indeed, the proto-scientific – humanist's belief in mankind's ability to do anything, to be its own master, to find knowledge and wisdom in the Light of Nature. In its place is the inexorable dominion of God:

> Some like to claim that we have free will. This is not true. We do not choose, God chooses . . . If you accept a good seed and revive the dead, heal lepers, drive out devils, you may not say, I can do it or not. You do not do it. God does it . . . If God is in you, you do God's works; if the devil is in you, you do his works . . . Man is not master, God is master. We are not free but subjects. Our freedom and our laws come from God and not from man.[14]

Here is the medievalist in Paracelsus; he has forsaken the future and can see only the safe and austere certainties of the past.

NATURAL SPIRITS

Whether or not he was in the grip of madness, Paracelsus saw insanity all around him on his aimless journeys. In his response is perhaps one of the most moving illustrations of his humanity in an age distinguished for its brutality.

Mental illness was a moral issue in the Middle Ages. Where people recognized a 'natural' form of madness at all, it was typically deplored as an indication of enthralment to one of the deadly sins, which caused humoral imbalances that upset the brain. More commonly, the idiot and the madman were considered to be invaded by demons, and that was reason enough to shun and mock them, and also to fear them. Epilepsy was a prime example of a mental affliction whose manifestation had all the signs of a demonic possession, and it seemed to confirm the view that people were vulnerable to the fearsome embrace of demons.

In the *Malleus maleficarum* (*Hammer of Witchcraft*, 1486), a tract written by the Dominican Inquisitor Heinrich Kramer* that was effectively an instruction manual for the persecution of alleged

* The *Malleus* was long thought to be the joint work of Kramer and Jakob Sprenger, another Inquisitor. It turns out that Sprenger and Kramer were in fact joined only by bitter rivalry.

devil worshippers, this theory of mental disease was portrayed in pseudo-medical terms that explain how demons can create hallucinations:

> Although to enter the soul is possible only to God Who created it, yet devils can, with God's permission, enter our bodies; and they can make impressions on the inner faculties corresponding to the bodily organs. And by these impressions the organs are affected in proportion as the inner perceptions are affected in the way which has been shown: that the devil can draw out some image retained in a faculty corresponding to one of the senses; as he draws from the memory, which is in the back part of the head, an image of a horse, and locally moves that phantasm to the middle part of the head, where are the cells of imaginative power; and finally to the sense of reason, which is in the front of the head. And he causes such a sudden change and confusion, that such objects are necessarily thought to be actual things seen with the eyes. This can be clearly exemplified by the natural defect in frantic men and other maniacs.[15]

When the English mystic Margery Kempe experienced episodes of mental illness in the late fourteenth century, she recalled how 'folk spat at her for horror of the sickness, and some scorned her and said that she howled as it had been a dog, and banned her and cursed her, and said that she did much harm among the people'.[16] Blessed with an understanding husband, she was luckier than most.

If Bosch and Bruegel are any authority, demons were truly terrifying, slicing off men's limbs, impaling them on skewers, forcing drink down their necks until they burst, fornicating with them, pouring molten lead over their heads and frying their dismembered parts in a pan. That many of these things were routinely perpetrated by men on other men gives us a fair idea of the role demons played as a repository for the violent fears and instincts of that tumultuous time.

Where did these supernatural beings come from? The idea that the world is filled with spirits is pre-Christian. Like many other aspects of paganism, the Church found it more expedient to incorporate this belief into Christian theology than to attempt the futile task of eliminating it. But whereas the spirits and monsters of myth and legend could be both good and bad, aiding as well as tormenting humankind, Christian dogma insisted that supernatural beings were universally evil. Demons were presented as the assistants of the devil – lesser angels that, like their master, had fallen from God's grace.

This orthodoxy was established in the ninth and tenth centuries in a Church document called the *Canon episcopi*, and it conditioned all subsequent views of demons until the Renaissance. Demons took many forms. Some were celestial, and responsible for bad omens such as comets. Others lived in the deep earth – these were the goblins and kobolds that Agricola blames for mining accidents: 'In some of our mines, though in very few, there are other pernicious pests. These are demons of ferocious aspect, [and] are expelled and put to flight by prayer and fasting.'[17] Demons could appear to men as monsters or giants, but they were incorporeal, compelling Kramer to explain at length how demons are nevertheless able to speak to witches, to see them and eat food with them and – of particular interest to the redoubtable Inquisitors – to 'practise such abominable coitus' with them.

The *Malleus* reawakened interest in demons and their wicked behaviour, and it encouraged efforts to eradicate the necromantic rituals in which they were invoked. It was through the agency of demons that witches were believed to exercise their powers. Indeed, as we have seen, some maintained that *all* magic was demonic – Paracelsus frequently had to refute accusations of sorcery. Agrippa even dared to imply that, by confusing natural and black magic, the Inquisition was sometimes guilty of ascribing to demons events that could be explained by the invisible agency of natural forces.

Agrippa also challenged the idea that all demons were evil: he shared the folk belief in harmless and even friendly spirits. Even Agricola, although in most respects a sceptic, attested to the relatively equable character of some subterranean gnomes in his book *On Subterranean Animals* (1549):

> Then there are the gentle kind which the Germans as well as the Greeks call *cobalos*, because they mimic men. They appear to laugh with glee and pretend to do much, but really do nothing. They are called little miners, because of their

> dwarfish stature, which is about two feet. They are venerable looking and are clothed like miners in a filleted garment with a leather apron about their loins. This kind does not often trouble the miners, but they idle about in the shafts and tunnels . . . Sometimes they throw pebbles at the workmen, but they rarely injure them unless the workmen first ridicule or curse them. They are not very dissimilar to Goblins, which occasionally appear to men when they go to or from their day's work, or when they attend their cattle . . . The mining gnomes are especially active in the workings where metal has already been found, or where there are hopes of discovering it, because of which they do not discourage the miners, but on the contrary stimulate them and cause them to labour more vigorously.[18]

Paracelsus went further. The prevailing view was that demons, being fallen angels, were not of this world. According to the Polish philosopher Witelo in the thirteenth century, demons are a kind of superior animal spirit: 'as much as an angel exceeds a man, so much does a demon exceed a brute.'[19] But Paracelsus insisted that some so-called demons are not supernatural in any sense, but merely 'natural' beings: a mixture of human, animal and spirit.

His book *On Nymphs, Sylphs, Pygmies and Salamanders* (a later work whose precise date of composition is uncertain) appears on one level to be a collection of almost whimsical fancies, a discussion of creatures straight out of the stories of wide-eyed peasants. Yet Paracelsus does not simply recount these fantastical tales but works determinedly to fit them into his cosmological and ontological scheme. Characteristically, this requires him to berate others for their ignorance, although perhaps with a little less vitriol than of old: 'There are more *superstitiones* in the Roman Church than in all these women and witches . . . Consider such things carefully, and be not blind with seeing eyes and dumb with good tongues, particularly since you will not let yourselves be called dumb and blind.' Paracelsus reminds us how the Scriptures command us 'to despise nothing, to ponder everything in mature understanding and judgement'.[20]

Like Agrippa, Paracelsus conceded that the strange beings that exist in the world are not all demons and agents of the devil. Like humans, they may be good or bad: 'They are witty, rich, clever, poor, dumb like we who are from Adam.'[21] They resemble men, although different in proportions, because God has seen fit to make these creatures in man's image just as man is made in God's image. 'They have man's diseases and his health,' he explains:

> Their customs and behaviour are human, as is their way of talking, with all virtues, better and coarser, more subtle and rougher. The same applies to their figures: they are very different, like men. In food they are like men, eat and enjoy the product of their labour, spin and weave their own clothing. They know how to make use of things, have wisdom to govern, justice to preserve and protect.[22]

Thus, to these elemental creatures – which Paracelsus never claims to have seen – he extends the same compassion and humanity as he does to the reviled and outcast members of humankind itself. They are all God's creations, even if not descended from Adam. 'Nobody should wonder that there are such creatures. For God is miraculous in His works which He often lets appear miraculously.'[23] As to the theologians, 'it becomes easily apparent that they have little understanding of these things. Making it short, they call them devils, although they know little enough about the devil himself.'[24]

These creatures differ from man in one essential aspect, however: they have no souls. That is to say, like men they have both an 'elemental' (terrestrial) and a 'sidereal' (astral) body, but not a spiritual one. In this they are like animals. Thus, while they are as capable as humans of noble and intelligent actions (as well as cunning and pernicious ones), they have no prospect of an eternal afterlife. When they die, they 'are dead like the beasts'.

And that is why they crave the company of men, because they long for the soul they do not possess. If one of these creatures is able to court and marry a human, then it too acquires a soul through the holy sacrament of marriage. 'When they enter into a union with man, then the union gives the soul. It is the same as with the union that man has with God and God with man, a union established by God, which makes it possible for us to enter the kingdom of God.'[25] Gaining a soul, the being rises above the level of the beasts and inherits moral responsibility. 'From this it follows', he continues, 'that they woo man, and that they seek him assiduously and in secret. A heathen begs for baptism and woos it in order to acquire his soul, and to become alive in Christ. In the same way, they seek love with man, so as to be in union with men.'[26] It is surely a profound Christian belief that admits the possibility of redemption even to the non-human!

Like a Victorian taxonomist, Paracelsus proceeds to describe the species of these wild creatures in some detail. There are four types, he says, and they correspond to each of the four Aristotelian elements. The water people are nymphs, the air people sylphs, the earth people pygmies, and the fire people salamanders. Sylphs are also known as sylvestres, 'forest people' or 'wind people'; the pygmies are 'mountain people', who some call gnomes. Nymphs are the legendary undines, and salamanders also go by the name of vulcani.

Each, says Paracelsus, inhabits its element just as we breathe and pass through the air. To the gnomes, earth is like a gas ('chaos') through which they can move with ease. And to each, a different element provides the 'soil' in which they grow their food: water is the 'soil' of gnomes. Although shaped essentially like men, they have their own characteristics. Nymphs look more or less identical to men and women, and it is for this reason that marriage is not uncommon between a female nymph and a man. But it is not always a happy union. A nymph once married a nobleman from Staufenberg, who then rejected her as a devil and took another wife. But the nymph turned up at his wedding ceremony and killed him by 'giving him the sign' through the ceiling. And justifiably so, Paracelsus argues, for this non-human creature would have no recourse to the justice of the magistrates.

Gnomes are small ('about two spans'), salamanders are 'long, narrow and lean', and the sylvestres are 'cruder, coarser, longer and stronger' than humans. Between themselves, each race occasionally gives birth to monsters, just as happens in human unions. These constitute some of the other creatures of myth. Monsters born of sylvestres are giants; those of the mountain folk are dwarfs; of the nymphs, sirens; and of the fire people, the will-o'-the-wisps. The birth of a monster, Paracelsus advises, is a portent of some calamity: a giant might herald an earthquake, for instance.

Why did God create these beings? They inhabit their respective elements, says Paracelsus, so that they might act as guardians of the treasures of the world, ensuring that men do not plunder them all at once. Most notably, the gnomes dwell in mines, just as countless miners assured Agricola, because it is precisely in those places that the ores and precious minerals of the earth are found.

Paracelsus' 'naturalistic' interpretation of apparent superstition is abundantly clear in his theory of incubi and succubae. According to the *Canon episcopi* these were among the worst of the demons. They come at night to women and men respectively and tempt them sexually in their sleep. Celibate medieval clerics presumably feared them because they felt their embraces all too often, and were powerless to resist. In the waking hours a man can oppose temptation; but who can control their passions in a dream? All the more reason, then, to blame these nocturnal impulses on some external fiend that preys on the weakness and vulnerability of the devout man. The element of denial and suppression is carried to almost comical extremes in the *Malleus maleficarum*, where Kramer recounts the kind of scene more familiar from Restoration farces:

> Husbands have actually seen Incubus devils swiving their wives, although they have thought that they were not devils but men. And when they have taken up a weapon and tried to run them through, the devil has suddenly disappeared, making himself invisible.[27]

Celibate Paracelsus must have been as susceptible as any priest to visitations from lascivious succubae. While he does not seem to have suffered great torments of guilt about it, he did have fears – and rather disturbing theories – about any 'seed' that was not properly used for the purposes of procreation (see also page 346). He did not question the demonic interpretation of night-time sexual fantasies, but he regarded this phenomenon, like any other, as one that can be accorded a 'scientific' explanation. This explanation has provoked much commentary. As he tells us:

> Incubi are male, succubae female creatures. They are the outgrowths of an intense and lewd imagination of men and women, formed of the semen of those who commit the unnatural sin of Onan. Such semen that does not come into

> the proper matrix [that is, the womb], will not produce anything good. Therefore the incubi and succubae grown out of corrupted seed are evil and useless . . . This semen, born from imagination, may be taken away by spirits that wander about at night, and that may carry it to a place where they may hatch it out. There are spirits that may perform an *actus* with it, as may also be done by witches, and, in consequence of that *actus*, many curious monsters of horrible shapes may come into existence.[28]

Products of the imagination? Is this Paracelsus as a sixteenth-century Freud? That anachronism, sometimes mooted, in truth makes little sense. By 'imagination' Paracelsus does not mean the workings of the unconscious but a mental force with motive power, able to act on matter and nature just as magnetic and electrical forces do. As Kramer implies in the *Malleus*, the brain was deemed to be composed of specific chambers where certain mental functions took root: memory, imagination, reason. Paracelsus believed that one person's imagination could have a direct effect on another's. He advanced this as a naturalistic explanation for some acts of supposed witchcraft that others would ascribe to the mediation of demons.

Jung discerns something still more profound in Paracelsus' attitude. While it would certainly be unwise to ascribe to him a presentiment of the psychoanalytic notion of projection, Jung thinks that Paracelsus might have subconsciously grasped the psychic function of evil spirits. 'The Church', says Jung, 'might exorcise demons and banish them, but that only alienated man from his own nature, which, unconscious of itself, had clothed itself in these spectral forms.'[29] To Jung, the whole Paracelsian concept of the 'invisible body' is a primitive representation of the psychic component of human life.

Paracelsus' belief in the efficacious power of the imagination leads him to conclude that some diseases are products of the mind. In that case, the cure must come from the same source. 'Nobody can be cured by faith', he said, 'unless the disease was imaginary in the first place.'[30] Even when we accept his particular concept of 'imaginary', it is an extraordinary claim, and one that appears to undermine any belief in faith healing. 'It is not the curse or the blessing that works,' he said, 'but the idea. The imagination produces the effect.'[31] In an age undoubtedly full of psychosomatic ailments, this is not only a profound insight but possibly a very useful one.

WITCH-HUNTING

Paracelsus never denied the reality of the demons that, in the eyes of witch-hunters like Heinrich Kramer, did the devil's work and brought misery, suffering and corruption on humankind. Yet Paracelsus argued that their powers were limited, for they were subject to God's will and could not act outside it. Moreover, he considered that they rarely acted in the terrestrial sphere, to cause storms and the like. When witchcraft was used to produce such phenomena, the demon did not act with arbitrary powers but merely helped to bring about those natural processes that set the events in train. If the skies rained frogs, demons did not conjure the hapless creatures out of thin air, but helped to initiate the physical influences (in this case a kind of magnetic force) needed to draw real, pond-dwelling frogs into the clouds.

Such views went some way towards tempering the persecutory mania that swept through Europe in the later sixteenth century. Although the Inquisition had the formal powers to try people for witchcraft ever since its inception in the early thirteenth century, such trials were rare before 1500. But for nearly two centuries after the publication of the *Malleus maleficarum*, thousands of people – mostly women, but a significant number of men too – were accused, tortured and burned on suspicion of consorting with demons and making pacts with the devil. At the height of the witch hunts, between 1580 and 1650, there were around 100,000 to 200,000 prosecutions brought in Europe.

It would be wrong to construe this wholly as a means by which the Church sought to control society; for while the execution of witches was officially sanctioned, it was often instigated and conducted at a local level that had more to do with the settling of grudges and with superstitious paranoia. Agrippa and Paracelsus were brave but lonely voices. While their scepticism echoed the calls for moderation in human affairs made by the humanists, few were listening, and Protestants were no less enthusiastic than Catholics about rooting out the agents of Satan with the rack and hot irons.

The most substantial and most admirable of challenges to witch-mania in the sixteenth century was that of Agrippa's disciple Johann Weyer. Weyer came to Agrippa as a fifteen-year-old student, and studied with him for four years, receiving a thorough grounding in Neoplatonism and the traditions of natural magic. But while Weyer retained a great affection for his 'former master and revered teacher', he turned aside from Agrippa's mysticism and adopted a temperate and rational outlook more reminiscent of that of Agricola. He was taught by the great French doctors Jean Fernel and Jacques Dubois (Sylvius), as well as by the discriminating Paracelsian Guinter von Andernach.

Weyer's attitude to tales of demonic magic and possession is typified by his impatient response to accusations that Agrippa's dog was in fact a pet demon: 'if anyone knew him well, I did, since I often walked him on a rope leash when I was studying under Agrippa . . . It never ceases to amaze me that men of such repute sometimes speak, think, and write so foolishly on the basis of an idle rumour.'[32]

In 1550 Weyer became physician to Duke William V of Jülich-Berg-Cleves in Düsseldorf, a centre of Erasmian humanism. He held this post for the rest of his life, and it was here that he wrote *De praestigiis daemonum* (*On the Activities of Demons*), which was published, to great controversy, in 1563. It was, in Sigmund Freud's opinion, one of the ten most important books ever written – an endorsement that has encouraged the tendency to see Weyer's book as a precursor of psychology and psychoanalysis.

Weyer is not particularly kind to so-called witches, referring to them as toothless, ignorant and vile old women. But he has no time for the idea that they have made pacts with the devil and are conduits for his malign forces. They are, said Weyer, simply mad or disturbed, or suffering from drug-induced hallucinations – or are the innocent victims of fraudulent charges. Weyer did not doubt the reality of the devil, nor his malign influence on individuals; but he considered it absurd to suppose that human agents can themselves cause possession or summon demons to do their bidding. Thus, he says, witches are not really dangerous, nor is witchcraft genuine. To persecute witches is to oppress people who, however deranged, are no more wicked than anyone else.

That was the case, Weyer insisted, of a woman from the town

of Büderich, who, it was popularly alleged, 'was being harassed by an evil spirit'. He explained:

> For weeks on end she lingered by night about the tombs in the cemeteries. Sometimes, too, she would run into the streets, breaking down one person's doors or smashing another's windows; and occasionally she would run away to more remote and wooded places.[33]

He diagnosed her condition to be due merely to an excess of melancholia that came over her at a certain time of the year.

It is not hard to see why Freud admired Weyer. His sensitivity to mental disturbances was sometimes remarkable, and might have been considered enlightened in the nineteenth century, let alone the sixteenth. Weyer was once summoned to treat a young girl called Barbara Kremers, who had come to be regarded as something of a miracle by her local community. It was said that she lived without taking any food or drink, sustained by God's grace alone. It seems clear that Barbara suffered from an eating disorder, which left her so weak that she had to use crutches. Noting that there was nothing more to the case than wilfulness, Weyer knew that the poor girl was liable to be strictly punished when 'exposed'. But he prevailed on his patron Duke William to prevent this, and instead he and his wife took Barbara into their household and patiently helped her to recover an ability to eat properly again.

A NEW VIEW OF MADNESS

Weyer's attitude towards witchcraft and mental disturbance is so close to that of Paracelsus that one might imagine the latter to have been an influence on Agrippa's pupil. But it isn't so. Weyer's disdain for mysticism inevitably led him into conflict with the followers of Paracelsus, for whom Weyer expressed more or less open contempt. It was Weyer who was the recipient of Oporinus' infamous letter portraying Paracelsus as a drunken maniac. Weyer's education as a physician was traditional, and he was representative of those doctors who observed essentially Galenic principles while acceding to some of the more useful measures of Paracelsian medicine. He fits the archetype of the early scientific empiric, possessed of a Baconian frame of mind that seeks for what works and not for some grand explanatory theory.

Paracelsus' interest in mental illness and 'diseases of the mind' was evident from his earliest days as a physician. If there is any truth in the tales of his mother's suicide, perhaps there lies the source of his interest. Certainly, and rather remarkably, he includes 'those who kill themselves' among a list of new diseases 'never before described' which he took it upon himself to investigate. 'It seems unjust to me', he said, 'that these diseases should never have been described by medicine, that they should have been forgotten.'[34] And despite his efforts, they remained so: 'For the diseases of the mind,' wrote the physician Robert Burton a century later, 'we take no notice of them.'[35]

Paracelsus' book *The Diseases That Deprive Man of His Reason*, which may have been written in the early 1520s, is a genuinely pioneering work, especially when one considers that until then the *Malleus maleficarum* summarized pretty much what was known and thought about mental illness. Here was a wholly different vision: one in which these conditions 'develop out of man's disposition'. Paracelsus is fully aware that he is challenging long-standing prejudice: 'The present-day clergy of Europe attribute such diseases to ghostly beings and threefold spirits; we are not inclined to believe them . . . nature is the sole origin of diseases.'[36]

By 'nature' we are to understand that he means his chemical cosmology: mental afflictions can be rationalized, just like bodily afflictions, in terms of alchemical processes. Epilepsy (the 'falling sickness', a term that also included catalepsy and analepsy) was attributed to 'the boiling of vapours in the *spiritus vitae*', and was the microcosmic analogue of the great earthquakes 'which make the whole earth tremble'.[37] Mania – characterized by 'frantic behaviour, unreasonableness, constant restlessness, and mischievousness'[38] – arose from the distillation, sublimation or coagulation of humours in the body, caused by heat.

This sounds like a thoroughly materialistic view. Yet Paracelsus describes these processes not as material but as spiritual, as 'invisible diseases' – though, as we have seen, this was not to deny them a chemical aspect. After all, 'spirit' was a concept indistinguishably theological and alchemical. Recall that when Paracelsus spoke about spirits distilled from wine or herbs, it was no metaphor: they were the direct homologues of the spirit or quintessence in humans, the intangible vapour that quickens us into life.

So when he maintains that spiritual diseases need a spiritual cure, we should not anticipate a regime of prayer or meditation, but rather a prescription of spirits and arcana extracted alchemically from nature. Paracelsus does seem to vacillate on this issue, however, conceding that 'there are some material remedies which cure spiritual diseases'.[39] And so his book on mental illnesses lists distinctly Paracelsian cures, with ingredients such as camphor, mandrake, verbena, willow bark, tinctures of gold and *unicornu*, ground up and served in a soft-boiled egg. These 'material' remedies help only in new cases, he advises, 'where the disease has not yet got the upper hand' – but the distinction between material and spiritual remedies is sometimes hard to discern in any case, except that the latter are graced with a heavier dose of Paracelsus' Latinate neologisms: *aurum potabile*, *magisterium antimonii*, *extractum sulphuris*. The book makes a particularly urgent case for the virtues of vitriol – here meaning not sulphuric acid but sulphate salts, which were mined in Germany, Hungary and elsewhere. Paracelsus recommends them for preventative medicines against mental illness:

> This vitriol is a special salt, different from all others, and it has more virtues and qualities than other salts and such high and great virtue that it is appropriate to mention it in this book . . . In medicine it is a miraculous remedy and in alchemy it is excellent and useful for many other things . . . It puts all Italian and German apothecaries and all their writings to shame, for it is a medicine which in itself is sufficient to fill one fourth of all dispensaries and is satisfactory against one fourth of all diseases[40]

– including jaundice, stones, fevers, worms and falling sickness.

Although this chemical diagnosis of madness was somewhat original, there was also a materialistic aspect in the traditionalists' view of mental illness, whereby it was attributed to humoral imbalances. In other words, the alternative to Paracelsus' explanation was not necessarily a moralistic one that adduced demonic possession; even Galenists could reject this as superstitious, as Weyer illustrates. Madness, they argued, was the result of an excess of melancholy or black bile. But traditional medicine and demonology were by no means incompatible, for it was commonly thought that it was through the agency of black bile that the devil gripped the mind and soul; Luther called this humour 'the devil's bath'.[41] Even progressive physicians such as Ambroise Paré, Jean Fernel and Felix Platter allowed that the devil could produce physical changes and symptoms. So the idea that insanity and other mental afflictions have naturalistic explanations and physical cures was not inconsistent with a belief in their ultimately supernatural origin.

The diseases of the mind, Paracelsus advises, can be given a systematic aetiology just like those of the body. They comprise one of the five *entia* of the *Paragranum*, and can be further subdivided into those that afflict people acutely and chronically: recurrent but temporary loss of reason, and permanent insanity. 'For as fools are of many kinds,' he tells us, 'so also are there many kinds of crazy people, not of one sort, nor in one way, but in many ways, of many sorts, in many patterns and forms.'[42]

There are four manifestations of insanity, says Paracelsus. *Lunatici* are driven to unreason by the influence of the moon, which exerts a kind of magnetism on the bodily humours and 'tears reason out of man's head'.[43] This occult (but completely natural) influence can be counteracted by medicines. The *Insani* have inherited madness or acquired it in the womb. They are hard to treat, though sedatives can keep the madness at bay. The illness can be prevented, however, if the parents conceive the child 'artificially' rather than by 'natural coitus'. This rather extraordinary process demands that the natural lusts be suppressed by cold baths and that the coitus be instead 'induced and stimulated by medicine'[44] – although Paracelsus does not specify exactly how this is achieved.

Then there are the *Vesani*, whose insanity comes from poisons in food and drink. Although excessive amounts of alcoholic drink were a likely cause of mental dysfunction at that time, Paracelsus seems more concerned with the ill effects of food. He says that some people are induced to love others beyond reason by being given food: 'If a person offers something to eat to another, whether man or woman, unbreakable, eternal love is the result; for this reason some servants give food to their masters so as to flatter them and to make love spring up in them, with the result that servants are above the masters.'[45] Others, especially if they are of a choleric nature, may be driven by particular foods into becoming obsessed by war. Some *Vesani*, he writes, 'who jump up and run about all the time, have received their insanity from eating the kind of thing that gives them an urge to mount and climb'.[46] The fourth category of insane people are the *Melancholi*, who have madness in their disposition owing to too much *spiritus vitae* in the brain.

Among the insane, Paracelsus also briefly mentions the *Obsessi*, who are 'obsessed' (possessed) by the devil. He did not deny such a supernatural cause of madness, potentially the result of witchcraft; but his belief in natural laws that bind all of creation left him convinced that even these cases were susceptible to chemical cures: 'Now just as heaven [the stars, not God] has to yield to the physician, so too must the devil yield through the right ordering of medicine.'[47] In other words, even demonic possession did not really carry *moral* overtones: if a curse could be effected by mechanical (albeit illicit) manipulation of occult forces, then natural magic could work a remedy.

This humanistic attitude does not always guarantee what we would consider a humane response, however. In his treatments for St Vitus's Dance, Paracelsus' recommendations are draconian. This condition can be induced by bacterial infections, which produce involuntary jerky movements of the limbs. It seems likely that the affliction was also equated with other kinds of apparently licentious behaviour presumably indicative of more strictly mental illnesses, since St Vitus's Dance was regarded as giving rise to actions that are 'voluptuous, lewd and impertinent'. The assumption was that it was directed by thoughts and intentions of a similar nature: the disease arises in the imagination, says Paracelsus, in 'a mere opinion and idea'. As a result, he sounds abruptly punitive in his prescriptions:

> Shut the patients into a dark, unpleasant place and let them fast on water and bread for some time, without mercy. Thus hunger will compel them to adopt a different nature and different thoughts, so that the lasciviousness is driven out by abstinence . . . Some think they would die if they could not act in such a way (singing, dancing, etc.) but it is not so. It is better to take a good stick and give the patients a good beating and lock them in.[48]

Throwing them into cold water was another option. Yet all of this is offered without any indication that Paracelsus regards the condition as reprehensible; the treatment is simply administered for the patient's own good. And he is keen to divest the disease of religious connotations, challenging its popular name: 'In our opinion, such diseases have nothing to do with the works of the saints.'[49] He calls it instead chorea lasciva, and indeed the condition is known as chorea today.

Most strikingly of all, Paracelsus suggests another remedy for St Vitus's Dance that psychotherapists will recognize as a form of displacement. One manifestation of the condition is in unprovoked rage and swearing, reminiscent of Tourette's syndrome. Paracelsus recommends that the patient make a wax image of his own person, direct all his venom and curses onto it, and cast it into the fire. For the rage, he says, is ultimately self-directed: 'curses work against those who utter them and not against the men at whom they are aimed.' By destroying the image, 'the thoughts are destroyed with it'.[50] Only by keeping in mind Paracelsus' belief in real, essentially 'physical' occult forces that operated on mind as they did on matter can we still see (as we should) the gulf between his sixteenth-century model of mankind's mental world and that of today.

Even people with congenital mental disabilities – the fools and idiots of the medieval lexicon – could be understood through Paracelsus' alchemical conception of humankind's inner world. These people posed a dilemma for theologians: why would God make men imperfectly? Broadly, Paracelsus' explanation is the fairly conventional one that 'fools' are a product of the corruption of the Fall. But the proximate cause, he says, stems from the way that man is fashioned by the alchemical craftsmen, the Vulcani (not to be confused with the fire spirits or salamanders). Occasionally their place is taken by 'heavenly apprentices and immature master craftsmen' who execute flawed work and mis-carve their creations. If that happens to the mind, says Paracelsus, it might also be evident in the body, so that fools 'sometimes also carry misgrowths on their body'.[51]

Here, as often elsewhere, it is not always clear how literal Paracelsus intends to be. The Vulcanus, the 'chemically efficient principle of nature'[52] as it has been called, is a wonderful metaphor for the self-organizing power of biochemistry, or for its versatile molecular agents, the enzymes. But Paracelsus seems to ascribe a real, physical existence to these beings. They never grow old, he says, but die when their skill reaches its peak. 'Each has his special hammer and chisel, and [carves] for himself a separate statue, so that no one becomes like another one.'[53]

Paracelsus' compassion for 'fools' contrasts with the prevailing vulgar attitude to mental affliction. When he speaks of the 'wisdom that is also in the fools',[54] he is expressing a belief that was widely held; but in sixteenth-century society mentally disabled people were more likely to become objects of derision than to be viewed as oracles. Paracelsus' notion of the 'wise fool' reveals a moving commitment to Christian charity:

> Because he is a fool, [his talk] is mocked and derided, for they go forth silly, foolish, and disrespectable. Also nobody suspects withal in such people a man to lie hidden, as if both body and soul were a fool. They also forget that Christ says we are not to call our brother a fool. Because, even if the animal body is a fool, yet the soul, his spirit, etc. is no fool . . . that which is eternal within him is without any folly and simplicity. Only that it might not come out and be agreeable to the people . . . Even if the nature [that is, the body] went wrong, yet nothing has been wrong with the soul and with the spirit . . . Therefore also no one ought to be regarded as simpleton or fool, or be called so.[55]

Christ, after all, died for all men, 'fools' as much as those of sound mind and body. Indeed, because fools are incapable of the guile and deception with which other men clothe their words in seeking after good opinion, the fool may come closer to the truth: 'Know ye withal that the fools reveal greater judgement, more shrewdness, more wisdom than the wise.'[56] The fool, says Paracelsus, cannot disguise his 'animal body' with the kind of artifice that, in normal men, is liable to suppress one's inner wisdom. He is, moreover, incapable of sin, just as the animals are; he is the 'holy fool'.

PROFESSIONAL HAZARDS

As Paracelsus passed along the valley of the Inn in 1533–4, the smelting works of the Fuggers swallowed up the diverted waters

The mining industry was polluting the air even in the sixteenth century.

of rivers and streams while their chimneys belched out smoke into the sky. Paracelsus visited the Tyrolean mines and was struck by how the miners suffered from ailments peculiar to their profession. It may have been at this time that he began to work on the first manual of occupational health, *On the Miner's Sickness and Other Miners' Diseases*.

The miner's lot was always a perilous one. The work was terribly hard, the conditions filthy, the air bad, and there was a constant danger of accidents, especially as mine shafts were sunk ever deeper. When Sigismund, cousin to Maximilian I, governed the Tyrolean mines in the early sixteenth century, conditions were so grim that the miners went on strike. They complained of long hours, slow payment of wages, and high prices for the food from the company stores. The Schwaz miners marched on Innsbruck, forcing the authorities to concede to some of their demands: underground shifts, for instance, were limited to eight hours a day. But when Maximilian took over the Tyrol and Jakob Fugger acquired the region's silver mines, the regime became harsher. Another strike was quelled by force: Maximilian's troops arrested the miners' leaders (after they were promised a hearing by the imperial officials) and any who persisted with the strike were banned from working in any mines within the Habsburg empire. In such a climate, it is not surprising that the miners took part in the revolts of the mid-1520s, which, however, did nothing to improve their situation.

But one of their worst concerns was a problem the authorities could do little about even if they had been so inclined. The 'miners' sickness' was a life-threatening respiratory disease – probably a mixture of all manner of ailments ranging from pneumonia to tuberculosis, fibrosis and bronchitis, and caused by hours spent in stale, dusty air. Paracelsus identified this sickness as an affliction of the lungs, which could be incurred by anyone involved in the collection, preparation and smelting of ores:

> Those who work at washing, in silver or gold ore, in salt ore, in alum and sulphur ore, or in vitriol boiling, in lead, copper, mixed ores, iron or mercury ores, those who dig such ores succumb to lung sickness, to consumption of the body, and to stomach ulcers; these are known to be affected by the miners' sickness.[57]

No one, he says, had previously identified these diseases or written about them. He realized that they are caused primarily by the corruption of the air – a consideration that, in Paracelsus' scheme of things, should be regarded as an 'astral' influence. This does not imply that the stars themselves are involved, but simply that the heat and moisture of the atmosphere are implicated, just as they are when the sun warms the earth and causes plagues.

In addition to these airborne diseases, Paracelsus recognized that miners were at risk from the toxicity of the metals they extracted – particularly mercury, which could turn their teeth black before eventually making them fall out. But the remedies that Paracelsus suggested, even had they been effective, can have brought little relief to the Tyrolean miners during his lifetime, since his book remained unpublished until Samuel Zimmermann (known as Architectus) brought it out in 1567.

Unlauded and unwelcome, Paracelsus appeared at the gates of Innsbruck in the summer of 1534 and petitioned the authorities for permission to practise as a physician. The response was predictable:

> Because I did not appear in the garnishry of the doctors, I was dispatched with contempt and was forced to clear out. The Burgomeister of Innsbruck had been used to doctors clad in silken robes at the court of princes, not in shabby rags grilled by the sun.[58]

And so he continued wandering southwards, over the Brenner Pass towards the plague-stricken city of Sterzing.

17
The Little Man

Animating the Earth

> Incubation offered a more appropriate metaphor than parturition for the particular transformation of matter at which [alchemists] aimed: alchemical texts reverberate with encomia to yolks and albumen and their potentiality. The eggshell itself becomes the metonymic vehicle of the alchemical process, the gourd-shaped vessel in which the raw matter is cooked and changed.
>
> Marina Warner, *Fantastic Metamorphoses, Other Worlds* (2002)

Perhaps it was Sterzing's predicament that rallied Paracelsus' spirit. For, to judge from his later letter of protest to the town authorities, he was soon back on form – even if his targets were now theological rather than medical:

> You daily denounce me because I have spoken the truth at times in taverns and inns, against useless churchgoing, luxurious festivals, vain praying and fasting, giving alms, offering tithes, confession, sacraments, and priestly rules and observances. You accuse me of drunkenness because this has taken place in taverns, and you call me an agitator because you say taverns are not appropriate for truth. But you were silent and well pleased when in the taverns I advised people to give you offerings. If that was proper in inns, then admit that the truth is proper in inns.[1]

But he did not only preach at Sterzing. When he arrived in the high summer of 1534, many of the townspeople were departing. This alpine town (now in Italy, where it is called Vipiteno) was visited by the plague that June. Cities had taken to confining plague victims in hospitals (called *lazaretti* in Italy) outside the walls, which were little more than the most basic of barracks where the afflicted would languish and die. Even such quarantining had limited effect in containing the disease, however, and able citizens commonly fled a plague-ridden city, seeking air free from the taint of poison.

By resolutely making his way to this endangered and perilous town, Paracelsus showed a degree of recklessness in which courage is hard to distinguish from obsession. In Sterzing he treated the sick and studied the disease, although as in Innsbruck his disreputable appearance brought him scant welcome. He administered pills of rolled bread tainted with infected faeces, a primitive (not to say dubious) form of inoculation supposedly learnt from the Turks in Constantinople. In this period he wrote a book on the plague and dedicated it to the beleaguered town itself. The contagion eventually abated – as it always did in the end, though we cannot know if Paracelsus deserves any of the credit – and the outlandish doctor was once again asked to move on.

In Merano his reception was warmer: he wrote his treatise on the miners' diseases, he found patients eager for his attentions, and enjoyed 'honour, happiness and fortune'. But not for long – soon he was on the road once more, crossing the mountains into the Tyrol, reaching St Veltlin ('a most healthy land') and then passing on to St Moritz.

The spa towns of southern Switzerland gave him another opportunity to pursue his fascination with mineral waters, agents of the earth's alchemy. He studied the acidic spring water of St Moritz, 'which drives away gout, and makes the stomach as strong in digestion as that of a bird that digests tartar and iron'.[2] He was invited in August 1535 to the Benedictine monastery at Pfäfers-Ragatz, where the abbot Johann Jakob Russinger was his host and his patient.

The great gorge of Bad Pfäfers remains a striking example of subterranean natural alchemy at work. As the path climbs upwards into the foothills of the Alps, the walls loom ever closer overhead until finally they lean up against one another. Then the stream rushes through a resonating geological cathedral, the gloom broken here and there by patches of sky far overhead. The path, clinging to the rocky ledge above the torrent, takes the traveller to a warren of caves incubating in the earth's heat. In this steamy seclusion, the mineral waters are cooked until they coat the rocks with bright deposits of ferrous red and cupric green.

In Paracelsus' time the gorge was reached from above, along a

The gorge and spa at Bad Pfäfers.

path from the abbey which still looks out over the Swiss pastures. People coming to seek a cure by bathing in the warm spa waters were let down into the chasm by ropes; the monks received the invalids and bore them to the warm springs in the rock. These waters, according to Paracelsus, have curative powers akin to those of the herbs melissa and hellebore. There is still a hostelry by the entrance to the enclosed gorge – but it is now a cultural centre and a restaurant, which commemorates Paracelsus' visit in a small and attractive museum.

His attendance on Russinger was just one of many professional calls he made on his endless journeys. But it is notable for having yielded a rare surviving document written in Paracelsus' own hand: a medical *consilium* (prescription) to the abbot himself, covering just three pages of yellowed paper. The writing is bold and confident, but the document surprises with a recommenda-

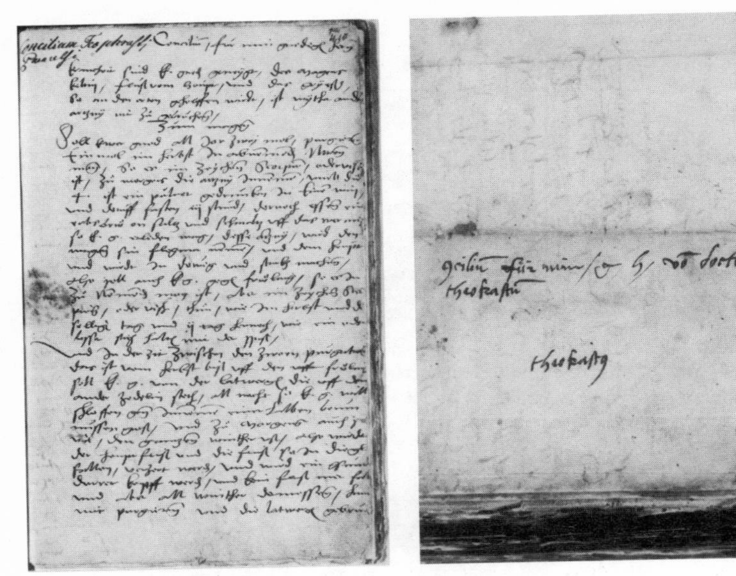

The first and signature pages ('Theofrastus') of Paracelsus' consilium for the abbot of Bad Pfäfers.

tion for bloodletting.

By September of 1535 Paracelsus was heading northwards, passing close by Einsiedeln and reaching Memmingen, Ulm and Mindelheim early in the following year. He treated patients wherever he went, wrote frantically, and according to a secretary's account was very much his old self – to the inevitable discomfort of those around him:

> He mumbles to himself for hours. When he speaks to others they can hardly understand him. He spends much time before his oven brewing powders but he does not suffer anyone to help him. He gets angry when spoken to. Suddenly he has a fit and yells like a wounded animal. He gets impatient with the slightest mistake of his amanuensis.[3]

At Memmingen he argued bitterly with an innkeeper; elsewhere he got into fights with patients who would not pay their bills. He petitioned the courts. He was ready once more to take up arms and tilt at windmills.

A LIFE'S WORK

Paracelsus' fortunes and reputation began to wax again in the late 1530s. The most notable sign of this was the publication in 1536 of his *Great Surgery*, one of the few major books on medicine that he ever saw through the printing press in his lifetime. It was written the previous year, and Paracelsus travelled to Ulm with the manuscript to take his chances with the publishers. The first one he tried, Hans Varnier, apparently did an execrable job, making such a mess of the proofs that Paracelsus gave up on him and proceeded instead to Augsburg, where the book was printed in two volumes in the late summer by Heinrich Steiner. It was dedicated to Charles V's brother Ferdinand, 'King of Rome and Archduke of Austria'. But the canny Varnier did not let his efforts go for nothing: he went ahead with an 'unauthorized' printing, prompting Paracelsus to specify in the Steiner edition that this was the only valid version of the book. In any event, it was a considerable success, becoming the most frequently reprinted of all his works.

The *Great Surgery* is a summary of a lifetime's experience. It sets out clearly the key merit of Paracelsian medical care: less is more. 'Surgery', he says, 'consists in protecting Nature from suffering and accident from without, that she may proceed unchecked in her operations.'[4] The surgeon's job is simply to create the best conditions for the healing powers of nature to take effect, which may mean nothing much more than keeping the wound clean. For all his neologisms and mysticism and complex chemical recipes, Paracelsus' most fruitful contribution to medicine may have been the principle of minimal interference, a stripping away of centuries of advice on useless or positively harmful tampering with the body of the patient.

The book marks the beginning of a productive and energetic time for the wandering physician, even a return to something of his former glories. His religious convictions, seemingly strengthened during the years in the wilderness, found voice in the eschatological almanac *Prognostication for the Next Twenty-Four Years*, published in 1536. But Paracelsus also began to regain his faith in his medicine – and concomitantly, to re-engage with his many opponents. He began the trenchant *Seven Defensiones* in early 1537, while lodged at Kromau in attendance on an eminent patient.

After the *Great Surgery* was published in Augsburg in late 1536, Paracelsus left for Munich and then Eferdingen, near Linz on the Danube, where he visited an old friend, the cleric and jurist Johann von Brandt. Here a summons reached him from the Chief Hereditary Marshall of Bohemia, Johann von der Leipnik, who dwelt in a castle at Kromau, near Brno in Moravia. Paracelsus may have had the Marshall in mind when he admitted in the *Defensiones* that he could not work miracles – that if a patient were too far gone, no doctor could save his life. Leipnik suffered from dropsy and paralysis, and was virtually on his deathbed when Paracelsus arrived to treat him. There was little to be done save to make up a prescription to ease the Marshall's suffering, and meanwhile Paracelsus, as a respected guest, was at liberty to start, complete or extend several works. He wrote a third volume of the *Greater Surgery*, worked both on the *Defensiones* and another polemical medical book entitled *The Labyrinth of Lost Physicians*, and started (but never completed) his final great work, the *Astronomia magna, or the Whole Philosophia Sagax of the Great and Little World*. Once he had done all he could for the Marshall, he left Kromau to look for a publisher.

Paracelsus' fortunes peaked at this time in Pressburg (now Bratislava), and in September 1537 he was received with a ceremonial dinner held in his honour, 'spread over two full tables'.[5] In Vienna he even found the ear of King Ferdinand, who availed himself of this near-legendary physician and awarded him a gold chain for his services. But for the orthodox medical community, little had changed: the Viennese doctors castigated him, and Paracelsus declined Ferdinand's offer of introduction to the court physicians. In some eyes he was still a sorcerer who dabbled in the black arts; rumours of his profane activities spread through Vienna.

The Viennese publishers proved no more amenable than the doctors. Presumably influenced by the latter, they refused to publish the confrontational *Labyrinth*. The book was another robust defence of alchemical medicine ('no physician should be ashamed of alchemy'[6]) and an attack on the physician who sits 'behind the stove and [treats] his patients with sophistical logic'.[7] Paracelsus soon fell out with the Austrian authorities too. He claimed that Ferdinand never paid him the 100 florins offered for the printing of a book on tartaric diseases. The Austrian treasury countered that Paracelsus had already squandered this sum in advance. In the end the king concluded that the great doctor was nothing better than an insolent mountebank. According to the anti-Paracelsian Johannes Crato von Krafftheim, Ferdinand saw through Paracelsus' postures and denounced him as a 'lying and impudent impostor'.[8] And so the old pattern played itself out once again: honours and adulation followed by a self-inflicted fall from grace. It is no surprise that he was on the move again at the end of 1537, or by early in the new year.

It is not clear whether his visit to Villach was made because of or in ignorance of the news that his father was dead. But either way, in May 1538 Paracelsus acquired a deed to his father's modest estate from the city magistrates. We have no evidence of how he responded to his father's death; but although he had not seen Wilhelm for twelve years, there is no reason to suspect any lessening of the deep affection that is evident from his earlier writings.

Perhaps because of the enduring suspicions of the Villach doctors, Paracelsus did not linger in the city but retired in the summer to nearby St Veit, where he continued writing and attempting to publish his books. Here he finished the pugnacious *Seven Defensiones* (1538). Subtitled 'A Reply to Certain Calumniations of His Enemies', it is an outspoken riposte to his critics and detractors – his Nietzschian *Ecce Homo*, complete with chapters such as 'To Excuse His Strange Manner and Wrathful Ways' and 'Concerning the Rejection of False Physicians and False Company'. If it is true that the *Defensiones* displays less of the spluttering rage and choleric of his earlier defiant outpourings, he now fires his bolts with a surer aim.

How can we imagine, he asks, that today's diseases can be cured by reading books written a thousand years ago and more? For just as every land and climate has its own diseases and its own cures, so too does every age. 'Of what avail is the rain that fell a thousand years ago? That which falls at present prevails . . . concerning the present should we trouble ourselves, not concerning the past.'[9] There are now new diseases abroad, such as the French pox, and new remedies are needed. 'There are never again the same *Causae*; things are more biting now, as both philosophies of heaven and the elements sufficiently prove.' In consequence, the doctors should recognize that 'an unlettered peasant heals more than all of them with all their books and red gowns'.[10]

Paracelsus defends not only his system of medicine but also his own demeanour. He makes no pretences of being cultivated, he admits – he comes from a rough land, and that has shaped his appearance and his words. 'I am said to be a strange fellow with an uncivil answer, I do not wash to the satisfaction of everyone, I do not answer everyone's contention with humility.'[11] But the other doctors are afraid of such plain speaking because they need to cover their incompetence with flattery and fawning:

> [They] know little of the arts; they resort to friendly, pleasing, charming words; they advise people with breeding and fine words; they set forth all things at length, delightfully, with distinct differentiations, and say: Come again soon, my dear sir; my dear wife, go and accompany the gentleman, etc. I say thus: What wilt thou? I have no time now; it is not so urgent . . . My intention is to gain nothing with my tongue, but only with works.[12]

And worse still, the people fall for it. For the physicians:

> have made such fools of the patients that they are completely of the belief that a friendly, affectionate manner, ceremony, ingratiating ways, much ado, constitute art and medicine . . . Medicine should be such, that the physician may answer according to his flesh and blood, his country's customs and his own nature: rough, rude, stern, gentle, mild, virtuous, friendly, delightful – according to how he is by nature and by acquired habit.[13]

Yet Paracelsus also now acknowledges his limits. I cannot cure everything, he admits, 'for I cannot master nor overpower God, but He me and all others'.[14] Neither can he be expected always to know the right cure immediately, but only after carrying out tests and experiments. Some diseases are, in any case, 'impossible'. One could say he is simply making excuses; but nonetheless he seems clearly to have revised his earlier view that any condition is curable if one searches hard enough in nature for the remedy.

In St Veit he was employed by the Fuggers to search for gold veins in Carinthia, which prompted him to write a history of the region, the *Chronicle of Carinthia*. He may well have hoped that this patriotic offering would help to secure publication of his other works, for he sent the *Chronicle* along with the *Defensiones*, the *Labyrinth* and the book on tartaric diseases to the authorities of the States of the Archduchy of Carinthia.

In September of 1538 Paracelsus received a message indicating that they planned to publish these works. But time passed, and nothing came of it. It seemed that his manuscripts had been forgotten, or lost. Whatever the reason, they were never printed in his lifetime.

THE MAKING OF A MAN

There is no clearer indication of Paracelsus' optimistic spirit in 1537–8 than the *Astronomia magna*. 'This', according to Karl Sudhoff, 'is undoubtedly the authentic indisputable kernel of the mature work of the ripened man – Paracelsus at his height.'[15] While the *Greater Surgery* is a practical manual and the *Defensiones* are rooted in mundane disputes, the *Astronomia magna* returns to the grand cosmic vision, the overarching theory of the universe and all life within it. 'The sagacious philosophy', Paracelsus explains in the preface, 'describes the whole religion of all creatures and the classification of their fundamentals and art.'[16]

Not the 'science' or 'philosophy', note, but the 'religion'. This book, perhaps more than any other of his works on natural philosophy, lays emphasis on the Christian basis for the mechanisms of nature. It is the finest exposition of his concept of a philosophy derived from the wisdom of God. 'Natural reason and eternal wisdom belong together,' he says:

> For Holy Scripture represents the beginning of all philosophy and natural science; without this beginning all philosophy would be used and applied in vain. Consequently, if a philosopher is not born out of theology, he has no cornerstone upon which to base his philosophy. For truth springs from theology, and cannot be discovered without its help.[17]

All of which might seem profoundly antithetical to the development of science. But crucially, Paracelsus implies that God's creation can be explored and understood more or less independently from God Himself. It is the beginning of the idea, essential to the seventeenth-century mechanists, that God built a clockwork universe, wound it up and then let it run of its own accord.

'There are two kinds of wisdom in this world,' says Paracelsus, 'one eternal and one temporal. The eternal wisdom springs directly from the light of the Holy Ghost, the other wisdom directly from the light of Nature.'[18] Although the latter has many aspects, man may study each of them in isolation, to master them and to use them to noble and beneficial ends. It could almost be Francis Bacon writing:

> In God there is only one way, one art, one teaching, one manner, one essence . . . but in the light of Nature there are many ways of working, as craftsman, in the arts, in other faculties, and in the religions . . . Every art should be complete in itself, the astrologer should know his art, the magician his, likewise the diviner, the necromancer, the reader of signs, the practitioner of uncertain arts, the instrument-maker, and the generators.[19]

Thus, even while religion apparently underpins all possible understanding of the world, Paracelsus is here beginning to distinguish the scientist – the magus – from the cleric. 'The difference between a saint and a magus', he says, 'is that the saint works through God, the magus through Nature.'[20] Nor should it be seen any longer as arrogant blasphemy to enquire into the workings of nature – of the 'world machine', as Paracelsus expresses it on one occasion. After all, he says, new phenomena and new laws are constantly coming into being, for 'not all the stars have yet completed their action and imprinted their influences'.[21] Do we not then have a duty to seek out and comprehend these new things? Indeed, 'it is a divine gift to investigate in the light of Nature.' In consequence:

> no one who has discovered something new or who undertakes to explore some unknown field should be held back . . . give heed to those who each day seek something new and also each day find something new, whatever it may be – whether in natural wisdom, arts or custom. For the heavens are responsible for it. Thus new teachings, new arts,

> new orders, new diseases, new medicine proceed from this, for the heavens are constantly at work. And it remains for man to decide what portion of these things he should take to himself and what he should not.[22]

It is fair enough to accuse Paracelsus sometimes of medievalism, but passages like this reveal the distance between medieval scholasticism and the new, questing spirit that gave rise to the great scientific institutions of subsequent centuries.

The *Astronomia magna* contains the clearest expression of Paracelsus' views on astrology, the relationship between the macrocosm and the microcosm, and a good deal else. Here he writes on magical beliefs, occult sciences such as phrenology, physiognomy and necromancy, as well as on geometry and meteorology. And he treats technology in a Baconian spirit, predicting techno-utopias where 'pipes and crystals' will 'carry the human voice over a distance of a hundred miles'.[23]

And in another book from this productive era, *On the Nature of Things*,* there appears one of his most intriguing digressions. Here he intimates that he has discovered the very secret of life itself, and knows how to make a living human-like being from scratch. This is the 'little man' – the homunculus.

Artificial procreation is arguably one of the few remaining taboos for natural science, which is why *Frankenstein* is still regarded as much more than a quaint Gothic fantasy. Today the idea of creating life provokes moral outrage even in an increasingly secular society. It is striking to note that the pious Paracelsus seems to find nothing blasphemous in such a project. This is not so much because he holds life as any less sacred and God-given than do our contemporary bishops and priests; but rather, to Paracelsus there is no real dividing line between the life of a human being and the life of the universe. His whole world is animistic, permeated by a *spiritum vitae*, a life-giving agency that can be extracted from metals and herbs. The homunculus is not so

* The authenticity of *De natura rerum* is disputed. Nicholas Goodrick-Clarke assigns it a date of 1537, but the first publication was by Adam von Bodenstein in 1572. It has been suggested that this book might have been a reworking of another, genuinely Paracelsian, text. In any event, it seems agreed that the work represents the true views of Paracelsus.

much an embodiment of life made from nothing, as a concentration and redirection of the life-giving potential in the world.

Paracelsus prepares us by first giving a prescription for resurrecting a dead chicken. You must burn it 'to dust and ashes', then seal up these remains and allow them to putrefy 'to the highest degree' until they become 'mucilaginous phlegm'. The magical arts can 'renovate and restore' this disgusting residue so that it becomes once again a living bird. 'By this process all birds can be killed and again made to live.' This, says Paracelsus, 'is the very greatest and highest miracle and mystery of God'.[24]

And what can be done for chickens works too for humankind: 'For you must know that in this way men can be generated without natural father or mother; that is to say, not in the natural way from the woman, but by the art and industry of a skilled Spagyrist [alchemist] a man can be born and grow.'[25]

The recipe begins with sperm, the very essence of procreativity. Paracelsus considered both sperm and egg to have the full potential for life already within them. In a procedure that we can hardly read today without being put in mind of the new sciences of *in vitro* fertilization and cloning, Paracelsus proposes that the life of the sperm can be brought to fruition 'without the female body and the natural womb'. Again he attributes this to the fecund alchemical process of putrefaction, acting in some sense via the occult force of 'magnetism'. 'This is in no way opposed to Spagyric art and to Nature,' he assures us, 'nay, it is perfectly possible.' This is how it was done:

> Let the semen of a man putrefy by itself in a sealed cucurbite [round glass vessel] with the highest putrefaction of the *venter equinus* [that is, horse manure] for forty days, or until it begins at last to live, move, and be agitated, which can easily be seen. After this time it will be in some degree like a human being, but, nevertheless, transparent and without body. If now, after this, it be every day nourished and fed cautiously and prudently with the arcanum of human blood, and kept for forty weeks in the perpetual and equal heat of a *venter equinus*, it becomes, thenceforth, a true and living infant, having all the members of a child that is born from a woman, but much smaller. This we call a homunculus; and it should be afterwards educated with the greatest care and

> zeal, until it grows up and begins to display intelligence. Now, this is one of the greatest secrets which God has revealed to mortal and fallible man. It is a miracle and marvel of God, an arcanum above all arcana, and deserves to be kept secret until the last times, when there shall be nothing hidden, but all things shall be made manifest.[26]

This fantastical idea was not entirely a product of Paracelsus' imagination. The legend of the artificial man is an ancient one. Simon Magus, the biblical wizard of Samaria, allegedly made a young boy by condensing human breath in a glass. The myth appears in Jewish tradition in the form of the golem, a being of animated clay. But it seems likely that Paracelsus' homunculus owes more to Islamic tradition. A fourteenth-century text called *De essentiis essentarium*, which was spuriously attributed to Thomas Aquinas, tells how the Arabic physician Rhazes made an artificial human. And Paracelsus' recipe sounds similar to prescriptions that appear in the Jabiran corpus for making human bodies from 'essences' or 'sperm' in a sealed vessel. Jabir refers to a work known in the Middle Ages as the *Book of Laws*, wrongly thought to be written by Plato, which explains how to make a 'rational animal' in the womb of a cow.

The importance that Paracelsus attributes to putrefaction in this process echoes the common belief in spontaneous generation: the formation of flies, maggots and other small creatures in putrescent waste. And the heat generated by fermentation (of manure, for instance) was regarded by alchemists as a source of potency and vitality. Even Avicenna argued that man can be generated from putrefaction. But Paracelsus also connects to the magical belief that the root of the mandrake plant could take on the form of a tiny human-like creature, the mandragora. He writes in *De vita longa* (1526–7) that this diminutive being is in fact the homunculus. 'It is Paracelsus', says the science historian William Newman, 'who made the generation of the *homunculus*, or artificial human . . . dear to Western civilization.'[27] This, Newman suggests, is the source of the manikin, the 'dapper form, that lives and breathes and moves',[28] which Wagner, the pupil of Goethe's Faust, conjures up in the laboratory to serve his master.

Whatever the origins of these ideas, in Paracelsus they seem to have fed some truly disturbing notions about the powers that reside in sperm. While he believed that male seed shed anywhere except in the female womb could grow to form monsters, he felt it was just as hazardous for a man to retain it. Sperm that putrefied internally, he said, could cause the growth of 'flesh, decay and lumps'.[29] Thus, for the unmarried man with no proper use for his seed, chastity was not enough: he needed to be castrated for his own good. The legends and rumours of Paracelsus' own castration or 'eunuchoid' sexual characteristics seem, in this light, perhaps to merit closer consideration than at first they might appear to, especially in view of Paracelsus' troubling conclusion that God gave men external genitalia in order to make this kind of self-mutilation easier to conduct. '[This] complex of ideas concerning sexual pollution, unnatural generation, disease, and religious purification by castration is, even by sixteenth-century standards, bizarre,'[30] Newman says.

LAST CALL

Paracelsus' disappointment at the fate of his 'Carinthian' books was sharpened by the continuing antipathy of the doctors. While he stayed at St Veit in 1538 it is said that one day he attended church in Villach and was abused as he passed through the churchyard by physicians who had gathered from miles around. By the end of the year he was travelling again. At Laibach in Carniola he met Augustin Hirschvogel, who painted his portrait and made the most striking engraving of him. Forty-five years old, he looks a weary sixty, and his strangely shaped pate has lost its hair. By the end of 1538 he reached Wolfsberg in Lavanttal, and in the following year details of his movements through southern Germany and Austria are somewhat uncertain and apparently aimless: Augsburg, Munich, Villach again, Graz, Breslau, Vienna. By the spring of 1540 we find him sickening at Klagenfurt.

Here Paracelsus explains to a patient that he cannot attend him, partly because he is ill himself but also because he is awaiting 'a certain written message of importance'.[31] Apparently he knew that the Bishop of Salzburg was about to invite him back to that city, the scene of such disputes fifteen years previously. And why not go? Perhaps there he will get some more of his books published.

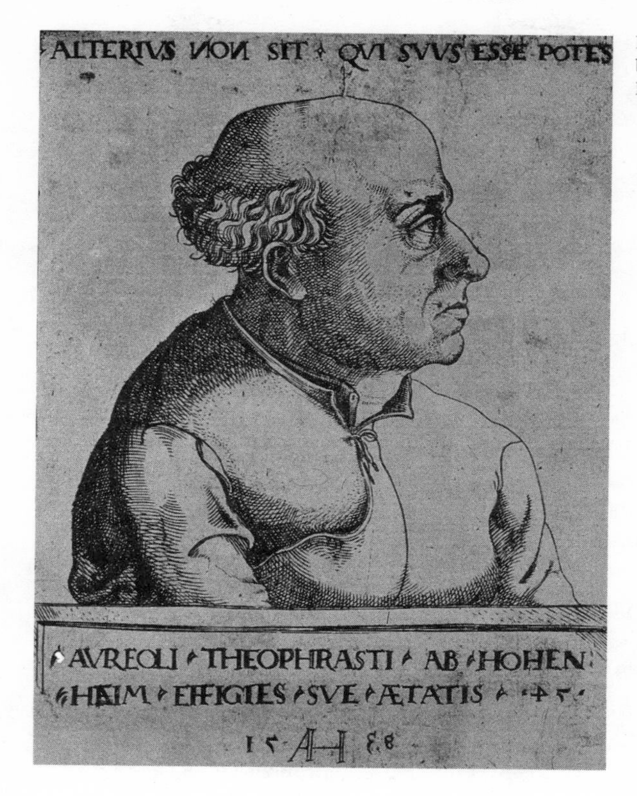

Profile of Paracelsus by Augustin Hirschvogel (1538).

This time, however, it was no revolutionary firebrand who arrived at the Austrian city, but an enfeebled man aged beyond his years. And he knew it. 'The snow of my misery has come,' he wrote to a friend. 'Summer is over.'[32] A year later he was dead.

18
The White Horse

Death in Salzburg

> And so I'll say goodbye. Clap your hands, live well, and drink, distinguished initiates of folly.
>
> Erasmus, *In Praise of Folly* (1511)

Off the Linzergasse, a cobbled alley leads behind the Sebastianskirche. Through an old wooden door, stone steps descend into a cloistered cemetery. Everywhere there are baroque visions of death: weeping cherubs, painted pietàs and carved skulls crawling with worms. On the right, in the entranceway to the church, a marble sepulchre dominates an alcove, bearing a familiar square-domed profile and an inscription:

Philippi Theophrasti Paracelsi
qui tantam orbis famam
ex auro chimico ADEPTUS est
effigies et ossa
donec rursus circumdabitur pelle sua.
Sub reparatione ecclesiae MDCCLII
ex sepulchrali tabe eruta
haec locata sunt.

Here are the effigy and the bones of Philip Theophrastus Paracelsus, who has won such fame in all the world through his alchemy; until they are again clad in flesh.

When this church was repaired in 1752 they were lifted from their mouldering grave and interred at this spot.

The records of Paracelsus' movements towards the end of his life are sketchy. Some say he set out for Salzburg in early 1541, stopping often along the way, and did not pass through the city gates until May. According to others, he had already taken up his lodgings at Platzl 3, on the right bank of the river Salzach – where a plaque and portrait now commemorate his visit – in 1540. Whether it was because of poison in his veins, the attrition of decades of exhausting travel, or simply the burdens of a time when old age came at forty, he was nearing his end. When the Pole Franz Boner wrote from Cracow to ask for his medical aid, Paracelsus replied that he was too ill to be of help.

As his condition deteriorated, he moved to a more comfortable situation at the Wirtshaus zum Weiszen Ross (the White Horse Inn) across the river in the Kaigasse, under the shadow of the city castle that the townspeople had besieged sixteen years earlier. Here he was attended by his servant Clauss Frachmaier, and in circumstances rather less straitened than usual – his wardrobe was full – he worked as best he could on his religious writings. But on St Matthew's Day, 21 September 1541, he summoned the public notary Hans Kalbsohr to record his will. Also present were six witnesses: Melchior Späch, Andree Setznagel, Hans Mülberger,

The sepulchre of Paracelsus in St Sebastianskirche, Salzburg.

The romanticized and highly unlikely depiction of Paracelsus' death in John Henry Pepper's *Playbook of Metals* (1861). Here the fatal elixir is said to be distilled alcohol.

Ruprecht Strobl, Sebastian Gross and Steffan Waginger, all good burghers of the district. This is what the will records:

> The most learned and honoured Master Theophrastus von Hohenheim, Doctor of Art and Medicine, weak in body and sitting in a camp-bed, but clear in mind and of upright heart . . . commits his life, death, and soul to the case and protection of Almighty God, in steadfast hope that the Eternal Merciful God will not allow the bitter sufferings, martyrdom, and death of his only begotten Son, our Saviour Jesus Christ, to be fruitless and of no avail for him, a miserable man.[1]

Was he, perhaps, wondering how far his own sufferings and death would be of avail to the sick and the oppressed, once he was gone?

He bequeathed his books, his medical equipment and his medications to a Salzburg doctor named Andree Wendl. As for the rest of his goods and possessions, except for a few small monetary bequests these were to be distributed 'to his heirs, the poor, miserable, needy people, those who have neither money nor provision, without favour or disfavour: poverty and want are the only qualifications'.[2] These belongings turned up subsequently in many places: Augsburg, Kromau, Leoben, possibly Villach and St Veit. To some they were precious relics, to others potent secrets or useless baubles. They were whatever legend required of them. But one item in particular was never traced: the great broadsword, a devil or the secret of life reputedly hidden in its pommel.

It is unlikely that Paracelsus was able to leave his bed before his death three days later, on Saturday, 24 September, the festival of St Rupert. But rumours nevertheless spread of a more dramatic demise. He took an overdose of his vital elixir. He perished when an assistant roused him prematurely from a magic sleep. He was killed in a drunken brawl. Or he tumbled down some stairs, blinded by wine, on his way back to the inn. His enemies put poison in his wine, or powdered glass or diamond in his beer, or hired thugs to beat him in an alleyway, or to fling him from high rocks after luring him to a banquet. It is surprising that the devil himself did not materialize to claim him that night, or that he did not, in death, adopt the posture of a damned man, as poor Agrippa did. But that did not for long hold the Faustian stories at bay. It did not seem credible that such a man could pass away quietly in his bed, drained and wearied by the world.

No indeed: for had not Paracelsus asked that his dead body be quartered and buried in manure, from the warm ferment of which it would regain life, much as he himself had quickened the limbs of the homunculus? And was it not the case that, when the body was exhumed thirty years later, the parts had grown back together again as if on their way to resurrection? His credulous biographer Franz Hartmann tells us ('and those who are supposed to know confirm the tale') that Paracelsus may not be dead at all but reincarnated in an 'astral body', living in 'a certain place in Asia, from whence he still – invisibly, but nevertheless effectually – influences the minds of his followers, appearing to them occasionally even in visible and tangible shape'.[3] The Hartmanns of this world will not let him go that easily.

Quartered or whole, Paracelsus was buried, according to his request, in the graveyard of St Sebastian's church, close to the city bridge in Salzburg. He asked that the first, seventh and thirteenth psalms be sung around his grave and that a penny be given to each and every poor man who stood in front of the church. But his bones did not find peace. In 1572 they were dug up when a new chapel was added to the church, and reburied by the graveyard wall. In 1752 Archbishop Andreas von Dietrichstein ordered that Paracelsus' remains be moved again, this time to the marble tomb in the entrance porch of the church. The old tombstone was inserted into this new monument, a testament to the (temporarily) elevated status of the man it commemorated. The inscription is written in Latin, although Paracelsus might have preferred German. It says:

> Here lies buried Philip Theophrastus, doctor of medicine of great renown, whose art most wonderfully healed even the most terrible wounds, leprosy, podagra [gout], dropsy, and other seemingly incurable diseases; and who honoured himself by having all his possessions distributed among the poor. He passed from life to death on 24th September in the year 1541.
>
> Peace to the living, eternal rest to the dead.

The bones were placed behind an iron door, onto which Paracelsus' portrait was painted. At least, that was the intention. But by some misunderstanding, it was his father's face that graced the monument – taken, along with the alleged Hohenheim coat of arms, from the portrait that once hung in the Carolino-Augustem Museum in Salzburg – until the error was recognized

Paracelsus' tombstone, now set into the sepulchre in Salzburg.

years later and a bas-relief of Paracelsus substituted in its place. The confusion is echoed in a late-sixteenth-century manuscript at the University of Leiden which speaks of 'a magician, Wilhelm Bombast, boldly called Theophrastus Paracelsus'.[4]

There is, at least, no mistaking the curiously shaped skull, which was examined by the eighteenth-century German physician Samuel Thomas von Soemmering. He pronounced that a fracture, once thought to be the mark of a fatal blow struck by his enemies, was probably caused by some milder violence earlier in Paracelsus' life and had simply become enlarged over time after his death.

The 'magnetic attraction' that Paracelsus believed drew supplicants to the graves and shrines of saints was still strongly felt in later times. For even into the nineteenth century, pilgrims came from all over Austria to the tomb at St Sebastian, hoping that the power of this fabled doctor's bones would grant them a cure. In times of epidemic, such as when cholera struck the region in 1830, crowds filled the churchyard.

BY WILL ALONE

Science is supposed to pay no heed to the contingencies of personality. However much Isaac Newton and Albert Einstein fascinate us as individuals, we are asked to imagine that the course of science would have been barely altered had Newton been the warmest, most collegial member of the Royal Society, or Einstein a misanthropic hermit, so long as they still had the same ideas. And even if they had never existed, science would have gone much the same way, for other geniuses (or cohorts of lesser minds) would soon have come up with the same theories. For every Darwin, there is always an Alfred Russel Wallace.

Maybe so. But Paracelsus, as he told us, was different. Single-handedly, he started a medical revolution and founded a chemical tradition. From this tradition came some of the greatest of the early chemists – not always directly, nor uncritically, but each with an evident debt to Paracelsus. For at least a hundred years after his death, medical practitioners had to define their position relative to two opposed poles: Galen and Paracelsus.

It is true that the ancient medicine of Galen and Hippocrates could not have long outlasted the Renaissance in any case. Vesalius opened up Galenic anatomy for inspection (although he thereby challenged only that aspect of ancient medicine), and William Harvey picked apart the corpse. The same with chemistry: it would surely have waxed in the Enlightenment regardless of Paracelsus, who was as much a target as Aristotle was in Robert Boyle's chemical reformation. Likewise, the tradition of experimental enquiry advocated by Francis Bacon did not draw on Paracelsus alone, or even primarily. Since there seems to be no real *need* for Paracelsus, his continuing presence cannot be explained by any deep validity or novelty in his ideas.

Yet whereas most scientific revolutions have relied on the force of logic and practical demonstration, confronting traditionalists with arguments, measurements and experiments that they could deny only by refuting reason and experience, Paracelsianism did not. Its impact was surely due in large part to the sheer force of his personality and the power of his literary style – and the allure of the mistaken vision it described. Paracelsus was never logically irrefutable, and more often than not he was wildly wrong, yet people listened. They listened, and they felt forced to take a position: Paracelsian or anti-Paracelsian.

Nicolaus Copernicus put forward his thesis quietly, even timidly, from his deathbed. It was meticulously argued, which is more than can be said for anything Paracelsus wrote. But astronomy was not shattered by it. The modest initial print run of 400 copies of *De revolutionibus* didn't even sell out. It was read in Rome without alarm (Cardinal Nicholas von Schönberg even wrote to Copernicus advising him to publish the book). As we have seen, some of the strongest interest came from astrologers, who recognized how the new scheme could improve the accuracy of their forecasts. Copernicus' ideas were introduced to England by the astronomer and astrologer Thomas Digges, who in 1576 described the Copernican system in a revised edition of his father Leonard's popular almanac, retitled *A Prognostication Everlasting*.

Paracelsus, on the other hand, wrote copiously and in an energetic, even frantic, style. He was verbose and often lapsed into incoherence, but there could be no doubt about his intention to overturn the old system and replace it with a new one. Much of what he said about chemical medicine and natural magic was not really new at all; but no one had said it before in quite the same way. John of Rupescissa had seemed obscure and narrow; Ficino,

Reuchlin and Pico della Mirandola wrote for their erudite peers. Agrippa wrote rather little, and that seemed at face value to be riven with contradiction. Paracelsus was less systematic and calculating, and indeed far less successful, than Luther at exploiting the new printing technology to mass-market his ideas in the vernacular; but his colourful, violent language worked a comparable magic. Moreover, he spread the word directly: his endless wanderings had, by the time of his death, given rise to legends across all of Europe.

What if someone with a Copernican temperament had conceived of the entire chemical cosmology of Paracelsus? Could it ever have become a movement on the scale of Paracelsianism? I rather think not; and I believe that is so partly because, unlike Copernicism, Newtonian mechanics, relativity and the rest, Paracelsianism was not *necessary*.

The earth goes round the sun. The earth is a giant magnet. Bodies remain in constant motion when no forces act on them. Gravity follows an inverse square law. We remember the authors of these statements – Copernicus, William Gilbert, Galileo and Newton – because what they said is true. But there is nothing equivalent in the many volumes that Paracelsus penned. We can find many fine ideas therein (as well as much gibberish): that an 'alchemist' works inside us, for example, and that diseases have chemical cures, that biochemistry, geochemistry and industrial chemistry all operate according to the same rules. But these things are either false as stated by Paracelsus, or too general to be regarded as theorems, or restatements of older wisdom. Paracelsus made no major discovery that is still a recognized part of modern science. For every useful and 'modern' aspect of Paracelsus' system, we can find another that is spurious, nonsensical or merely incomprehensible. We cannot suppose that, but for Paracelsus, Antoine Lavoisier would not have constructed the rudiments of modern chemistry in the late eighteenth century and the periodic table would not have crystallized a century after that.

No – we can imagine a slow and steady withering of Galen and Aristotle taking place over the two hundred years or so that followed the Renaissance, without the furious ructions that attended Paracelsianism. That Paracelsus improved on the ancients in a robust and lasting way was largely incidental, and owes much to the patience of the careful men among his successors, who sorted the pearls from the dross. Paracelsus would have had little time for their reconstituted versions of his work. As far as he was concerned, his was the final word, and you could either take it or (if you were a fool) leave it.

In other words, Paracelsus did not do science. Walter Pagel says that he 'produced scientific results from a non-scientific world of motives and thoughts';[5] but even that, really, goes too far. There is arguably more of the modern scientific spirit in the work of medieval scholars such as Roger Bacon and Robert Grosseteste, who were searching for a rational philosophy based on experimental fact-finding. However bound they were to the religious codes of their time, they were in some sense looking, like scientists, for knowledge that was objective and universal.

Paracelsus' philosophy strove to be no less overarching, but it was intensely personal. His infuriatingly obtuse neologisms can be seen on the one hand as evidence of an unconscious sleight of hand, an attempt to impress when in truth he does not really know what he is trying to say; but they are also a symptom of an inner vision that cannot really be articulated, like someone striving to explain the ambience of a dream. It is precisely this deeply personal nature of Paracelsus' vision that led to its being interpreted and used in so many different ways by his successors. There was no such latitude in Newtonian mechanics.

There are ample examples of scientists being misled by wrong ideas. But such ideas do not last long unless the errors are seductive. We have seen what seduced people to believe in the Aristotelian elements, the Galenic humours, the tall tales of astrology, and the Philosopher's Stone. Such 'aberrations' persisted not because of the personalities who advocated them but because, like pre-quantum physics, they made a lot of sense at the time.

But Paracelsianism only ever really made sense to one man: Philipp Theophrastus Aureolus Paracelsus Bombastus von Hohenheim. What we can see of it now is a kaleidoscopic mixture, sometimes beautiful, sometimes grotesque or terrifying or inchoate, always profoundly human. That is, in the end, why it is still valuable and captivating. It tells us – and both scientists and artists must believe it – that we can look outside ourselves and commune with the universe.

19
Work With Fire

The Chemical Legacy of Paracelsus

> It was indeed a discouraging contrast in intellectual history, and a sad exhibition of human inability to detect the true and sober from the false and fantastic, that the same century which refused to digest and accept solid demonstrations of *De revolutionibus* should have swallowed so eagerly the innumerable half-baked and incoherent tomes of Paracelsus and his followers . . . Why should so few be ready to accept the Copernican modifications of the Ptolemaic system, when so many were ready to accept the Paracelsan revolt against Hippocratic and Galenic medicine?
>
> Lynn Thorndike, *History of Magic and Experimental Science* (1941)

If a man is revealed by his library, it is no wonder that John Maynard Keynes considered Isaac Newton a sorcerer. Almost one in ten of Newton's books were about alchemy, and he also owned works on palmistry and astrology. A quarter of the library was devoted to theological tracts. To judge from his books alone, Newton's scientific achievements stemmed from just a subset of his many interests.

No serious scholar of alchemy in the seventeenth century would lack the writings of Paracelsus, and Newton possessed a major edition of his works. By that time there was an entire *Corpus Paracelsia* – volumes written by followers who revised, expanded and clarified the chemical philosophy. Newton's copy of *A New Light of Alchymie* (1650), a translation of the *Novum lumen chymicum* (1614) by the Paracelsian chemist Michael Sendivogius, is dog-eared and clearly well used; likewise *On the Transmutations of Metals* by Paracelsus himself. Newton may not have been a disciple of Paracelsus, but he was clearly familiar with his ideas.

There is nothing particularly unusual or surprising about that, for the legacy of Paracelsus pervaded the natural philosophy of the seventeenth and eighteenth centuries. Self-confessed Paracelsians held prestigious positions in the universities and the chemical and medical colleges, and his medical teachings were promoted by reputable physicians everywhere.

Yet this was a man who, by the time of his death, had published rather little beyond his works on prognostication, which in themselves were fairly unremarkable for their times. Paracelsus' name had entered local legend across Europe – but folk tales do not confer academic respectability. He was reputed to be a wondrous doctor, but Europe was full of itinerant wonder-workers peddling miracle cures. The first question that arises in assessing Paracelsus' considerable legacy is: why him? Why not John of Rupescissa, Pico della Mirandola, Trithemius, Agrippa, or one of the many other alchemists, natural magicians, unorthodox doctors and eccentrics of the Renaissance?

This is perhaps the greatest mystery of Paracelsus' story. Had he made some vital discovery, like Copernicus or Newton, it would be far easier to defend his place in the pantheon of science and the history of ideas. It is not even that he was simply in the right place at the right time – often quite the opposite. His was instead the thankless task of preparing for the hegemony of science, lacking the tools he needed, derided by onlookers and having no real notion of what it was he was doing. At face value his ambitious programme was doomed from the outset.

Yet by the end of the sixteenth century, Paracelsus' works provided the defining framework for a good deal of the practice and theory that was evolving into modern science. Somehow he spoke to men in a language they could comprehend, and what he said seemed to them to provide the key to the secrets of the world.

THE EARLY REVIVALISTS

There is a sense in which Paracelsus was ahead of his time. That is certainly what his early-twentieth-century biographers would have us believe, although what they meant by it is highly questionable: Paracelsus certainly did not anticipate nuclear theory or quantum physics (yes, such claims have been made). Rather, he rode the wave of Neoplatonism a little too early. Interest in the occult and Hermetic arts may have gathered pace from the support of Ficino, Pico and Reuchlin, but it was not until the late sixteenth century that the stream became a flood – one that unleashed a second wave a century later. Paracelsus' works provided a focal point for this interest, and they were attuned to the burgeoning appeal of chemistry, previously the lowly art of druggists and artisans but now fast becoming an esteemed science. This interest was stimulated by the efforts of Guglielmo Gratarolo, who in the 1560s published a string of important alchemical works including those of Arnald of Villanova, Raymond Lull, John of Rupescissa and Michael Savonarola.

For almost twenty years after his death, however, it seemed that Paracelsus had become consigned to one of the more obscure corners of the Renaissance. In 1550 Cyriacus Jacobus referred to a 'Theophrastus Transsilvanus' who had found a secret key to medicine and used it to unlock the terrible diseases of gout, leprosy and epilepsy. But these were rumours – nothing more.

Then from around 1560 various publishers and editors began to translate and disseminate the manuscripts that Paracelsus had left behind. Chief among them were Michael Toxites, Adam von Bodenstein (the son of Luther's estranged follower Andreas Karlstadt), Gerard Dorn, Peter Perna of Basle, and Johannes Huser, who produced the first complete edition of Paracelsus' known works. These men were energetic champions of Paracelsian ideas; Dorn in particular produced extensive commentaries, interpretations and defences of Paracelsus, as well as several original works on alchemy and Hermeticism; the most notable of these is his *Key to All Chemical Philosophy* (1566).

Dorn illuminates the uneasy relationship between Paracelsianism and humanism. The humanists were never quite sure what to make of Paracelsus: he shared their dislike of scholastic dogma and artifice and their openness to new ideas, but he was too coarse, too wild and irrational, to sit easily among them. In this much, at least, he resembled Luther. But there was nothing Lutheran, and much that was humanist, in Paracelsus' belief in the dignity and merit of humankind, for whose benefit God had filled the world with traces of his Grand Design. Dorn, however, argued that both old-fashioned Aristotelianism and the new humanism were un-Christian, cold and empty, and that a better philosophy – a philosophy of love and spirituality – was to be found in the writings of Paracelsus.

Leo Suavius (Jacques Gohory) was more explicit in his *Compendium* (1567) about Paracelsus' debts to earlier alchemists, including Roger Bacon, the medieval Geber, Arnald, Lull and Nicolas Flamel. Nonetheless, his book advertises many Paracelsian remedies, including laudanum, mummy, antimony and mercury. Suavius praises Paracelsus for using natural magic in ways that Ficino was too timid to risk. Yet he confesses to being occasionally confused by the great doctor, admitting – and who can blame him? – that sometimes he isn't sure if Paracelsus is writing about alchemy, religion, astrology or medicine. Suavius was one of Paracelsus' first supporters in France, and he founded the Lycium Philosophal in Paris, a private academy that anticipated the great Jardin du Roi, the medicinal herbarium established by Gui de la Brosse in 1635.

Paracelsus' unpublished manuscripts might have vanished had it not been for the efforts of men like Bodenstein and Toxites. Adam von Bodenstein, physician to the Elector Palatine Prince Ottheinrich von Wittelsbach, was a conventional Galenist until shortly before he joined the Basle medical faculty in 1558. Two years previously he was cured of a fever by a Paracelsian remedy prepared by a doctor named Cyriacus Legher. Bodenstein later admitted:

> In my distress I accepted it despite the fact that I was taught to consider Paracelsus, the author of the prescription, an impostor, and I should have hesitated. As I was at the height of distress and pain I took the medicine as a last resort, using it as much as possible. By the grace of God I recovered from all pain and illness in thirty-four days.[1]

Bodenstein become a convert, and he subsequently used Paracelsian remedies with such success that he 'was suspected to have conjured the devil'.[2] At Basle, he was delighted to discover many of Paracelsus' manuscripts still in the possession of Oporinus, who did not consider them worth publishing. But when Bodenstein printed them, he was expelled from the university in 1564 by the medical faculty, which regarded them as 'heretical and scandalous books' and disowned Bodenstein for being 'an adherent of the false teaching of Theophrastus'.[3] Undeterred, Bodenstein remained at Basle and eventually published forty of Paracelsus' works.

Michael Toxites studied with Bodenstein and subsequently became a physician in Strasbourg and Alsace. Here he tracked down and published many of Paracelsus' books, including the *Archidoxa*. In Cologne, the town physician Theodor Birckmann persuaded Archduke Ferdinand of Austria to issue an appeal for Paracelsus' manuscripts to be collected, whereupon they accumulated at the court of the Elector Palatine Ottheinrich (fast becoming a centre of Neoplatonism), at Neuburg castle on the Danube. Paracelsus himself had found a warm welcome here in 1525, and amazingly his old colleague Hans Kilian was still there four decades later, now the Elector's chief alchemist. He curated the manuscripts, which eventually found their way to Johannes Huser, physician to Prince Ernst von Wittelsbach, Elector of Cologne, who published them in Basle in 1589–90.

Once Paracelsus was in print, his ideas spread rapidly. A sign of his growing reputation is the alchemical dialogue in Gratarolo's *Chrysorrhoas* (1561) that features Theophrastus ab Hohenheim, 'a man, if ever Germany bore one, who is absolute artificer of this art'.[4] Among the twenty-seven medical books in the collection of Joachim Wittenheder, a barber-surgeon of Braunschweig who died in 1567, there were eight by Paracelsus.

One of the most influential Paracelsians of the late sixteenth century was the Dane Peder Soerensen (Petrus Severinus), a doctor who established himself in Venice before becoming royal physician to Christian IV, king of Denmark. Severinus' *Idea of Philosophic Medicine, Containing the Foundations of the Entire Doctrine, Paracelsic, Hippocratic and Galenic* (1571) was, despite its even-handed title, dominated by the 'Paracelsic'. He laid stress on the Paracelsian tradition of experiment, inquiry and exploration. The true doctor was not a rich man who read dusty books and gazed at flasks of piss, but a pilgrim who renounced worldly goods. As Severinus wrote,

> Sell your lands, your houses, your clothes and your jewellery; burn up your books. Instead buy yourselves stout shoes, travel to the mountains, search the valleys, the deserts, the sea shores and the deepest recesses of the earth; mark the distinctions between several kinds of animals, plants, minerals . . . be not ashamed to study the heavenly and the earthly lore of peasants. Lastly, buy coal, build furnaces, work with fire. Thus and thus only will you attain the knowledge of things and their properties.[5]

Severinus was no crude propagandist. He has been called the Melanchthon to Paracelsus' Luther: a cultivated and sober man who stripped Paracelsus' words of their rant and rhetoric and presented the central ideas in calm and orderly Latin. According to Gui de la Brosse, Severinus understood Paracelsus better than Paracelsus did himself. Francis Bacon, no great fan of Paracelsus, admitted that 'Only one of your followers do I envy you, Paracelsus, and that is Peter Severinus, a man who does not deserve to perish with your absurdities.'[6]

He was fortunate also to elicit the discerning support of Guinter von Andernach (Johann Winter), a humanist scholar in Strasbourg who taught both Vesalius and the ill-fated heretic Michael Servetus. A medical traditionalist for most of his life, Andernach was introduced to Paracelsus' works in his seventieth year by Michael Toxites, and although he regarded Paracelsus himself as arrogant, he became a convert to chemical remedies. In the same year that Severinus' *Idea of Philosophic Medicine* appeared, the Italian exile Peter (Pietro) Perna in Basle published Andernach's *Of the Old and the New Medicine*, which eased Paracelsus' acceptance by reconciling his ideas with conventional medicine and presenting him as a rediscoverer of an ancient medicine that had been lost.

PARACELSUS ACCUSED

As Paracelsianism emerged in the late sixteenth century, so did resistance to it. Even after Vesalius' *De fabrica*, medicine was slow to change, and the anti-Galenic recipes of Paracelsus tasted bitter to many. His cause was not helped by the Counter-Reformation, arising out of the Council of Trent in 1564 at which the Roman Church attempted to stem the tides of Lutheranism and Calvinism. Neoplatonism was considered to be dangerously allied to the Protestant cause – and even if this association was inaccurate in theological terms, it was commonly true in practice. The works of Paracelsus fell foul of this new Catholic intolerance.

Yet it was a humanist and Protestant who did the most damage. The title of the tract *Disputations Concerning the New Medicine of Paracelsus* was an unambiguous indication of the author's intentions – an author unusually dogmatic and vituperative even by the standards of the time. The Swiss physician Thomas Liebler, known as Erastus, manifests, in Lynn Thorndike's words, 'an expressed conviction that whoever fails to agree with him is either very stupid or very wicked, and an inclination to attribute as much as possible to God on the one hand and to the demons on the other, and as little as possible to either natural law or human initiative'.[7] This misanthrope was a professor of medicine at Heidelberg; his writings, Thorndike continues, 'display that controversial temper, that crabbedness and want of good taste and tact, that lack of any sense of humour, that dry and narrow-minded erudition, that unwillingness to see the least merit in or excuse for another's point of view, which so generally characterize the literary and learned controversies of the sixteenth century'.[8]

Erastus' attack on Paracelsian medicine was no spontaneous outburst but a premeditated campaign, possibly with a somewhat self-serving goal. In the 1560s he fell out with the Calvinists over his assertion that the Church should be subservient to the state (a position now known as Erastianism). By denouncing Paracelsus as the most wicked and impious of men, Erastus can be seen to be polishing his own tarnished halo. But his dislike of Neoplatonic magic and mysticism seems unfeigned. His *Disputations* excoriated the entire Paracelsian project:

> Certainly there have never lived under the sun (I am speaking of Philosophers) more diligent worshippers of demons than the Platonists. And are we to say that the words of truth are confirmed by their execrable lies? Ficino was so addicted to the loathsome and clearly diabolical fables, that he preferred to lick up the stinking spittle of the Platonists, rather than taste the most sweet honey of truth.[9]

No corner of Paracelsian philosophy escaped Erastus' flail: the new theory of the elements, the value of chemistry in treating disease, the 'magnetic' attraction between disease and remedy, natural magic in general, the idea that man is a microcosm of the universe ('why can he not fly, lay eggs, live in the sea, grow fruit or remedial arcana, if he really contains all other objects of nature?'[10]), the idea that disease is an agent entering the body from outside (as opposed to an internal imbalance of humours), the use of metals in medicine, and so on. The originator of these ideas, according to Erastus, was a 'grunting swine', a blasphemer who would revive the 'pestilence' of Gnostic heresy (which denied that the Creation was the work of God) adduced by the first Faustian man, Simon Magus.

To discredit Paracelsus, Erastus assembled a formidable range of witnesses, in particular the Zwinglian Heinrich Bullinger, who related his unfavourable impressions on meeting Paracelsus in Zurich forty years earlier (Chapter 11). In October 1570 Erastus wrote to this powerful Protestant leader explaining that 'I am completely absorbed in reflection regarding confuting the most monstrous and absurd dogma of Paracelsus.'[11] Joining Bullinger on Erastus' panel for the prosecution were Johannes Crato von Krafftheim, Conrad Gesner, Joachim Camerarius the Younger of Nuremberg, Johann Jakob Grynaeus, a pastor of Rötelen (near Basle), Sigmund Melanchthon (the nephew of Luther's associate), Wilhelm Xylander, a professor at Heidelberg – and finally, Johannes Oporinus, who became Erastus' publisher and even wrote his biography.

From these men and others, Erastus gathered and enthusiastically recounted every damning story about Paracelsus he could find. Crato, for instance, related what he had heard from a Moravian nobleman, Berthold von Leipa, of Paracelsus' exploits in the region of Moravsky Krumlov, a city near Brno. Berthold said that his father, Johann von Leipa, summoned Paracelsus to cure his arthritis. The doctor treated Johann for two years without effecting the slightest improvement, succeeding only in damaging the sight in one of his patient's eyes. It could have been worse, Berthold claimed, for the treatment Paracelsus administered to the wife of Baron Johann von Zerotín gave her seizures of which she eventually died, forcing Paracelsus to flee from Hungary. Berthold adds that while Paracelsus was writing the *Astronomia magna*, which dates from his time in Moravia, he was usually drunk.

It was Crato, too, who contributed the unfavourable account of Paracelsus' time in the court of Ferdinand in Vienna (page 339). Another physician, Markus Recklau, furnished a further set of tales of Paracelsus' incompetence and untrustworthiness. For example, in Munich Paracelsus was summoned to attend a man known as Monachus ('monk'), an official in the court of the Duke of Bavaria. The man was already receiving apparently effective treatment from the duke's own doctors, but Paracelsus insisted on prescribing mercury. Seeing that Monachus was reduced to a moribund state, the doctor fled to Austria just before the man died. Erastus alleges that in fact all those who Paracelsus treated in Basle died within a year.

Moreover, Erastus found another eye witness to indict Paracelsus for wicked and demonic practices. This man, one Georg Vetter, claimed to have attended Paracelsus' lectures in Basle and to have accompanied him on his journeys in Austria and Transylvania in the later years of his life. Erastus is at pains to establish Vetter's good character and reliability. 'Recently', he says, 'a most pious, learned and industrious man Georg Vetter – loving and devoted to Paracelsus – most strongly brought to my attention that [Paracelsus] was extremely devoted to impious magic and that he was accustomed to call an evil spirit his friend.' According to Vetter:

> I feared nothing more, as often as he was drunk (which he was, however, frequently), that he would summon a troop of devils: which again and again he wanted to do (produced by the type of his art), but at my request he abandoned it. When I warned Paracelsus (when sober) that he made these utterances, because he gravely offended God, and because the devil would be the customary one to pay his students at the final sad stipend, he responded that not much later he would be singing for refuge.[12]

It is Erastus who began the rumour that Paracelsus was castrated by a soldier as a young boy. He asserts that Paracelsus learnt magic and chemistry in Spain, perhaps because of the obvious echoes of the works of John of Rupescissa and Arnald of Villanova. Explicit charges of plagiarism were made by Crato, who said that Paracelsus' medicines could all be found in a book written two hundred years earlier by 'a monk of Ulm'.

How much of all this should one believe? It appears that Erastus was generally accurate in reporting the words of others, but he was clearly selective in his choices, eager for defamations and unhesitant in transcribing them. We can question both his motives and his methods, but if there was falsity or distortion involved, it seems more likely to have come from his informants.

The *Disputations* cannot, in any event, be wholly dismissed as the empty invective of a reactionary. In claiming that all we receive from the heavens are heat and light, not astral 'virtues', Erastus sounds decidedly modern. And he is sympathetic to the idea of using distillation to prepare medicines. Indeed, his concept of medicine sometimes seems surprisingly Paracelsian in its emphasis on empiricism. But at root Erastus is a medical traditionalist. He sees no merit in Paracelsus' classification of diseases. He rejects the idea that imagination plays a role in health and illness. He criticizes Paracelsus' failure to explain the relationship of the *tria prima* to the four Aristotelian elements. He charges (rightly, although not for the right reasons) that antimony, a favourite remedy of the Paracelsians, is poisonous and that metals in general are not good to ingest. Erastus' classical scholasticism is revealed in the way that he often opposes Paracelsian remedies for philosophical reasons, as though they could be discredited by logic alone. His opposition to magic is not on account of its absurdity, but its impiety: Erastus' own world was still inhabited by demons and by witches, whose severe punishment he applauds.

Thus Erastus devotes a considerable part of the *Disputations* to a critique of Paracelsus' theology, focusing in particular on the issue of human materiality. Paracelsus' view that the human body was, after death, a useless husk fit only for the worms conflicted (said Erastus) with the claim in the New Testament that the resurrection of the Last Judgement embraced both body and soul, and would thus have a material aspect. Then there was the question of the flesh of Christ and the Virgin Mary. By asserting that Mary is descended not from Adam, like other humans, but from Abraham, born 'without male seed' and so free of Adam's original sin, Paracelsus seemed to be saying that Mary did not possess normal human flesh. This, to Erastus, bore the taint of Catholicism. Christ, in turn, was said by Paracelsus not to have received his flesh from Mary in the womb, but to have been embodied directly in 'holy flesh'. This again implies that Christ was not truly human, in which case the sacrifice of the Crucifixion is undermined. In short, Paracelsus was clearly a heretic – a more serious charge than being merely a bad doctor.

Erastus completed his first anti-Paracelsian diatribe in 1571, and it was published, curiously enough, by Peter Perna, an enthusiastic editor of Paracelsus' works. Between 1570 and 1603 Perna and his son-in-law, Konrad Waldkirch, published sixty books and tracts by Paracelsus, along with other important works by Johann

Weyer, Niccolò Machiavelli and others. Maybe Perna reasoned that, by stirring up debate, he would invigorate the market for his Paracelsian books. Erastus followed this initial broadside with several other anti-Paracelsian volumes. In *Disputation Concerning Potable Gold* (1578) he calls his fellow countryman 'an evil magician, atheist, pig' whose followers are 'either most ignorant or indubitably wicked'.[13]

Although perhaps unmatched in his loathing of Paracelsus, Erastus was not alone in opposing the 'new medicine'. John Donne, whose interest in alchemy ran deep, accused Paracelsus of leading a 'legion of homicide physicians'.[14] Jacques Fontaine of Provence argued in *Dispute of the Elements against the Paracelsists* (1581) that these people 'fill pages with useless words',[15] and he accused Paracelsus of black magic. Likewise, the Swiss physician and humanist Theodor Zwinger (Oporinus' nephew) asserted that Paracelsus believed 'If God will not help, the devil will.'[16] All this was refuted vigorously by Dorn, citing the distinction made by Ficino and Pico between natural magic and sorcery.

Conrad Gesner's attitude to Paracelsus is ambivalent: Charles Webster characterizes it as 'a mixture of admiration and fright'.[17] Gesner was a mild and irenic soul, an encyclopaedist in the tradition of Pliny, and ill suited to controversy. He gave ammunition to future critics by suggesting that Paracelsus wrote and taught in German only because his Latin was so poor (and the criticism was not entirely unwarranted). But his main concerns were theological: he thought he identified the Arian heresy in the beliefs of the Paracelsians, the suggestion that Christ was 'a quite ordinary man, [in whom] was no other spirit than in us'.[18] (This is, of course, effectively the opposite of Erastus' charge that Paracelsus rendered Christ supernatural.) Nevertheless, Gesner admitted that he learnt a great deal about metals from Paracelsus' writings, and sometimes used antimony remedies himself.

Johann Weyer was, as we have seen, highly sceptical of Paracelsian medicine. He poured scorn on Paracelsus' blind and ignorant followers, who, he says, apply all sorts of worthless potions, charms, pills and powders. They were infatuated with 'the confused heap of useless words with which Paracelsus filled his writings', and as a result they have become 'special slaves of arrogance, self-love and vainglory, who can accomplish all things whatsoever by very loud cries, and by promises and very long words, in perfect imitation of their master'.[19] Yet Weyer, like other broadminded physicians who came after him, appreciated the chemical dimension that Paracelsus had added to medicine – while remaining reluctant to credit the innovation to him alone:

> I do not here make light of chemistry, which is an important part of medicine, and which I value highly, as do all my fellow-practitioners of the ancient [that is, Galenic] medicine . . . And I freely acknowledge that by the power of chemistry spirits are extracted to combat all sorts of illnesses, and oils and powders and salts are compounded from sulphur, vitriol, antimony, and similar minerals, as well as metals. I have such things in supply myself and use them with success.[20]

This does not sound very Galenic at all; if not Paracelsus, then at least Arnald of Villanova and John of Rupescissa deserve credit for such remedies. If a Galenist could happily embrace them, clearly medicine was changing.

THE POLITICS OF ALCHEMY

These critics were, by and large, confounded, for by the turn of the century Paracelsianism was in the ascendant. Influential advocates began to appear all over Europe, encouraged and supported by Paracelsus' busy (and successful) publishers. At the conservative University of Basle, Theodor Zwinger executed a *volte-face* in his appraisal of his countryman Adam von Bodenstein. As an anti-Paracelsian he had helped to expel Bodenstein from the city; but by the late 1570s Zwinger was full of praise for him, referring to Paracelsus as the 'German Hippocrates'. He collaborated with Perna in editing Paracelsus' works, and would have worked on Huser's complete edition had he not died. In 1575 Perna published the Paracelsian tract *On the Secrets of Antimony* by Alexander von Suchten, a Pole who spread the new medicine in his native country after serving the Elector Palatine in Germany.

It was also in Basle that the Anglo-Scot Thomas Moffett (or Muffet) converted to Paracelsian medicine in 1578. In his doctoral thesis at the university he attacked Galen and Erastus; but the faculty was not yet ready for such unconventional boldness, and

Moffett was forced to retract his remarks before he was awarded his doctorate. He pursued the cause, however, in his book *On the Justice and Excellence of Chemical Medicine* (1584), a Platonic dialogue between a Paracelsian protagonist and a fierce opponent of the new medicine pointedly named Philerastus ('lover of Erastus'). In Copenhagen Moffett befriended Severinus, and when he returned to England his position as a member of parliament, a Fellow of the Royal College of Physicians, and physician to such luminaries as Philip Sidney, Francis Walsingham and Francis Drake gave him many opportunities to promote the chemical medicine. A keen entymologist, Moffett's interest in spiders led to the immortalization of his young daughter, who encountered one of her father's pets while 'eating her curds and whey'.

Paracelsian ideas were accepted in England with less acrimony than elsewhere; but he had his critics there too. Richard Bostocke is generally regarded as the first to popularize his ideas in England in *The Difference betwene the Auncient Physicke and the Latter Physicke* (1585).* He became familiar with the works of Paracelsus through his friendship with the Elizabethan alchemist John Dee, with whom he had attended St John's College in Cambridge in the 1540s. Dee's famous library in Mortlake, near London, included many of Paracelsus' books, and Bostocke is recorded as having borrowed from it in the 1580s. Moffett's was an influential voice, but a fellow member of the Royal College of Physicians, Francis Herring, was to write in 1604:

> I have often marvelled how any man of wisdome and modestie, seeing the incredible insolencie and impudencie, the intollerable vanitie and folloie, the ridiculous and childish crakings and vantings of Paracelsus, should once commend him without noting his contrary vices, and giving him a dash with a blacke coale.[21]

The Galenic traditionalists remained contemptuous of the brewers

* The book's author is indicated only as 'R. B. Esq.', which has prompted much discussion over his identity. Some have associated 'R. B.' with the London printer Robert Bostock; but Richard, a gentleman and Member of Parliament from Tandridge in Surrey, seems a far better candidate.

of chemical medicine: one of them, James Hart of Northampton, argued that the Paracelsians would 'feed us with smoky promises'.[22]

Michael Sendivogius' *Novum lumen chymicum* did much to popularize Paracelsianism across Europe. Its author, Michal Sedziwój, was a Moravian gentleman of Cracow, and he gained wide renown as an alchemist who possessed the secret of transmutation. It has been often claimed that he never truly understood the mysteries of the Hermetic art but was merely bequeathed the Philosopher's Stone by a Scottish adept named Alexander Seton. But the Seton tale seems to be pure fabrication: Sendivogius' reputation appears to rest on his own diligent labours. Studying at Leipzig in the 1590s, he made contact with the arch-Paracelsian Johann Thölde, and after his arrival at the court of Rudolph II of Prague, a centre of Hermeticism, in 1593, his renown as an alchemist became known through Bohemia and Poland. Rudolph sent him on diplomatic missions to his homeland, where Sendivogius' successful 'projection' of base metals into gold was conducted before the Polish king and a number of witnesses. From around 1607 he conducted alchemical experiments at the castle of the industrial metallurgist Marshal Wolski in Krzepice, Poland, and in 1616 he went to Marburg to work on Paracelsian medical chemistry with Johann Hartmann, professor of chemistry at the university there. Among those whom Sendivogius impressed was the great mystic and alchemist Michael Maier, who testified to a transmutation performed by the Polish adept before his own eyes.

Another of the pre-eminent figures of the Paracelsian movement was Oswald Croll, physician to Prince Christian of Anhalt-Bernberg. The title page of his popular pharmacopoeia *Basilica chymica* (*The Royal Chemistry*, 1609) awards Paracelsus a place in a stellar gallery of great alchemists that includes Hermes Trismegistus, Geber, Roger Bacon and Raymond Lull. The famous astronomer Tycho Brahe, a colleague of Severinus, called Paracelsus an 'incomparable' philosopher and doctor and defended him against the attacks of Erastus.

It was not long before the battlelines began to be defined by new criteria. The disputes over the 'new medicine' were not simply a matter of Paracelsianism versus Galenism, nor even of new versus old, or radicals versus conservatives. The boundaries soon became aligned with political and religious differences. In England the Paracelsians were associated with Puritanism, while the Galenists were generally Anglicans. During the reign of James I there was a degree of mutual tolerance, but this benign atmosphere evaporated as the country approached civil war under Charles I. To the king and his Catholic queen, Henrietta Maria, the chemical doctors smelled of hard-line Protestantism, and Charles I preferred the services of William Harvey, a follower of Aristotle and Avicenna, who called the Paracelsians 'shit-breeches'. When the war came, Harvey went with the Royalists, the Paracelsians with the Parliamentarians.

In Germany the divisions were even more explicitly linked to religious affiliations. Under the pressure of the Counter-Reformation, Protestantism began to seem vulnerable and became increasingly radicalized. The Neoplatonic philosophers who allied themselves with (or had merely been associated with) the Lutherans and Calvinists became ever more mystical, utopian and secretive. In this climate, Rosicrucianism and other secret brotherhoods flourished, and Ficinian magic resurfaced in Prague at the court of the Emperor Rudolph II, who became known as the 'alchemical emperor'. It was here in the early seventeenth century that Paracelsians, as well as many vagrant mystics and alchemical charlatans, flocked from all over Europe, assured of the sympathetic ear of the emperor. Oswald Croll came to Prague prophesying the imminent arrival of the prophet 'Elias Artista' whom Paracelsus had mentioned – he 'who is to restore all things' and to usher in a chemical golden age. John Dee paid an ill-fated visit in 1584 with his angel-scryer Edward Kelley. The director of Rudolph's alchemical experiments was Michael Maier, a deeply enigmatic figure. Having graduated in medicine at Basle, Maier came to Rudolph's court in the 1590s, where he was dubbed 'Imperial Count Palatine' and served as a secret agent for the emperor. Like Croll, Maier befriended Christian of Anhalt, to whom he dedicated some of his writings.

After Rudolph died in 1612 his brother Matthias became Holy Roman Emperor. Matthias favoured religious tolerance, but his successor was the fervently Catholic archduke Ferdinand. A group of Protestant Bohemian nobles, having indecorously ejected the Habsburg governors in Prague and then fomented revolt in Austria, refused to acknowledge Ferdinand as King of

In the frontispiece of Oswald Croll's *Basilica Chymica*, Paracelsus takes his place alongside such lengendary chemists as Hermes Trismegistus, Jabir (Geber), Raymond Lull and Roger Bacon.

Bohemia and adopted as their monarch the Calvinist son-in-law of James I, the Elector Palatine Frederick V, who was an enthusiast of chemical medicine and Paracelsian cosmology. To the Paracelsians in Prague, this seemed like the dawn of the promised new age.

But it was a very different dawn that lit up the skies of Bohemia. The army of the Elector Palatine, the so-called 'Winter King', was defeated in battle at Weissenberg, the White Mountain, in 1620, and the imperial forces brutally suppressed Calvinism in Bohemia. The repercussions ignited the Thirty Years War, which decimated the population of Germany and scattered the radical Paracelsians throughout Europe.

In France, too, Paracelsianism was politicized. The conservative Parisian medical faculty scored the first victory in the war of medicine when they succeeded in expelling from the city the Norman Paracelsian empiric Roch le Baillif. In 1578 the faculty denied Baillif the right to practise medicine, saying that he lacked the proper qualifications. When Baillif defied them, he was tried before a parliamentary court and banished.

The Parisian doctors' dislike of Baillif stemmed not only from his semi-literate and a bombastic self-promotion, but also from his Protestantism. To their alarm, however, his unorthodox methods had been well received at court. The situation seemed to them even graver when, in 1589, Henri, a Huguenot prince of Navarre, was crowned King Henri IV, because the new king brought to Paris a number of doctors with wayward ideas. In 1594 Henri appointed as his *premier médecin* (royal doctor) the Paracelsian Jean Ribit, a Huguenot from Geneva. He also summoned to court the Gascon Joseph Duchesne (Quercetanus), who had made a name for himself as a Paracelsian physician among the German princes.

Duchesne helped to bind chemistry and biology closer together by considering respiration as a form of distillation. Breathing introduced into the body substances inherent in the air, such as the explosive components of lightning and thunder: 'aerial nitre' and sulphur. These substances, said Duchesne, react in the body to create diseases that induce hotness. Yet aerial nitre came also to be seen as a source of vitality, the *spiritus mundi*, which was ultimately the repository of the human life force. It was thought that the lungs separate this substance from the air and turn it into new blood – which gave the Paracelsians even more reason to reject Galenic blood-letting, revealed thus to be simply a draining of the patient's vitality. The vital role of 'nitro-aerial' seemed to be confirmed in 1674 when Robert Boyle's assistant John Mayow found that the gas released when nitre (potassium nitrate) is heated turns arterial blood from dark to bright red in the lungs. Mayow's nitro-aerial is indeed a vital essence: oxygen, the element whose identification in the late eighteenth century marks the birth of modern chemistry. Once again, Paracelsianism sowed seeds that later blossomed into science.

At the court of Henri IV, Paracelsian chemical medicine (commonly known as iatrochemistry, from the Greek *iatros*, physician) reached its first apogee. This was due in considerable measure to the efforts of another Genevan Huguenot, Theodore Turquet de Mayerne, a friend and protégé of Duchesne and an irrepressible figure of seemingly limitless curiosity, who came to Paris in 1599 after Ribit secured him a court position.

His arrival meant that there were now three Huguenot doctors surrounding the supposedly Catholic king, and the Paris medical faculty, led by the conservative Jean Riolan, was outraged. In 1603 it issued a defence of tradition, *Apology for the Medicine of Hippocrates and Galen, against Mayerne and Quercetanus* – a condemnation not just of these named individuals but of Paracelsians in general. In their defence, the two doctors enlisted the assistance of the famous Saxon physician Andreas Libavius, and eventually a compromise was reached whereby Duchesne and Mayerne agreed to practice medicine 'according to the rules of Hippocrates and Galen . . . and not otherwise'[23] in return for a retraction of the condemnation.

When Mayerne cured an English ambassador, he was invited to London and appointed physician by James I. Mayerne subsequently returned to France, but soon had reason to seek James's patronage again. In 1609 Duchesne died, and so did Ribit's successor as *premier médecin*, André du Laurens. Henri offered Mayerne Laurens's post, but under pressure from the Paris medical faculty the king attached the condition that Mayerne must convert to Catholicism. The physician refused. In the following year Henri was assassinated. Now lacking royal protection, Mayerne fled to England escorted by the agents of James I, and there he flourished. In 1616 Mayerne was elected a Fellow of the Royal College of Physicians, which two years later published a new official pharmacopoeia containing many Paracelsian remedies. Mayerne's royal patronage was withdrawn by Charles I, who preferred the orthodox William Harvey; but when he died in Chelsea in 1655 it was as a wealthy man.

Around 1604, Ribit and Mayerne helped a fellow Huguenot and iatrochemist named Jean Béguin establish a pharmaceutical laboratory in Paris, where he offered public lectures on chemical medicine. In 1608 Béguin sponsored the publication of a new edition of Sendivogius' *Novum lumen chymicum*, and two years later he published his own exposition of Paracelsian chemistry, a textbook entitled *Tyrocinium chymicium* (*The Chemical Beginner*). It was the first of its kind – neither a 'book of secrets' nor an obscure academic treatise, but a manual for the would-be iatrochemist, or 'chymist', as this transitional discipline began to style itself. The book's practical nature may have stemmed in part from Béguin's expedient avoidance of the more radical and controversial aspects of Paracelsian speculative theory in the reactionary climate that followed Henri's assassination; but whatever the reason for its emphasis, the *Tyrocinium* assisted the separation of Paracelsus' chemistry from his philosophy. In 1615 Béguin's book was published in French, with revisions and additions, as *Les Elemens de chymie*, heralding the seminal works of the later French chemists Pierre Joseph Macquer (*Elémens de chymie théorique*, 1749) and Antoine Lavoisier (*Traité élémentaire de chemie*, 1789, subsequently published in English as *Elements of Chemistry*).

THE ANTIMONY WARS

Béguin's didactic example was perpetuated at the Jardin du Roi. The traditionalist Jean Riolan drew up plans for a royal herb garden in 1618, but the Paris medical faculty could not have been pleased when the institution established almost twenty years later by Gui de la Brosse turned out to look favourably on iatrochemistry. In 1647 the Jardin appointed the first professor of chemistry in France:* William Davidson, a Scot from Aberdeen. Davidson, who gallicized his name to Guillaume d'Avissone, was a committed Paracelsian. He had been lecturing in Paris since 1618, when, with Ribit's support, he took over Béguin's position.

No sooner was he appointed than Davidson offended the anti-Paracelsian physicians with his defence of the medicinal value of antimony, a classic Paracelsian cure. A so-called 'antimony war' had been waged between French physicians and iatrochemists since the beginning of the seventeenth century. What it lacked in bloodletting, this war made up for in bile.

The arguments over antimony also serve to remind us that, alongside the assimilation of chemical Paracelsianism into academia, the mysticism and obscurantism of sixteenth-century alchemy still survived. For one of the most enthusiastic supporters of antimoniacal iatrochemistry was the mysterious Basil Valentine, apparently another of the wholly fictitious 'great names' of alchemy.

'Antimony, you affirm, is a poison,' Basil Valentine rages in his combatively titled *The Triumphal Chariot of Antimony* (1604):

* The first university chair in chemistry within Europe was awarded in 1609 in Marburg, to the iatrochemist Johann Hartmann, physician to the Landgrave of Hesse-Cassel.

> Therefore [you say] let everyone beware of using it. But this conclusion is not logical, Sir Doctor, Magister, or Baccalaureas; it is not logical, Sir Doctor, however much you may plume yourself on your red cap . . . Antimony can be so freed of its poison by our Spagyric Art as to become a most salutary Medicine.[24]

Is this the voice of Paracelsus, ranting from beyond the grave? Well, it is not that of 'Basil Valentine', supposedly a Benedictine monk of Erfurt from the fifteenth century, yet for whose existence there is not a shred of evidence. His name means mighty or valiant king, a suitably grand and vague designation for a supposedly great alchemist. His 'works', full of references to the archeus, arcana and the *tria prima*, were published by Johann Thölde, a late-sixteenth-century saltmaker from Frankenhausen in Hesse, who was probably their true author. Some of Paracelsus' later critics were to charge him with plagiarizing 'Valentine', not realizing who really came first.

The Triumphal Chariot is full of archetypically cryptic alchemical jargon, but it also gives us a delightful (if sadly apocryphal) explanation of the origin of the name 'antimony'. Basil Valentine claims that he once threw the antimoniacal residue from an alchemical experiment into the pigs' trough at his monastery, whereupon the pigs' appetite increased and they became fatter. This, he decided, would be a good treatment for some of his thin and sickly Benedictine brothers. But he was sure that the over-cautious monks would refuse their 'medicine', and so he added it secretly to their food. To his alarm and distress, the effects were quite the opposite of those intended: the monks developed intestinal illnesses, attacks of fainting and violent coughs. Some of them even died. Valentine never revealed his part in all this, but thenceforth he designated the metal *anti-monachos* ('anti-monk'). The etymology is not far askew: the Greek *monos* (alone) has the same root as the Latin *monachos* (solitary, i.e. a monk). But the term *anti-monos*, 'not alone', was probably first used to denote the sulphide mineral of antimony (stibnite) by Constantine of Africa in his eleventh-century work *De gradibus simplicibus*.

Antimony is indeed fairly toxic. It has been suggested that Mozart may have died after taking excessive doses of antimony tartrate prescribed by his doctors. And poisoning with antimony was a notorious way for Victorian doctors to dispose of unwanted wives and relatives. The toxicity produces purgative effects: pills of pure antimony metal were often prescribed in the Middle Ages as laxatives. The pills, retrieved from the excrement, could be reused.

As 'Basil Valentine' indicates, the Paracelsians had no illusions about the poisonous nature of antimony, but they believed that, as with mercury, their chemical manipulations could separate the good from the bad, removing the poison and producing an antimony remedy. Oswald Croll listed no fewer than twenty-three recipes containing antimony. Anti-Paracelsians were not convinced, however – the common confusion of antimony with arsenic at that time scarcely helped matters – and the argument raged furiously during the early seventeenth century.

When Davidson became a professor at the Jardin du Roi, he announced his intention to 'devote much attention to the preparation and exaltation of antimony, because words are lacking by which to name so rich a substance . . . there is no more lofty medicine under heaven.'[25] His robust defence of antimony drew the admonition of the Parisian medical faculty, who added to their objections the fact that he was a foreigner and a Protestant. He was hounded from his post in 1651, the same year in which the ageing Jean Riolan, son of the staunch opponent of Duchesne and Mayerne, reprinted the slanderous condemnation of those two Paracelsians that his father helped to write in 1603. It could hardly have harmed Mayerne himself, now eighty years old and living comfortably in England, but it showed that the battle for Parisian medicine was as bitter as ever.

None of this, however, affected the loyalties of the Jardin du Roi, which continued to maintain a string of iatrochemists, alternatively known as 'spagyric' chymists (from the Greek *spao*, to divide, and *ageiro*, to bind, a reference to (al)chemical manipulations). Davidson was succeeded by Nicaise Lefebvre, who later became Royal Professor of Chemistry to Charles II of England in 1660. Lefebvre's *Cours de chymie*, published in that year, became a standard chemistry textbook for the next century. Samuel Pepys records how he visited Lefebvre's London laboratory in 1669, and 'there saw a great many chemical glasses and things, but understood none of them'.[26]

After Lefebvre came Christophe Glaser, who continued the textbook tradition with *Traité de la chymie* (1663); and then Nicolas

Lemery, whose *Cours de chimie* (1675) also proved popular. By this stage, academic chemistry or chymistry had begun to make a clean break with alchemy. Lefebvre divided chymistry into distinct sub-disciplines: philosophical (that is, theoretical) chymistry, iatro-chymy, and pharmaceutical chymistry, which refers to the decidedly 'applied' aspects of medicinal chemistry practised by apothecaries. Lemery's book, meanwhile, marks a shift in emphasis that was to prove crucial to the development of chemistry as a coherent science. Whereas Paracelsus and his earlier followers focused their attention on distillation processes, and specifically on the volatile substances they produced – the oils and spirits that formed the basis of many Paracelsian arcana – Lemery was more concerned with solutions: substances dissolved in a solvent such as water or alcohol. He described how to make medicines from extracts of plants, separated by dissolution rather than distillation. Dissolving vegetable substances in alcohol is a far gentler extraction process than using heat to distil them, which can destroy delicate medicinal compounds. Lemery also investigated solutions of minerals and of metal salts, like those produced when an acid eats away at a pure metal. To Paracelsus this kind of dissolution was often merely the start of a process that culminated in purification of a 'quintessence' by distillation. But by investigating the solutions themselves, Lemery widened the scope of chemistry and directed attention towards the salts, acids and alkalis that were eventually to lead chemists to a clear understanding of their art.

To Lemery at least, the alchemical 'puffers' had had their day – he wrote and lectured in transparent terms designed to appeal to fashionable Parisians. Yet the very fact that Lemery felt the need to condemn the charlatans whose 'subtle inventions . . . they too often impose on such as have plenty of money, to make them become fellow-partners with them in their operations'[27] is a reminder that such people were still very much around.

The Jardin du Roi continued to attract the leading French chemists for two hundred years after Davidson held its first chemical post: among its luminaries were Macquer, the great chemist of gases Joseph Louis Gay-Lussac, and the master of colour science Michel-Eugène Chevreul. The Jardin helped to make Paris arguably the birthplace of modern chemistry: a provenance secured, in the face of rancorous opposition, by a few practically minded and discerning followers of Paracelsus.

The demonstrator of chemistry at the Jardin in the mid eighteenth century was the eccentric Guillaume-François Rouelle, who lectured to Macquer and Lavoisier as well as to Jean-Jacques Rousseau and Denis Diderot. When Diderot prepared his famous *Encyclopédie*, the epitome of French Enlightenment philosophy, he asked another of Rouelle's students, Guillaume-François Venel, to write the entry for 'Chymie'. Venel complained that chemistry was still an under-appreciated science, misunderstood by 'incurious' persons as a puffers' manual art. What it needed in order to win the respect of philosophers, he said, was a 'new Paracelsus, who will make of chemistry the science that understands nature and displaces geometry from that pretension'.[28] The old Paracelsus would surely have approved of this idea – although the person often considered to have filled the role, Antoine Lavoisier, has been described as an 'anti-Paracelsus': 'an insider rather than an outsider, a physicalist rather than a chemist, and a man of algebraic sensibility rather than a mystic'.[29] By the time chemistry was ready to take its final steps towards modernity, Paracelsianism had run its course; and the man had been, like his ideas, discarded.

It is in the transitional discipline of iatrochemistry that we can find Paracelsus' most substantial legacy to the development of science. To say that modern chemotherapy can be attributed to him is inaccurate in many ways. He was not, as we have seen, by any means the first to commend and use chemical medicine. And to attend to his potions and pills while discarding the mystical framework within which they were embedded is to mistake the appearance of Paracelsianism for its true substance. But without the translations of Paracelsus' works in the 1560s and 1570s, it is hard to see what later medical chemistry would have looked like, or what path it would have taken. In his own terms, and certainly during his own lifetime, Paracelsus failed. But what a valuable and indeed what a glorious failure it was.

20
Philosopher's Gold

The Last of the Chemical Magicians

Those who become practically versed in nature are, the mechanic, the mathematician, the physician, the alchemist, and the magician, but all (as matters now stand) with faint efforts and meagre success.

Francis Bacon, *Novum organum* (1620)

'There are not', wrote Seth Ward, astronomer and friend of Isaac Newton, 'two waies in the whole World more opposite, than those of the L. *Verulam* and D. *Fludd*, the one founded upon experiment, the other upon mysticall Ideal reasons.'[1] The L[ord] Verulam is Francis Bacon, whose vision of science informed and shaped the early Royal Society in London. D[octor] Fludd was Robert Fludd, a fellow of the Royal College of Physicians and a committed Paracelsian.

This was, to Ward and many of his scientific contemporaries, the choice they faced in the middle of the seventeenth century. Which path to take? The one founded on experiment, on the sober accumulation and practical application of facts unencumbered by the need to arrange and explain them according to any overarching theory; or the other that invoked 'mysticall reasons' for everything that happened in the world?

It is clear enough what Ward thought. But Fludd was not so easy to dismiss. If he was a rather marginal figure in the emerging scientific community in England, nonetheless he was also a respected doctor, an able self-publicist, and a knight who had studied at Oxford and had eminent contacts throughout Europe. While many Paracelsians shunned mathematics, Fludd extolled it as the most excellent way to understand the workings of nature. For him, geometry and numbers – a kind of Pythagorean 'spiritual mathematics' – offered a path to a quasi-Gnostic revelation of the secrets of the universe. For years Fludd sought to isolate the *spiritus*

The English Paracelsian Robert Fludd (1574–1637).

mundi, the substance in the air that (according to Paracelsian philosophy) could create new life in non-living matter through the process of putrefaction. This, Fludd explained in his *Philosophicall Key* (*c.*1619), is what gave rise to the spontaneous generation of creatures in decaying matter.

Fludd and his compatriot John Dee, sometime magician in the court of Elizabeth I, personify one pole of the Paracelsian movement: magical, Cabbalistic, Gnostic, a world apart from the practically oriented iatrochemists. The hard-headed Johannes Kepler attacked Fludd's ideas in the 1620s, characterizing them as 'enigmatic, emblematic and Hermetic'.[2] Fludd, in turn, dismissed Kepler as one of those who 'concern themselves with quantitative shadows'.[3]

Thus the twin strands of Paracelsus' works had, by the seventeenth century, become teased out and pulled to opposite extremes. On the one hand logic and reason, the tools of Aristotle, had become allied to careful experiment and observation of nature, producing the standard historical model for the Scientific Revolution. On the other hand a deeply teleological universe was deemed to be comprehensible through mystical revelation or by decoding the cryptic signs implanted by God in nature.

Which of these was the better route to knowledge and insight? It is tempting to see the first of them as progressive and 'modern', the latter as retrograde and medieval. But such distinctions are simplistic. Let us consider, for instance, the Royal Society itself, that bulwark of the Enlightenment, that rarefied and exclusive club for men of science and the model for subsequent scientific institutions. It was out of the mystical approach to natural philosophy that this concept of a 'scientific' brotherhood emerged: a confraternity dedicated to uncovering the secrets of nature. This vision was given its fullest expression by Francis Bacon, whose *New Atlantis* (1627) describes 'a college', in the words of Bacon's friend and posthumous publisher William Rawley, 'instituted for the interpretation of nature and the producing of great and marvellous works for the benefit of men'.[4] Bacon called it Salomon's House, a kind of scientific temple encountered in the fictitious land of Bensalem ('New Jerusalem') by a group of travellers crossing the Pacific Ocean from Peru to Japan.

The scholars of Salomon's House are scientists, but their role in the society of Bensalem resembles that of priests. They fulfil the ultimate promise of science, which is, according to Bacon, 'the production of wonderful operations'[5] – the fabrication of new and marvellous things, not the abstract and generalized theorizing of Aristotelians. Scientists, according to Bacon, should not be like ants, busy doing mindless practical tasks, nor like spiders, weaving tenuous philosophical webs, but like bees, mining nature for her goodness and using it to make useful things.

And what useful things the scholars of Salomon's House wrought! There are great artificial lakes, furnaces, engines, astronomical instruments and mechanical devices. There are caves where alchemists have succeeded in imitating the natural production of metals. And by means of their great art, these scientists can modify and 'improve' nature:

> We make, by art, . . . trees and flowers to come earlier or later than their seasons; and to come up and bear more speedily than by their natural course they do. We make

> them also by art greater much than their nature; and their fruit greater and sweeter and of differing taste, smell, colour and figure, from their nature . . . We have also parks and enclosures of all sorts of beasts and birds, which we use not only for view or rareness, but likewise for dissections and trials . . . We also try all poisons and other medicines upon them . . . By art likewise we make them greater or taller than their kind is, and contrariwise dwarf them and stay their growth. We make them more fruitful and bearing than their kind is, and contrariwise barren and not generative. And we also make them differ in colour, shape, activity, many ways. We find means to make commixtures and copulations of different kinds, which have produced many new kinds, and them not barren, as the general opinion is.[6]

The men who know how to do these things swear an oath of secrecy 'for the concealing of those [inventions] which we think fit to keep secret',[7] lest they fall into the wrong hands and be abused.

The focus of *New Atlantis*, then, is on *works*, not *theories*. Whereas the Aristotelian and medieval tradition of natural philosophy held that man can (within limits) understand the world by logic and reason, Baconian science exists for the purpose of *doing*: creating wondrous devices and using empiricism to discover how nature can be tamed, subjugated, altered. This was precisely the distinguishing feature of natural magic, which sought to harness and control the forces of nature for mankind's mundane, material ends. Bacon himself was hardly a mystic – he deplored the secretive ways of magic and alchemy – but he studied both subjects enthusiastically.

BROTHERHOODS OF SECRETS

Like the scientists of Salomon's House, the founders of the Royal Society felt themselves to be part of a closed circle of initiates bound together by privileged knowledge. Bacon's programme for accumulating reliable knowledge about nature by systematic experimentation, outlined in his *Great Instauration* (1620), was promoted in England in the 1640s by the Prussian exile Samuel Hartlib, a Puritan and Paracelsian. Hartlib harnessed the Paracelsian tide that washed through England in the wake of the Parliamentarians' victory over Charles I in 1646. The Puritans had backed Hermetic, chemical medicine against the Galenism that persisted in the Royal College of Physicians, and the end of the Civil War prompted the printing of many pamphlets vindicating this new medicine. Hartlib was one of a circle of progressive thinkers that included the politician and mathematician William Petty, the 'chymist' Robert Boyle and the American alchemist George Starkey. He petitioned the government to finance an institute designed to execute the scheme that Bacon had outlined in *New Atlantis*. Cromwell's Protectorate was not favourably inclined to the proposal, however, as it sounded to them a dangerously utopian, even millenarian idea. So it was not until the Restoration in 1660 that the Royal Society was established, with Petty, Boyle and later Isaac Newton among its leading lights. But this august institution was – unsurprisingly, given its provenance – not exactly the force for rationalism that it represents today. Among the close friends of its first Secretary, Henry Oldenburg, for example, was Petrus Serrarius, a mystic, astrologer, prophet – and Paracelsian. 'Somewhat inconveniently for standard interpretations of the Scientific Revolution,' says the historian Charles Webster, 'the decades following the foundation of the Royal Society witness a last outburst of judicial astrology, the continuing flourishing of Paracelsian medicine [and] undiminished appeal of alchemy and hermeticism.'[8]

Indeed, the image of a utopian society guided by 'scientific' wisdom was a familiar Neoplatonic fantasy. Hartlib outlined model societies of his own, called Antilia and Macaria. A 'scientific' utopia is described in *The City of the Sun* by Tommaso Campanella, written around 1602 but not published until 1623, the year before Bacon wrote *New Atlantis*. Campanella's city state admits an astonishing degree of equality: there is no private property and no slavery, and women are treated as equals with men. The seven-walled city, designed to mirror the Copernican universe, is dedicated to the instruction of its citizens in the sciences. To this end the walls are painted with pictures and diagrams that illustrate aspects of astronomy, geology, zoology, botany, metallurgy, geography, mathematics and other sciences. Campanella, who was punished for this bold and politically defiant vision with twenty-seven years of imprisonment and torture, was fundamentally a Neoplatonist, and at the heart of his utopia is a mystical sect of priests who use natural magic to direct the power of the stars for the benefit of humanity.

But the concept of a 'brotherhood' that guarded profound insights from the masses gained perhaps its most explicit realization in the Rosicrucian movement, which flourished in the early seventeenth century. Some say that the spiritual founder of the Rosicrucians was Paracelsus himself. In Huser's edition of his *Prognostication Concerning the Next Twenty-Four Years* there is a woodcut of a child looking towards a heap of Paracelsus' books, some inscribed with a capital R and one bearing the word *Rosa*. But the significance of this imagery for the Rosicrucians seems spurious.* The rose that the secret society chose as its symbol is in fact derived from the emblem of Martin Luther, in which a heart and cross spring from the centre of the flower. The movement began as a society of Protestant Paracelsians founded by the alchemist Johann Valentin Andreae of Herrenberg.

In the beginning, Rosicrucianism was merely one shadowy sect among many. It may have developed out of the German Orden der Unzertrennlichen (Order of the Inseparables), founded in 1577, which concerned itself with alchemy and mining technology. That would certainly account for the interest in the works of Paracelsus. The Inseparables became linked to the curiously named Fruchtbringende Gesellschaft (Fruit-Bringing Society), founded in 1617, of which Andreae was a member.

The Rosicrucians gained widespread notoriety after the publication in Kassel of two declarations, the *Fama fraternitatis* (*Report of the Brotherhood*, 1614) and the *Confessio fraternitatis* (*Confession of the Brotherhood*, 1615). The first of these, at least, is probably the work of Andreae, although they appeared anonymously. They announced that the original leader of the order was one Christian Rosencreutz, a Dutchman of the early fifteenth century. Rosencreutz supposedly gained occult knowledge in the Middle East and formed a group of nine disciples (all virgin bachelors), whose successors were now ready to spread the new learning. The *Fama* called for the

* The Paracelsus connection remains puzzling, however. In the first edition of the *Philosophia magna*, published by Birckmann in 1567, the Hirschvogel woodcut of Paracelsus appears in modified form with various strange images in the background that later became clearly associated with Rosicrucianism, such as a child's head emerging from a cleft in the ground. What is the significance of these symbols, fifty years before the Rosicrucian movement came into the open?

medicine of Galen and the philosophy of Aristotle to be replaced by wisdom gained directly from God and nature – as Paracelsus, whose works allegedly lay alongside those of Rosencreutz in a secret vault, had also proposed. The Paracelsian flavour was unmistakeable, which we can understand once we learn that the secretary of the Rosicrucian Order was reputed to be none other than Johann Thölde, the 'real' Basil Valentine.

This Rosicrucian manifesto maintained that many men throughout Europe now knew of Rosencreutz's legacy – for 'there is nowadays no want of learned men, magicians, cabalists, physicians, and philosophers'[9] – and it called on them to come forth and declare themselves, so that they might be admitted to the brotherhood. They alone knew who they were. (Or sometimes not: towards the end of his life Michael Sendivogius reputedly refused the entreaties of two Rosicrucians who travelled to his remote castle to enlist him.)

This was all a prescription for conspiracy, rumour and paranoia. And indeed the manifesto appeared to have just that effect, as an account from 1619 declares:

> What a confusion among learned men followed the report of this thing, what a conflict among the learned, what an unrest and commotion of impostors and swindlers, it is needless to say . . . there were some who in this blind terror wished to have their old, and out of date, and falsified affairs entirely retained and defended with force. Some hastened to surrender the strength of their opinions; and after they had made accusation against the severest yoke of their servitude, hastened to reach out after freedom.[10]

For a short time, the Rosicrucian influence was felt across Europe by the intelligentsia. Even the rationalist Descartes was among those who tried (and failed) to make contact with the supposed brotherhood. Francis Bacon's *New Atlantis* may have owed a debt to Andreae's own utopian vision, *Republicae Christianopolitanae descriptio* (1619), and was sometimes itself interpreted as a Rosicrucian fable. In England, prominent alchemists such as Thomas Vaughan and George Starkey were influenced by Rosicrucianism, and in the eighteenth century the movement was kept alive by the Freemasons, who developed their own interpretations of Rosicrucian symbolism. The Masonic Rosicrucian Society established colleges in Britain and the USA in the nineteenth century, and is still active today.

Robert Fludd was an enthusiastic supporter of the Rosicrucian declaration, of which he learnt from Michael Maier, who came to visit him in London. Fludd wrote a defence of the *Fama* and *Confessio* in 1616 that endorsed the proposal to replace Aristotle and Galen with magic, alchemy and chemical medicine in the schools and colleges.

Fludd's apology for Rosicrucianism was stimulated by an attack of the previous year by Andreas Libavius. Libavius had assimilated Paracelsian chemistry without accepting its mysticism. He was an iatrochemist and an opponent of Galen, but he had no time for magic: 'Certain devotees of chemistry differ little from magicians,'[11] he wrote with contempt. He was to iatrochemistry what Agricola was to metallurgy: a practical-minded and rational man with little interest in abstract theology and philosophy. He ridiculed the Paracelesian weapon salve, attributed all magic to the devil, criticized astrology, and derided superstitions such as the healing power of a king's touch. A conservative Lutheran, he inveighed against the 'abominable impiety' and 'madness'[12] of Paracelsus and his followers, spurred on perhaps by the suspicion that these Swiss notions were allied with Calvinism.

Libavius was not free of all magical thinking, believing for example that the wounds of a murdered body resume bleeding in the presence of the murderer. But he firmly rejected Paracelsus' flights of mystical fantasy and discounted stories of his supposedly miraculous healing powers. Libavius describes Paracelsus as an impious sorcerer, a 'limb of Satan', and the world's worst liar. Commenting on Paracelsus' works on mineral waters, Libavius charges that 'as in many other matters he is stupid and uncertain, so also here [he] writes like a madman'.[13] Although he presents a somewhat distorted picture of the ideas of Paracelsus and his followers, nonetheless his attack has more substance, and had more influence, than diatribes like those of Erastus. Libavius has been described as 'the first chemist of note in Germany who stood up manfully against the excesses of Paracelsus and who vigorously combated the defects in his doctrine, the obscurities in his writings, his phantasies and sophisms, and the employment of "secret remedies".'[14]

Libavius' *Alchemia* (1597) was a masterful synthesis of all that was known at that time about alchemy, pharmacy and metallurgy, presented in a clear and systematic manner that exerted a strong influence on Jean Béguin's *Tyrocinium chymicum*. In his *Syntagma* of 1611–13 he gives instructions on how to prepare a number of important chemical substances, including sulphuric and hydrochloric acids. His attempts to rid iatrochemistry of its mystical baggage were paralleled by those of Daniel Sennert, a professor of medicine at Wittenberg, who argued that 'there is a difference between Chymistry and Paracelsian Physick, for Chymistry was used before the time of Paracelsus. And though he used it, yet it is not necessary that all Chymists should be Paracelsians, and embrace his opinions.'[15] Sennert's *On the Consensus and Discord between the Chemists and the Followers of Aristotle and Galen* (1619) was essentially an attempt to reconcile the new medicine with the old, although it pulls no punches when dealing with the more fanciful of Paracelsian inventions:

> His works are full of incredible nonsense, for example that nightfall is not due to the setting of the sun but to the rising of the night stars, that some of the stars are shaped like cucurbits and phials containing salt, sulphur and mercury and emitting winds like man.[16]

It was certainly valuable for physicians to be told that by using chemical remedies they need not consider themselves to be endorsing Paracelsian cosmology. But in rejecting Paracelsus' originality Sennert cites Basil Valentine and Isaac Hollandus as precedents – both of them fictitious characters whose 'works' post-dated those of Paracelsus.

MAGIC AGAINST MECHANICS

Fludd of course lost his battles – the names we remember today are those of his enemies: Kepler, van Helmont and the disciples of the mechanist Descartes. Yet this was no simple conflict between mysticism and rationalism, for like Isaac Newton neither Kepler nor van Helmont was immune to Hermetic fabulation. The Cartesians were another matter.

In the Cartesian world-view, everything was made up of particles in motion. Descartes maintained that forces could be explained in terms of atom-like particles moving in vortices. The meetings of his supporters in Paris eventually led to the formation of the Académie des Sciences in 1666, the French equivalent of the Royal Society. The intellectual leader of this group was the monk and philosopher Marin Mersenne; throughout the 1630s Mersenne was just about the only person in Europe who knew how to contact the reclusive Descartes in the Netherlands.

These men were primarily mathematicians and physicists, and had a low opinion of chemistry. Mersenne's Cartesian vision of the human body as a system of pulleys and springs was a long way from the Paracelsian idea of an alchemical microcosm, and when the Parisian iatrochemist Étienne de Clave was arrested in 1624 for promoting anti-Aristotelian (and pro-Paracelsian) ideas about the nature of the elements, Mersenne expressed his approval at this suppression of alchemical heresies. The following year he published *La Verité des sciences*, in which he mocked what he saw as the false certainties of the alchemists and claimed that the Paracelsian chemical philosophy needed to be replaced by a mathematical description of nature.

Mersenne was not opposed to chemistry *per se*; rather, like Libavius and Boyle, he felt it needed purging of charlatans, allegorical ideas and obfuscating terminology. 'I do not deny that the metals have some power in medicines,' he said, 'but not because of the planets to which they are subject, for Saturn does not preside over lead, or Jupiter over tin, any more than does the Sun or Venus.'[17] Thus chemistry should limit itself to modest ambitions and refrain from making claims of a philosophical or a theological nature. One of the worst offenders, according to Mersenne, was Robert Fludd, whom he described as a magician and heretic.

Fludd responded to Mersenne's accusations by restating his Paracelsian position: true alchemy, he said, could furnish an explanation of all the universe, which was a 'chemical unfolding of nature'.[18] The English Paracelsian Thomas Tymme elaborated by saying that the Creation itself was an 'Halchymicall Extraction, Separation, Sublimation, and Conjunction'[19] – which was precisely the kind of view that Mersenne objected to. The Cartesians regarded this as wild, generalized, unquantifiable speculation that had nothing to do with their emerging concept of science.

In 1628 Mersenne, exasperated by Fludd, sought the support of his colleague Pierre Gassendi. Gassendi and Fludd had fallen out previously over William Harvey's theory of the circulation of the blood (and the dispute illustrates that scientists of this age cannot be divided neatly into the deluded mystics and the rational progressives, for it was Fludd who was in the right on that occasion, defending his friend Harvey against Gassendi's Galenic orthodoxy). Gassendi warned Mersenne that debating with Fludd was like arguing with a religious fanatic who would accept no truth but his own: he would make 'alchemy the sole Religion, the Alchemist the sole Religious person'.[20]

Fludd refused to give ground right up to his death in 1637; but Mersenne subsequently continued his attempts to discredit him. Among the scientists that he and Gassendi sought to win over to their cause was one who might have seemed an unlikely ally. The Flemish chemist and physician Jan Baptista van Helmont can be considered a Paracelsian himself, and few did more to make the name of Paracelsus respectable. Yet van Helmont had no hesitation in advising Gassendi that he should not waste his time over Fludd, whom he judged to be a poor physician and a worse alchemist.

A SCEPTICAL DISCIPLE

It is significant that van Helmont's critique was aimed at Fludd's competence rather than his beliefs. Although by the 1630s iatrochemistry was starting to abandon its mystical aspects, van Helmont nonetheless maintained a belief in magic, 'the most profound inbred knowledge of things'.[21] He was convinced that the Philosopher's Stone existed, and indeed claimed that he was once given it by 'a strange man, being a friend of one evening's acquaintance'[22] (precious knowledge transmitted by a mysterious stranger is a common theme in alchemical writings). 'It was of a colour such as is saffron in its powder,' said van Helmont, 'yet weighty and shining like unto powdered glass.'

> He who first gave me the gold-making powder had likewise also at least as much of it as might be sufficient to change two hundred thousand pounds of gold. For he gave me perhaps half a grain of that powder, and nine ounces and three quarters of quicksilver were thereby transchanged.[23]

So it always was with alchemy: a tantalizing taste of its power, while the real riches remain at the rainbow's end.

Van Helmont began his career as a Galenist. Like Adam von Bodenstein, he was converted to the new medicine by a successful Paracelsian cure, a sulphur ointment that he used on an itch. In 1621 van Helmont published a tract supporting the Paracelsian weapon salve, provoked by criticisms that a prominent Jesuit had directed at this remedy. He rather unwisely suggested that, as a purely natural phenomenon, it worked in a way akin to religious relics. He was denounced by the University of Louvain in 1623, and two years later van Helmont was declared a heretic by the Spanish Inquisition. After a spell in prison, he was forced to live under house arrest until his death. As a result of this persecution, his work was little known until it was published posthumously in a collected volume, *Ortus medicinae*, in 1648. The book was popularized in English by John Chandler, who translated it as *Oriatrike; or, Physick Refined* (1662); and by 1707 it had been reprinted twelve times in five languages.

Van Helmont displays a defiant self-reliance comparable to that of Paracelsus. He was offered employment by princes, including Rudolph II, but he preferred instead to devote himself to his research at home. He rejected the Aristotelian elements, but did

The Flemish physician and iatrochemist Jan Baptista van Helmont (1577–1644).

not wholly hold with the Paracelsian *tria prima* either – his criticisms informed Boyle's *Sceptical Chymist*. Van Helmont suggested that there were just *two* elements: air and water. Air was inert, however, so all tangible things were made from water. In a famous justification, he showed how a willow tree in a carefully weighed pot of soil gained 164 pounds over five years, supplied only (so it seemed) with pure water.

The experiment is notable not so much for its conclusions as for its methods: it depends on accurate quantification. Van Helmont applied the same quantitative principles to other experiments, such as his studies of the combustion of charcoal. These investigations led him to conclude that air was not the only vaporous substance, for some chemical transformations produced emanations to which he gave the name 'Gas', derived from Paracelsus' use of the word *chaos* to describe air. This was a profound insight that, alas, subsequent chemists failed to develop, persisting for more than a hundred years in the belief that all gases are varieties of 'air' in different states of purity.

Like Paracelsus, van Helmont had a proclivity for coining strangely named concepts: alongside his 'Gas', he added the emanations Blas (seemingly a kind of astral power that caused motion and change) and Magnall (a substance that fills 'empty' space). As non-material, or at least non-sensible, entities, these are clearly a part of the legacy of Neoplatonism. In other words, van Helmont had grander objectives than simply cataloguing chemical processes in the manner of Libavius or Béguin; he was still looking, like Paracelsus and Fludd but in his own way, for a 'chemical philosophy', a chymical explanation for everything.

By the same token, van Helmont revealed Paracelsian inclinations in his adherence to vitalism: the belief in a life force that pervaded all matter. In this he was quite divided from the French Cartesians, who supported their mentor's separation of body and soul. Like Fludd, van Helmont sought to isolate the *spiritus mundi*, which he pursued by distilling blood. And he elaborated Paracelsus' idea of the archeus, the internal alchemist that digests food to extract its goodness, by suggesting that digestion is a fermentation process involving an acid: the first clear statement of this important biochemical principle.

Van Helmont's appreciation of the role of fermentation in the body's chemistry stemmed in part from an episode that typifies this intensely curious and somewhat unworldly man. Once, when in London, he went with some noblewomen to visit the king's palace, and 'for civility's sake' he refrained from urinating for a full twelve hours. This led him to fear that his urine, 'having been long detained and cocted [boiled] beyond measure, would now be of a sandy grain'[24] – that is, it would contain fine particles that were the tell-tale warning of the much feared onset of kidney stones. Thus, once he had returned home he pissed through a napkin, but found no sign of gritty particles. The following day, however, he strained his urine in the same way and then left it to stand for twelve hours, whereupon he did indeed find 'sand' in it. He concluded that it needed to stand and ferment in order for the particles to appear: fermentation is here revealed as a process that brings about chemical alteration.

François de la Boe (Franciscus Sylvius), a disciple of van Helmont at the University of Leiden, expanded his theory of digestion by arguing that it requires several components: saliva, which provides the 'ferment', alkaline bile, and acidic pancreatic juices. This focus on acids and alkalis led to the almost Galenic idea that diseases were the result of an imbalance of the two: Franciscus Sylvius' pupil Otto Tachenius advertised as much in the title of his work *Hippocrates chemicus* (1666) – a title that seemed to promise a reconciliation of new and old. In such ideas we can see Paracelsus' fertile concept of an alchemical biology developing slowly towards modern biochemistry. Indeed, an interest in fermentation – an important industrial process for bakers and brewers, defined by the English Paracelsian Edward Jorden as a heat-producing phenomenon occurring in the absence of air – led eventually to the discovery of enzymes in the late nineteenth century.

Van Helmont is often called Paracelsus' greatest disciple, yet in the *Oriatricke* he writes, 'I have laughed at *Paracelsus* because he hath erected serious trifles into the principles of healing.'[25] Despite his faith in magic and chrysopoeian alchemy, van Helmont was a discerning thinker and intolerant of mystical chemical analogies. Paracelsus, he wrote:

> was deceived by the *Metaphor* of a *Microcosme* or little World. To wit, he translated the metaphor of a *Microcosme* into the truth it self . . . Away with thy trifles: For we have no

> fountains of Salt, no reducements of venal bloud into feigned and lurking metals. Neither are there minerals in us . . . Neither also are there microcosmical Lawes in us, any more than the humors of four Elements mutually agreeing in us, and the fights or grudges of these . . . Alas! with how sorrowful a pledge are all these things, and by how sporting a means hath that man [Paracelsus] invaded the principality of healing? to wit, that we are all little Worlds! for at how dear a rate doth he sell us this Idea or Image of the Macrocosme! and by what a scanty argument doth he found his dreams![26]

FROM PROJECTION TO COMBUSTION

Van Helmont comes at the end of the Paracelsian tradition of chemical mysticism, and there is little magic left in his version of a chemically based universe. A better indication of where chemistry was headed in the seventeenth century was to be found in Johann Rudolph Glauber, a practical man whom Paracelsus might have dismissed as an 'honest puffer'. Although his *Pharmacopeia spagyrica* (1654) provides many iatrochemical recipes, Glauber was more interested in the chemistry of the non-living world: what today we would call inorganic chemistry. He was a great technician and experimentalist, discovering several new salts (one of which, sodium sulphate, was long known as Glauber's salt) and improving on ways of synthesizing known compounds. Yet he pioneered the chemistry of organic (carbon-containing) compounds too, distilling volatile hydrocarbons from coal and wood. He used a tarry extract of wood as a repellent against insects that 'are wont to damnifie Fruit',[27] and claimed that one of his concoctions acted as a hair restorative, which he tested on his own balding pate. 'Had I more of the like Tincture', he said, 'it would have wholly renewed me.'[28] Like Paracelsus, Glauber worked with dangerous substances in poorly ventilated conditions, and metal poisoning blighted his final years, which, lacking a patron, he lived out in poverty.

But the extravagances of Paracelsian alchemy had one last contribution to make. Appropriately enough, perhaps, this contribution was confusing, misleading, sometimes couched in outrageous terms – and instigated by a brilliant and wayward

Johann Becher (1635–82): gold-maker or charlatan?

vagrant with a penchant for wild claims, noisy confrontations and ingenuous credulity.

That man was Johann Joachim Becher, a German from Speyer, and to judge from the shifty aspect of the figure depicted in his portrait we might expect a degree of scepticism to have greeted his claim that he could manufacture gold. But that was welcome news to the Dutch government in 1673, for Holland was at war with France and badly needed money. Give me a hundred pounds of silver, said Becher, and I will transform hundreds of tons of sand into gold.

Yet the Dutch were neither so desperate nor so trusting of alchemists to take up the offer without first demanding evidence of Becher's claims. So they gave him the money for a trial run, allowing him to construct a great water wheel and a furnace. Finally in 1679 the test began, and in front of a lone witness Becher turned sand to gold – a feat that sparked the interest of the great German mathematician Gottfried Wilhelm Leibniz. 'He claims that there will be almost as much net profit in it as in the mines of Hungary,'[29] Leibniz wrote. A month later Becher allegedly repeated his success before the mayor of Amsterdam and a government committee.

But while he was busy scaling up the process, Becher's enemies in the Dutch court began to conspire against him – or so he said (does this sound familiar?). Eventually he was forced to flee for his life. 'Mr Becher,' Leibniz's informant Christian Philipp of Hamburg told him, 'notwithstanding the cunning spirit of the Dutch, has succeeded in extracting a good sum of money and then disappeared afterward, but they are searching for him everywhere and will treat him most rigorously if he allows himself to be found.'[30] Weeks later, the sceptical Philipp told Leibniz that Becher had surfaced in England, 'and there he has found his dupes as always'.[31]

Becher saw many strange things on his restless travels – geese that lived in trees, stones of invisibility, bottles that captured spoken words so that they sounded again when the vessel was opened hours later. In a truly Paracelsian declaration, he said that he was 'one to whom neither a gorgeous home nor security of occupation, nor fame, nor health appeals to me; for me rather my chemicals amid the smoke, soot and flame of coals blown by bellows'. Poisoned by mercury fumes, 'deprived of the esteem and company of others, a beggar in things material',[32] he lived nevertheless as a Croesus in his mind.

He says all of this in the introduction to his *Physica subterranea* (1669), a treatise on minerals which lays out a 'new' theory of the elements. It was in fact a reincarnation of the familiar. Agreeing with van Helmont that air is inert and cannot take part in chemical processes, Becher admitted only water and earth into his elemental scheme. But water is a substance unique in itself, so that all the rest of matter acquires its diversity from *three* distinct types of earth. He called them *terra fluida* or *terra mercurialis*, mercurious earth, *terra pinguis* or fatty earth, and *terra lapida* or vitreous earth. The first is what makes (non-aqueous) substances fluid and volatile; the second makes them oily, sulphurous and combustible, the third gives them body and fusibility. We can recognize these earths at once, of course, for they are the *tria prima*: mercury, sulphur and salt, renamed but barely disguised. Had not Paracelsus, after all, identified sulphur as the source of that 'latent fatness' that is 'the life of metals'?[33]

But this *terra pinguis* soon had another incarnation. Georg Ernst

Stahl, a chemist and physician at the University of Halle, edited and published a new version of Becher's *Physica subterranea* in 1703. Stahl reiterated Becher's elemental system but proposed to call fatty earth by the name *phlogiston.* The word dates back to the early seventeenth century, although it had never previously gained wide usage; it comes from the Greek *phlogistos*, 'to set on fire'. Phlogiston, said Stahl, was the 'principle' of combustibility.

Stahl argued that phlogiston could explain combustion, a central concern of eighteenth-century chemistry. When a substance such as wood burns, it releases vapours (*terra fluida*) and phlogiston into the air, and the ashes that remain are the *terra lapida.* The same with metals, which, when heated in air, are converted to a salty residue or 'calx' as they release their phlogiston. Stahl claimed that the extraction of metals like iron from their ores by heating the ore with charcoal may be explained by the fact that charcoal is rich in phlogiston. By adding phlogiston to a calx or an ore, the metal is restored to its pure form.

Phlogiston is not a Paracelsian idea as such, but Paracelsus' role in its conception is easy to trace. Phlogiston is midway between the old notion of an abstract and ultimately inaccessible 'prime substance', be it an Aristotelian element or one of the *tria prima*, and a chemical element that can be extracted, isolated, weighed and measured. For Paracelsus such fundamental components (whether elements or principles) embodied general properties and tendencies of matter; for later chemists they were tangible realities, the ingredients of the world.

Phlogiston theory won support from most of the major chemists of the eighteenth century, but it was full of holes. Not the least of its difficulties was the evident fact that metals heated in air do not lose but gain weight. How can this be, if they are releasing phlogiston? Simple, said the phlogistonists – for phlogiston has negative weight, or an ability to confer buoyancy. It was not until the 1780s that these flaws began to tell, when Antoine Lavoisier argued that a newly discovered elemental component of air that he called oxygen is the real principle of combustibility. When things burn, said Lavoisier, they combine with oxygen rather than releasing phlogiston. This remnant of the alchemists' sulphur was nothing but oxygen's illusory shadow. Even if Lavoisier's arguments relied as much on artful propaganda (such as a complete revision of chemistry's nomenclature) as on the force of his practical demonstrations, in the end there was no disputing that the oxygen theory was correct. From that point, chemistry never looked back.

And so the last of the magicians died or were banished from science, and it became the business of sober professionals whose aspirations were typically modest and mundane. But we should hesitate before assuming that Paracelsus would have regretted this. It has given us, at last, a medicine that works, and a considerable understanding of the chemical composition of the macrocosm and the microcosm, and it has liberated us from the tyranny of the stars and put our fate into our own hands. Today's science is in many ways precisely that 'art' which Paracelsus, in his more lucid moments, wished to find:

> For the true art is reason, wisdom, and sense, and sets in order the truth which experience has won; but those who hold to fantasy have no ground to stand on: only formulae that are past and done with, as you know well enough.[34]

References

Introduction: Fool's Quest

Epigraph: M. Shelley, *Frankenstein* (1831), Penguin, Harmondsworth, 1994, p. 8.

1. F. Hoefer, *Histoire de la chimie*, vol. 2, ii, Paris, 1843, p. 9. Quoted in J. R. Partington, *A History of Chemistry*, vol. 2, Macmillan, London, 1961, p. 123.
2. R. E. Schlueter, 'Fact and fiction in the names and titles of Paracelsus', *Annals of Medical History*, **7** (1935), p. 274.
3. J. Read, *From Alchemy to Chemistry*, Dover, New York, 1995, p. 100.
4. J. Jacobi (ed.), *Paracelsus: Selected Writings*, Princeton University Press, Princeton, 1988, p. 3.
5. F. A. Yates, *Giordano Bruno*, University of Chicago Press, Chicago, 1964, p. 258.
6. C. Webster, *From Paracelsus to Newton*, Cambridge University Press, Cambridge, 1982, p. 58.
7. W. Shumaker (ed.), *Natural Magic and Modern Science: Four Treatise 1590–1657*, State University of New York at Binghamton, 1989, p. 3.
8. Thomas Aquinas, *Summa theologiae* (2a.2ae.96.2), in K. Hutchison, 'What happened to occult qualities in the Scientific Revolution?', *Isis* **73** (1982), p. 237.
9. ibid., p. 250.
10. ibid., p. 251.
11. H. P. Bayon, 'Paracelsus: personality, doctrines and his alleged influence in the reform of medicine', *Proceedings of the Royal Society of Medicine*, **35** (1942), p. 69.
12. J. Ferguson, *Encyclopaedia Britannica*, 14th ed. Quoted in Bayon.
13. Paracelsus (*c.* 1530). *De generatione stultorum*, trans. P. F. Cranefield and W. Federn, in *Bulletin of the History of Medicine* **41** (1967), pp. 56–74.
14. Partington, p. 127.
15. D. P. Walker, *Spiritual and Demonic Magic from Ficino to Campanella*, Warburg Institute, London, 1958, p. 96.
16. In H. E. Sigerist, 'Karl Sudhoff, the man and the historian', *Institute of the History of Medicine Bulletin* **II** (1934), p. 3.
17. C. Webster, 'The nineteenth-century afterlife of Paracelsus', in *Studies in the History of Alternative Medicine* (ed. R. Cooter), Macmillan, Basingstoke, 1988, p. 78.
18. W. Blake, *Complete Writings* (ed. G. Keynes), Nonesuch Press, London, 1957, p. 158.
19. A. S. Byatt, *Possession*, Vintage, London, 1990, p. 172.
20. I. S. Turgenev (1862), *Fathers and Sons*, trans. E. Schuyler, Leypoldt & Holt, New York, 1867, p. 141.
21. J. L. Borges, 'The rose of Paracelsus', in *Collected Fictions*, trans. A. Hurley, Penguin, Harmondsworth, 1999, p. 507.
22. J. K. Rowling, *Harry Potter and the Order of the Phoenix*, Bloomsbury, London, 2003, pp. 252–3.
23. Shelley, pp. 37–8.

24. W. Godwin, *Lives of the Necromancers*, Chatto and Windus, London, 1876, p. 218.
25. P. B. Shelley, letter of 3 June 1812, in F. L. Jones (ed.), *Letters*, vol. I, Clarendon Press, Oxford, 1964, p. 303.
26. M. Shelley, p. 8.

Chapter 1: Black Madonna

Epigraph: J. Winterson, *Gut Symmetries*, Granta, London, 1997, p. 1.

1. J. Jacobi, *Paracelsus: Selected Writings*, Princeton University Press, Princeton, 1988, p. 4.
2. C. G. Jung, 'Paracelsus as a spiritual phenomenon' (1942), in *Alchemical Studies*, trans. R. F. C. Hull, Princeton University Press, Princeton, 1983, p. 112.
3. H. M. Pachter, *Paracelsus: Magic Into Science*, Henry Schuman, New York, 1951, p. 20.
4. Jacobi, p. 3.
5. P. Amelung, *Das Bild des Deutschen in der Literatur der italienischen Renaissance, 1400–1559*, Hueber, Munich, 1964, p. 149.

Chapter 2: The Metal Makers

Epigraph: Georgius Agricola (1556), *De re metallica*, trans. H. C. Hoover and L. H. Hoover, Dover, New York, 1950, p. 24.

1. Paracelsus, *Chronicle of Carinthia*, quoted in A. M. Stoddart, *The Life of Paracelsus*, John Murray, London, 1911, p. 32.
2. Agricola, p. 17.
3. ibid., p. 19.
4. ibid., p. 19.
5. Tacitus, *Germania*, book V, quoted in J. Gimpel, *The Medieval Machine*, Pimlico, London, 1992, p. 69.
6. W. Manchester, *A World Lit Only by Fire*, Macmillan, London, 1993, p. 155.
7. Hoover and Hoover, in Agricola, p. xv.
8. D. C. W. Baumgarten-Crusius, *Georgii Fabricii chemnicensis epistolae ad W. Meurerum et alios aequales*, Leipzig, 1845, p. 139.
9. Paracelsus, *De mineralibus*, quoted in Agricola, p. 409.
10. W. Shakespeare, *Twelfth Night*, act II, scene V, 1600/1601.
11. J. Jacobi, *Paracelsus: Selected Writings*, Princeton University Press, Princeton, 1988, p. 115.

Chapter 3: The Universal Scholar

Epigraph: L. Olschki, 'The scientific personality of Galileo', *Bulletin of the History of Medicine*, **12** (1942), here p. 251.

1. H. M. Pachter, *Paracelsus: Magic Into Science*, Henry Schuman, New York, 1951, p. 25.
2. B. de Telepnef, *Paracelsus: A Genius amidst a Troubled World*, Banton Press, Largs, 1991, p. 24.
3. Pachter, p. 33. See also Telepnef, p. 25.
4. Telepnef, p. 26.
5. H. E. Sigerist, *Paracelsus: Four Treatises*, Johns Hopkins Press, Baltimore, 1941, p. 25.
6. Erasmus, *In Praise of Folly* (1511–21), trans. B. Radice, Penguin, Harmondsworth, 1993, pp. 86–7.
7. Telepnef, p. 23.
8. L. Thorndike, *A History of Magic and Experimental Science*, vol. IV, Columbia University Press, New York, 1934, p. 525.
9. ibid.
10. G. Mora (ed.), *Witches, Devils and Doctors in the Renaissance*, Medieval Renaissance Texts and Studies Vol. 73, State University of New York at Binghamton, 1991, p. 114.

11. J. R. Partington, 'Trithemius and alchemy', *Ambix*, **2** (1938), p. 58.
12. N. L. Brann, *Trithemius and Magical Theology*, State University of New York Press, Albany, 1999, p. 182.
13. Telepnef, p. 23.

Chapter 4: The Staff and the Snake

Epigraph: J. W. von Goethe, *Faust, Part I* (1773–1801), trans. P. Wayne, Penguin, Harmondsworth, 1949, p. 65.

1. J. Jacobi, *Paracelsus: Selected Writings*, Princeton University Press, Princeton, 1988, p. 50.
2. ibid., p. 68.
3. A. Cunningham and O. P. Grell, *The Four Horsemen of the Apocalypse*, Cambridge University Press, Cambridge, 2000, p. 274.
4. A. Weeks, *Paracelsus: Speculative Theory and the Crisis of the Early Reformation*, State University of New York Press, Albany, 1997, p. 53.
5. K. Park, 'Medicine in the Renaissance', in I. Loudon (ed.), *Western Medicine: An Illustrated History*, Oxford University Press, Oxford, 1997, p. 66.
6. A. Paré, *A Treatise of the Plague* (1630), trans. T. Johnson, London, p. 68.
7. ibid., p. 84.
8. D. J. Boorstin, *The Discoverers*, Vintage, New York, 1985, p. 346.
9. L. Thorndike, *A History of Magic and Experimental Science*, vol. I, Columbia University Press, New York, 1934, p. 165.
10. Pliny, *Natural History*, 29.9, quoted in W. Eamon, *Science and the Secrets of Nature*, Princeton University Press, Princeton, 1994, p. 22.
11. Jacobi, p. 52.
12. H. M. Pachter, *Paracelsus: Magic Into Science*, Henry Schuman, New York, 1951, p. 40.
13. W. Pagel, *Paracelsus: An Introduction to Philosophical Medicine in the Era of the Renaissance*, 2nd edn, Karger, Basle, 1982, p. 15.
14. G. Chaucer, *The Canterbury Tales* (*c.*1372–87), trans. N. Coghill, Penguin, Harmondsworth, 1951, p. 31.
15. Pachter, p. 57.
16. J. Aubrey, MS 10, f.113^{v}, Bodleian, quoted in K. Thomas, *Religion and the Decline of Magic*, Penguin, Harmondsworth, 1991, p. 17.
17. C. Rawcliffe, *Medicine and Society in Later Medieval England*, Sandpiper, London, 1999, pp. 125–6.
18. F. Guicciardini, *History of Italy*, ed. J. R. Hale, Washington Square Press, New York, 1964, p. 124.
19. T. More, *Utopia* (1516), trans. P. Turner, Penguin, Harmondsworth, 1965, pp. 57–8.
20. J. R. Hale, *War and Society in Renaissance Europe 1450–1620*, Fontana, London, 1985, p. 29.
21. J. R. Hale, *Renaissance Europe*, University of California Press, Berkeley, 1977, p. 95.
22. Hale, *War and Society*, p. 116.
23. Pagel, p. 44.
24. Cunningham and Grell, p. 138.
25. ibid., p. 140.
26. Hale, *War and Society*, p. 89.
27. A. Chastel, *The Sack of Rome, 1527*, trans. B. Archer, Princeton University Press, Princeton, 1983, p. 220.
28. Park, p. 76.
29. Thorndike, *History of Magic*, vol. V, 1941, p. 460.
30. Boorstin, p. 360.
31. F. Platter, *Beloved Son Felix: The Journal of Felix Platter, a Medical Student in Montpellier in the Sixteenth Century 1552–7*, trans. S. Jennett, Muller, London, 1961, p. 47.
32. Boorstin, p. 355.
33. C. Singer, *A Short History of Medicine*, Oxford University Press, Oxford, 1928, pp. 89–90.

34. Boorstin, p. 358.
35. I. Galdston, 'The psychiatry of Paracelsus', *Bulletin of the History of Medicine* **24** (1950), p. 207.
36. J. Hale, *The Civilization of Europe in the Renaissance*, Fontana, London, 1994, p. 557.
37. J. R. Mulryne and M. Shewring (eds.), *War, Literature and the Arts in Sixteenth-Century Europe*, Palgrave Macmillan, London, 1989, p. 113.
38. Hale, *Civilization of Europe*, p. 558.
39. D. T. Atkinson, *Magic, Myth and Medicine*, World Publishing Co., Cleveland and New York, 1956, p. 147.

Chapter 5: Intellectual Vagabonds

Epigraph: D. Englander, D. Norman, R. O'Day and W. R. Owens (eds.), *Culture and Belief in Europe 1450–1600: An Anthology of Sources*, Blackwell, Oxford, 1990, pp. 281–2.

1. Paracelsus, *Seven Defensiones*, in H. E. Sigerist, *Paracelsus: Four Treatises*, Johns Hopkins Press, Baltimore, 1941, p. 26.
2. ibid., pp. 24–5.
3. ibid., p. 25.
4. ibid., p. 29.
5. ibid., p. 27.
6. J. Jacobi (ed.), *Paracelsus: Selected Writings*, Princeton University Press, Princeton, 1988, p. 58.
7. Sigerist, p. 26.
8. ibid., p. 28.
9. ibid., p. 28.
10. Jacobi, p. 4.
11. Paracelsus, *Archidoxa*, I, iii (1526), in N. Goodrick-Clarke, *Paracelsus: Essential Readings*, Atlantic Books, Berkeley, 1999, p. 69.
12. J. Neusner, E. S. Frerichs and P. V. McFlesher, (eds.), *Religion, Science, and Magic: In Concert and In Conflict*, Oxford University Press, Oxford, 1989, p. 261.
13. B. Malinowski, 'Magic, science and religion', in J. Needham (ed.), *Science, Religion and Reality*. Macmillan, New York, 1925. See K. Thomas, *Religion and the Decline of Magic*, Penguin, Harmondsworth, 1991, p. 775.
14. L. Thorndike, *A History of Magic and Experimental Science*, vol. V, Columbia University Press, New York, 1941, p. 126.
15. H. C. Agrippa, *Three Books of Occult Philosophy*, ed. D. Tyson, Llewellyn Publications, St Paul, Minnesota, 1994, p. liii.
16. H. C. Agrippa, *De occulta philosophia*, I, 2, in *Opera*, vol. 1, Leiden, no date, p. 2. Quoted in C. Webster, *From Paracelsus to Newton*, Cambridge University Press, Cambridge, 1982, p. 59.
17. Hugh of St Victor, *Didascalicon* (*c.*1120), in *The Didascalicon of Hugh of St Victor*, trans. J. Taylor, Columbia University Press, New York, 1961, p. 154.
18. H. M. Pachter, *Paracelsus: Magic Into Science*, Henry Schuman, New York, 1951, p. 78.
19. F. A. Yates, *Ideas and Ideals in the North European Renaissance*, Routledge and Kegan Paul, London, 1984, p. 262.
20. H. C. Agrippa, *Of the Vanitie and Uncertaintie of Artes and Sciences* (1530), ed. C. M. Dunn, California State University, Northridge, California, 1974.
21. Agrippa, p. 15.
22. L. Thorndike, p. 130.
23. C. G. Nauert Jr, *Agrippa and the Crisis of Renaissance Thought*, University of Illinois Press, Urbana, 1965, p. 210.
24. Nauert, pp. 209–10.
25. H. C. Agrippa, *Three Books of Occult Philosophy*, trans. J. French, Moule, London, 1651, Introduction.
26. G. Mora (ed.), *Witches, Devils, and Doctors in the Renaissance*, Medieval Renaissance Texts and Studies Vol. 73, State University of New York at Binghamton, 1991, p. 113.
27. Agrippa, *Of the Vanitie and Uncertaintie of Arts and Sciences*, p. xxiii.

28. Tycho Brahe, *Astronomiae instauratae progymnasmata: Opera*, vol. III, Copenhagen, 1916, p. 116. Quoted in Thorndike, p. 138.
29. C. Gesner, *Epistolae Medicinales*, 16 January (1561). Quoted in P. M. Palmer and R. P. More, *Sources of the Faust Tradition from Simon Magus to Lessing*, Oxford University Press, New York, 1936, p. 100.
30. G. Ryga, *Two Plays: Paracelsus and Prometheus Bound*, Turnstone Press, Winnipeg, 1982, p. 15.
31. S. Palmer, *The General History of Printing*, London, pp. 31–2, 87–8, (1732), in A. Johns, *The Nature of the Book*, University of Chicago Press, Chicago, 1998, p. 351.
32. Palmer and More, p. 123.
33. Trithemius, letter to Johannes Yirdung, 1507, in Palmer and More, p. 86.
34. E. M. Butler, *The Myth of the Magus*, Cambridge University Press, Cambridge, 1948, p. 124.
35. Jacobi, p. lxxi.
36. B. G. Kohl and H. C. E. Midelfort, *On Witchcraft*, trans. J. Shea, Pegasus Press, Asheville, NC, 1998, p. 52.
37. Acts 8:9.
38. Palmer and More, p. 91.
39. Jacobi, p. 4.
40. Paracelsus, in *Astronomica et astrologica*, Arnoldi Byrchmans Erben, Cologne, 1567. Quoted in B. de Telepnef, *Paracelsus: A Genius amidst a Troubled World*, Banton Press, Largs, 1991, p. 32.
41. P. Ackroyd, *London: The Biography*, Chatto and Windus, London, 2000, p. 107.
42. Telepnef, p. 35.
43. ibid., p. 40.
44. ibid., p. 45. For Paracelsus' accounts of the Nile, see Paracelsus, *Sämtliche Werke*, ed. K. Sudhoff, Barth, Munich, 1922–33: vol. VIII, p. 298 and vol. VII, p. 129.
45. R. Mackenney, *Sixteenth Century Europe: Expansion and Conflict*, Macmillan, London, 1993, p. 249.

Chapter 6: A New Religion

Epigraph: F. Rabelais, *The Histories of Gargantua and Pantagruel (c.1534)*, trans. J. M. Cohen, Penguin, Harmondsworth, 1955, p. 556.

1. A. Weeks, *Paracelsus: Speculative Theory and the Crisis of the Early Reformation*, State University of New York Press, Albany, 1997, p. 148.
2. ibid, p. 128.
3. H. M. Pachter, *Paracelsus: Magic Into Science*, Henry Schuman, New York, 1951.
4. W. Manchester, *A World Lit Only by Fire*, Macmillan, London, 1993, p. 112.
5. V. H. H. Green, *Renaissance and Reformation*, Edward Arnold, London, 1952, p. 34.
6. D. Maland, *Europe in the Sixteenth Century*, 2nd edn, Macmillan, Basingstoke, 1982, p. 46.
7. A. Johns, *The Nature of the Book*, University of Chicago Press, Chicago, 1998, p. 369.
8. Manchester, p. 38.
9. ibid.
10. R. Mackenney, *Sixteenth Century Europe: Expansion and Conflict*, Macmillan, London, 1993, p. 141.
11. ibid, p. 141.
12. Manchester, pp. 128–9.
13. 'Humanism', in the *Catholic Encyclopedia*, Encyclopedia Press, 1913. See http://www.newadvent.org/cathen/07538b.htm
14. Maland, p. 81.
15. ibid, p. 69.
16. Manchester, p. 125.
17. K. Randell, *Luther and the German Reformation 1517–55*, 2nd edn, Hodder and Stoughton, London, 2000, p. 16.
18. R. H. Bainton, *Here I Stand: A Life of Martin Luther*, Penguin, London, 2002, p. 45.

19. Maland, p. 85.
20. Manchester, p. 139.
21. Bainton, p. 63.
22. ibid., p. 65.
23. Randell, p. 35.
24. Bainton, p. 185.
25. Maland, p. 249.
26. Stoddart, p. 265.
27. ibid., p. 267.
28. Pachter, p. 106.
29. A. M. Stoddart, *The Life of Paracelsus*, John Murray, London, 1911, p. 265.
30. Pachter, p. 156.
31. Stoddart, pp. 266–7.
32. Pachter, p. 267.
33. Weeks, p. 82.
34. Mark 14:22–24.
35. Mackenney, p. 156.
36. Maland, p. 250.
37. Manchester, p. 182.
38. Randell, p. 16.
39. ibid., p.16.

Chapter 7: Revolution under the Sign of the Shoe

Epigraph: M. Luther, *Against the Robbing and Murdering Hordes of Peasants* (1525), in D. Englander, D. Norman, R. O'Day and W. R. Owens (eds.), *Culture and Belief in Europe 1450–1600: An Anthology of Sources*, Blackwell, Oxford, 1990, p. 191.

1. H. M. Pachter, *Paracelsus: Magic Into Science*, Henry Schuman, New York, 1951, p. 106.
2. N. Goodrick-Clarke, *Paracelsus: Essential Readings*, North Atlantic, Berkeley, 1999, pp. 156–7.
3. ibid., p. 160.
4. ibid., p. 164.
5. ibid., p. 165.
6. ibid., p. 165.
7. ibid., p. 166.
8. ibid., pp. 167–8.
9. ibid., p. 169.
10. A. G. Dickens, *Reformation and Society*, Thames & Hudson, London, 1966, p. 135.
11. Pachter, p. 102.
12. D. Maland, *Europe in the Sixteenth Century*, 2nd ed, Macmillan, Basingstoke, 1982, p. 237.
13. Englander *et al.* (eds), *Culture and Belief*, pp. 192–3.
14. A. Weeks, *Paracelsus: Speculative Theory and the Crisis of the Early Reformation*, State University of New York Press, Albany, 1997, p. 95.
15. ibid., pp. 95–6.

Chapter 8: Transmutation at Ingolstadt

Epigraph: C. G. Jung, 'Paracelsus as a spiritual phenomenon' (1942), in *Alchemical Studies*, trans. R. F. C. Hull, Princeton University Press, Princeton, 1983, p. 159.

1. A. M. Stoddart, *The Life of Paracelsus*, John Murray, London, 1911, pp. 252–3.
2. ibid., p. 254.
3. J. Harris, *Lexicon Technicum: or, an Universal English Dictionary* (1704), in K. Thomas, *Religion and the Decline of Magic*, Penguin, Harmondsworth, 1991, p. 772.
4. J. Read, *From Alchemy to Chemistry*, Dover, New York, 1995, p. 29.
5. Paracelsus, *Paragranum* (1565), ed. A. von Bodenstein, f.50^{v}, Frankfurt, in W. Pagel, *Paracelsus: An Introduction to Philosophical Medicine in the Era of the Renaissance*, 2nd edn, Karger, Basle, 1982, p. 207.

6. Stoddart, p. 216.
7. H. Boynton (ed.), *Beginnings of Modern Science*, Walter J. Black, Roslyn, NY, 1948, p. 341.
8. Paracelsus, *Elf Traktat; Eleven Treatises on the Origin, Causes, Signs and Cure of Specific Diseases* (*c.*1520).
9. A. Weeks, *Paracelsus: Speculative Theory and the Crisis of the Early Reformation*, State University of New York Press, Albany, 1997, p. 121.
10. J. Jacobi, *Paracelsus: Selected Writings*, Princeton University Press, Princeton, 1988, pp. 196–7. See Paracelsus, *De imaginibus*, in *Sämtliche Werke*, ed. K. Sudhoff, vol. XIII, Barth, Munich, 1922–33, p. 378.
11. O. Temkin, 'The elusiveness of Paracelsus', *Bulletin of the History of Medicine*, **26** (1952), p. 201.
12. Jacobi, p. 122.
13. G. Chaucer, *The Canterbury Tales* (*c.*1372–87), trans. N. Coghill, Penguin, Harmondsworth, 1951, pp. 472, 484, 492.
14. E. J. Holmyard, *Alchemy*, Dover, Mineola, NY, 1990, p. 149.
15. L. Thorndike, *A History of Magic and Experimental Science*, vol. IV, Columbia University Press, New York, 1934, p. 395.
16. B. Jonson, *The Alchemist* (1610), in M. Jamieson (ed.), *Three Comedies*, Penguin, Harmondsworth, 1966, p. 228.
17. Jacobi, p. 84.
18. Weeks, p. 71.
19. Plotinus, *Enneads*, V, 3, 14, trans. S. McKenna, in B. Russell, *A History of Western Philosophy*, Unwin, London, 1984, p. 294.
20. G. Pico della Mirandola, *On the Dignity of Man* (1496), in D. Englander, D. Norman, R. O'Day and W. R. Owens, *Culture and Belief in Europe 1450–1600: An Anthology of Sources*, Basil Blackwell, Oxford, 1990, p. 25.
21. ibid., p. 26.
22. G. Dorn, *Physica trismegisti*, in *Theatrum chemicum*, Argentorati, Strasbourg, p. 371 (1659), in R. Patai, *The Jewish Alchemists*, Princeton University Press Princeton, 1994, p. 158.
23. *The Table Talk of Martin Luther*, trans. W. Hazlitt, George Bell and Sons, London, 1883, p. 326.
24. H. M. Leicester, *The Historical Background of Chemistry*, Dover, New York, 1971, pp. 44–5.
25. T. Burckhardt, *Alchemy*, trans. W. Stoddart, Penguin, Baltimore, 1971, p. 196.
26. Read, p. 28.
27. R. P. Multhauf, *The Origins of Chemistry*, Gordon and Breach, Langhorne, Penn., 1993, p. 183.
28. R. Bacon, *Opus tertium*, in *Opera quaedam hactenus inedita*, ed. J. S. Brewer, vol. I, Longman, Green, Longman and Roberts, London, 1859, pp. 39–40.
29. Thorndike, vol. IV, p. 501.
30. A. G. Debus, *Man and Nature in the Renaissance*, Cambridge University Press, Cambridge, 1978, pp. 17–18.
31. William of Auvergne, *De legibus*, 1.24, p. 69 (13th century), in *Opera omnia*, vol. I, Paris, 1674.
32. St Augustine, *Confessions*, 10.30, 10.35, in *The Confessions of St Augustine*, trans. R. Warner, Mentor Books, New York, 1963.
33. C. A. Brown, *Scientific Monthly*, x (1920), p. 202. See P. M. Dawson, 'The heritage of Paracelsus', *Annals of Medical History* **10** (1928), p. 258.
34. Cennini Cennino, *Il libro dell'Arte* (*c.*1390), trans. D. V. Thompson Jr., Dover, New York, 1954, pp. 101–2.
35. H. M. Pachter, *Paracelsus: Magic Into Science*, Henry Schuman, New York, 1951, p. 114.
36. ibid., p. 116.
37. A. E. Waite (ed.), *Hermetic and Alchemical Writings of Paracelsus the Great*, Part I, Alchemical Press, Edmonds, Washington, 1992, p. 157.
38. ibid., Part II, p. 41.

Chapter 9: Elixir and Quintessence

Epigraph: D. J. Boorstin, *The Discoverers*, Vintage, New York, 1985, p. 339.

1. A. E. Waite (ed.), *Hermetic and Alchemical Writings of Paracelsus the Great*, part II, Alchemical Press, Edmonds, Washington, 1992, p. 3.
2. ibid., p. 5.
3. ibid., p. 4.
4. ibid., p. 3.
5. H. M. Pachter, *Paracelsus: Magic Into Science*, Henry Schuman, New York, 1951, p. 119.
6. Waite, *Hermetic and Alchemical Writings*, p. 3.
7. ibid., p. 5.
8. ibid., p. 82.
9. Heraclius, *De coloribus et artibus Romanorum* (10th century AD), ed. and trans. M. P. Merrifield, in *Original Treatises on the Arts of Painting* (1849), vol I Dover, New York, 1967, p. 182.
10. Waite, p. 81.
11. ibid., p. 32.
12. C. Rawcliffe, *Medicine and Society in Later Medieval England*, (translated to modern English), Sandpiper, London, 1999, p. 148.
13. H. C. Agrippa, *Of the Vanitie and Uncertaintie of Artes and Sciences* (1530), ed. C. M. Dunn, California State University, Northridge, Calif., 1974, p. 314.
14. F. Rabelais, *The Histories of Gargantua and Pantagruel* (*c.*1534), trans. J. M. Cohen, Penguin, Harmondsworth, 1955, pp. 92–3.
15. G. Chaucer, *The Canterbury Tales* (*c.*1372–87), trans. N. Coghill, Penguin, Harmondsworth, 1951, p. 30.
16. G. C. Macaulay (ed.), *The Complete Works of John Gower*, vol. I, Clarendon Press, Oxford, 1899, pp. 283–4.
17. H. R. Lemay, 'Anthonius Guainerius and medieval gynecology', in J. Kirshner and S. F. Wemple (eds.), *Women of the Medieval World*, Blackwell, Oxford, 1988, p. 324.
18. Waite, p. 22.
19. ibid., p. 24.
20. ibid., p. 25.
21. W. Pagel, *Paracelsus: An Introduction to Philosophical Medicine in the Era of the Renaissance*, 2nd edn, Karger, Basle, 1982, p. 252.
22. L. Thorndike, *History of Magic and Experimental Science*, vol. III, Columbia University Press, New York, 1934, p. 363.
23. J. R. Partington, *A History of Chemistry*, vol. 1, Macmillan, London, 1970, p. 108.
24. Thorndike, p. 360.
25. Paracelsus, *The Great Surgery*, Book II, Ch. 13. In Waite, p. 22, footnote.
26. Waite, p. 37.
27. Paracelsus, *Paragranum*, Tract II, 'De astronomia', in Waite, p. 38, footnote.
28. Waite, p. 48.
29. ibid., p. 48.
30. ibid., p. 69.
31. ibid., p. 71.
32. J. C. Cooper, *Chinese Alchemy*, Sterling, New York, 1990, p. 66.
33. Pachter, p. 5.
34. Cooper, p. 57.
35. J. Jacobi, *Paracelsus: Selected Writings*, Princeton University Press, Princeton, 1988, p. 93.
36. Paracelsus, *On the Diseases That Deprive Man of His Reason* (*c.*1525), in R. P. Multhauf, *The Origins of Chemistry*, Gordon and Breach, Langhorne, Pennsylvania, 1993, p. 220, n. 60.
37. Pachter, p. 137.
38. Jacques Bongars, *J. Bongarsii epistolae ad Joach. Camerarium medicum*, Leiden (1647), in H. Trevor-Roper, *Renaissance Essays*, Secker and Warburg, London, 1985, p. 437.
39. Pachter, p. 133.

Chapter 10: Bitter Medicine

Epigraph: W. Caxton, *The Game and Playe of the Chesse* (1474), ed. W. E. A. Axon, London, 1883, p. 120.

1. J. Huizinga, *Erasmus of Rotterdam*, Phaidon, London, 1952, p. 252.
2. H. M. Pachter, *Magic Into Science*, Henry Schuman, New York, 1951, p. 146.
3. K. Randell, *Luther and the German Reformation 1517–55*, Hodder and Stoughton, London, 2000, p. 67.
4. A. G. Dickens, *Reformation and Society in Sixteenth-Century Europe*, Thames and Hudson, London, 1966, p. 113.
5. Randell, p. 70.
6. ibid.
7. Pachter, pp. 152–3.
8. A. M. Stoddart, *The Life of Paracelsus*, John Murray, London, 1911, p. 90.
9. Paracelsus, *Paragranum*, Adam von Bodenstein, Frankfurt (1565). See Pachter, pp. 158–9.
10. H. E. Sigerist, *Paracelsus: Four Treatises*, Johns Hopkins Press, Baltimore, 1941, p. 35.
11. Pachter, p. 157.
12. Stoddart, p. 101.
13. W. Pagel and P. Rattansi, 'Vesalius and Paracelsus', *Medical History*, **8** (1964), p. 309.
14. Pachter, pp. 154–6.
15. ibid., p. 156.
16. Pachter, p. 297.
17. R. B[ostocke], *The difference between the auncient Phisicke . . . and the later Phisicke*, Liii (1585), London. Quoted in C. D. Gunnoe Jr, *Thomas Erastus in Heidelberg: A Renaissance Physician during the Second Reformation, 1558–1580*, Ph.D. dissertation, University of Virginia, 1998, p. 276.
18. Pachter, p. 160.
19. Stoddart, p. 106.
20. J. Jacobi, *Paracelsus: Selected Writings*, Princeton University Press, Princeton, 1988, pp. 52–3.
21. Stoddart, p. 110.
22. Jacobi, p. 64.
23. ibid.
24. ibid., p. 50.
25. ibid., p. 26.
26. ibid., p. 57.

Chapter 11: The Battle of Basle

Epigraph: Miguel Cervantes, *Don Quixote* (1605), trans. J. M. Cohen, Penguin, Harmondsworth, 1950, p. 44.

1. C. G. Jung, 'Paracelsus as a spiritual phenomenon' (1942), in *Alchemical Studies*, trans. R. F. C. Hull, Princeton University Press, Princeton, 1983, p. 120.
2. A. E. Waite, *Alchemists through the Ages*, Banton Press, Largs, 1990, p. 139.
3. J. Jacobi, *Paracelsus: Selected Writings*, Princeton University Press, Princeton, 1988, p. 143.
4. H. M. Pachter, *Magic Into Science*, Henry Schuman, New York, 1951, p. 210.
5. Jacobi, p. 144.
6. Paracelsus, *Paragranum* (1529–30), I, viii, 62–5, in N. Goodrick-Clarke, *Paracelsus: Essential Readings*, North Atlantic, Berkeley, 1999, p. 74.
7. Jacobi, p. 91.
8. A. M. Stoddart, *The Life of Paracelsus*, John Murray, London, 1911, p. 180.
9. ibid., p. 130.
10. Pachter, pp. 162–3.
11. ibid., p. 164.
12. Goodrick-Clarke, p. 72.
13. Stoddart, p. 135.

14. ibid., pp. 134–5.
15. ibid., p. 135.
16. Pachter, p. 168.
17. J. Hargrave, *The Life and Soul of Paracelsus*, Victor Gollancz, London, 1951, pp. 130-31.
18. Stoddart, p. 137.
19. ibid., p. 152.

Chapter 12: Against the Grain

Epigraph: C. Quétel, *History of Syphilis*, Polity Press, Cambridge, 1990, p. 17.

1. Paracelsus, *Paragranum* (1529–30), I, viii, 53–4, in N. Goodrick-Clarke, *Paracelsus: Essential Writings*, North Atlantic, Berkeley, 1999, pp. 71–2.
2. ibid., p. 72.
3. W. Pagel, *Paracelsus: An Introduction to Philosophical Medicine in the Era of the Renaissance*, 2nd edn, Karger, Basle, 1982, pp. 110–11.
4. Goodrick-Clarke, pp. 73–4.
5. ibid., p. 74.
6. A. M. Stoddart, *The Life of Paracelsus*, John Murray, London, 1911, p. 143.
7. ibid., p. 150.
8. L. Thorndike, *A History of Magic and Experimental Science*, vol. V, Columbia University Press, New York, 1941, p. 435.
9. In J. Telle, 'Wissenschaft und Öffentlichkeit im Spiegel der deutschen Arzneibuchliteratur', *Medizinhistorisches Journal*, **14** (1979), p. 40. Quoted in W. Eamon, *Science and the Secrets of Nature*, Princeton University Press, Princeton, 1994, pp. 102-3.
10. Thorndike, vol. V, p. 432.
11. ibid., p. 435.
12. P. H. Smith, *The Business of Alchemy*, Princeton University Press, Princeton, 1994, p. 230.
13. H. M. Pachter, *Magic Into Science*, Henry Schuman, New York, 1951, p. 175.
14. ibid.
15. S. Franck, *Paradoxon* LXV (1534), in *Paradoxa*, ed. H. Ziegler, Jena, 1909, p. 93. See Pagel, p. 42, n. 122.
16. A. G. Dickens, *Reformation and Society in Sixteenth-Century Europe*, Thames and Hudson, London, 1966, p. 147.
17. B. McGinn (ed.), *Apocalyptic Spirituality: Treatises and Letters of Lactantius, Adso of Montier-En-Der, Joachim of Fiore, the Franciscan Spirituals, Savonarola*, Paulist Press, New York, 1979, p. 198.
18. Quétel, p. 10.
19. U. von Hutten, *Of the Wood Called Guaiacum, that Healeth the French Pockes, and also Helpeth the Goute in the Feete, the Stone, Palse, Lepre, Dropsy, Fallynge Evyll, and Other Diseases*, trans. T. Paynel, f.2^{r}. Berthelet, London (1540).
20. In J. R. Hale, *Renaissance Europe*, University of California Press, Berkeley, 1977, p. 24.
21. G. Fracastoro, *Syphilis: or, A Poetical History of the French Disease*, trans. N. Tate, book 3, J. Tonson, London, 1686, lines 327–30.
22. J. Astruc, *A Treatise of the Venereal Disease, in Six Books*, trans. W. Barrowby, vol. II, London, 1737, p. 229.
23. Hutten, 2^{v}–3^{r}.
24. ibid., 4^{r}–4^{v}.
25. R. Mackenney, *Sixteenth Century Europe: Expansion and Conflict*, Macmillan, London, 1993, p. 148.
26. Hutten f. 5v.
27. R. P. Multhauf, *The Origins of Chemistry*, Gordon and Breach, Langhorne, Pennsylvania, 1993, pp. 217–18, n. 52.
28. J. Jacobi, *Paracelsus: Selected Writings*, Princeton University Press, Princeton, 1988, p. 95.

29. A. Cunningham and O. P. Grell, *The Four Horsemen of the Apocalypse*, Cambridge University Press, Cambridge, 2000, p. 348, n. 117.
30. Pachter, p. 181.
31. Stoddart, p. 168.
32. ibid., p. 169.

Chapter 13: The Alchemist Inside

Epigraph: J. J. Berzelius, *A View of the Progress and Present State of Animal Chemistry*, trans. G. Brunnmark, London (1818), in M. Teich (ed.), *A Documentary History of Biochemistry 1770–1940*, Leicester University Press, Leicester, 1992, p. 445.

1. Paracelsus, *Spital-Buch* (1529), in A. Weeks, *Paracelsus: Speculative Theory and the Crisis of the Early Reformation*, State University of New York Press, Albany, 1997, p. 139.
2. Galatians 5:22.
3. Corinthians I, 13:7.
4. Corinthians I, 13:4–5.
5. H. M. Pachter, *Magic Into Science*, Henry Schuman, New York, 1951, p. 167.
6. A. M. Stoddart, *The Life of Paracelsus*, John Murray, London, 1911, p. 144.
7. Pachter, p. 191; Jacobi, p. 68.
8. Pachter, p. 191.
9. ibid., p. 192.
10. J. d'Indagine, *Briefe Introductions . . . unto the Art of Chiromancy, or Manuel Divination, and Phisiognomy*, trans. F. Withers, sig. Jv (1575), in K. Thomas, *Religion and the Decline of Magic*, Penguin, Harmondsworth, 1991, pp. 375–6.
11. Paracelsus, *Astronomia magna* (1537–8), I, xii, 3, in N. Goodrick-Clarke, *Paracelsus: Essential Readings*, North Atlantic, Berkeley, 1999, p. 109.
12. R. Olson, *Science Deified and Science Defied*, vol. 2, University of California Press, Berkeley, 1990, p. 32.
13. J. Jacobi (ed.), *Paracelsus: Selected Writings*, Princeton University Press, Princeton, 1988, p. 143.
14. Paracelsus, *Volumen medicinae paramirum* (*c.*1520), I, i, 189, in Goodrick-Clarke, p. 50.
15. ibid.
16. ibid.
17. Pachter, p. 136.
18. Paracelsus, *Opus paramirum* (1530–31), I, ix, 71–4, in Goodrick-Clarke, pp. 80–82.
19. Stoddart, p. 192.
20. Goodrick-Clarke, p. 50.
21. Paracelsus, *De Natura Rerum*, I, xi, 400 (1537), in Goodrick-Clarke, p. 191.
22. ibid., I, xi, 384, in Goodrick-Clarke, p. 188.
23. ibid., I, xi, 400, in Goodrick-Clarke, p. 191.
24. Stoddart, p. 192.
25. ibid., pp. 208–9.
26. Paracelsus, *Volumen medicinae paramirum* (*c.*1520) I, i, 207, in Goodrick-Clarke, p. 53.
27. ibid.
28. ibid., p. 52.
29. ibid., p. 53.
30. ibid.
31. ibid., p. 54.
32. ibid., pp. 54–5.
33. ibid., p. 55.
34. M. Luther, 'Whether one may Flee from a Deadly Plague' (1527), in G. K. Wiencke (ed.), *Luther's Works. Volume 43: Devotional Writings II*, Fortress Press, Philadelphia, 1968, p. 127.
35. C. Rawcliffe, *Medicine and Society in Later Medieval England*, Sandpiper, London, 1999, p. 1.
36. Luther, in Wiencke, p. 132.
37. A. Cunningham and O. P. Grell, *The Four Horsemen of the Apocalypse*, Cambridge University Press, Cambridge, 2000, p. 286.

38. Goodrick-Clarke, p. 56.
39. ibid.
40. Paracelsus, *De morbis ex incantationibus et impressionibus*, in C. Webster, 'Paracelsus confronts the saints: miracles, healing and the secularization of magic', *Social History of Medicine*, **8** (1995), p. 403.

Chapter 14: Beyond Wonders

Epigraph: Pablo Neruda, 'Poetry', in *Isla Negra: A notebook*, trans. A. Reid, Farrar, Straus & Giroux, New York, 1982.

1. H. M. Pachter, *Magic Into Science*, Henry Schuman, New York, 1951, pp. 196–7.
2. ibid., p. 197.
3. ibid., p. 199.
4. ibid.
5. B. de Telepnef, *Paracelsus: A Genius amidst a Troubled World*, Banton Press, Largs, 1991.
6. Pachter, p. 206.
7. N. Goodrick-Clarke, *Paracelsus: Essential Readings*, North Atlantic, Berkeley, 1991, p. 83.
8. ibid.
9. ibid., pp. 76, 78.
10. J. Jacobi, *Paracelsus: Selected Writings*, Princeton University Press, Princeton, 1988, p. 13.
11. A. Weeks, *Paracelsus: Speculative Theory and the Crisis of the Early Reformation*, State University of New York Press, Albany, 1997, p. 65.
12. W. Pagel, *Paracelsus: An Introduction to Philosophical Medicine in the Era of the Renaissance*, 2nd edn, Karger, Basle, 1982, p. 91.
13. Jacobi, p. 15.
14. ibid.
15. C. G. Jung, 'Paracelsus as a spiritual phenomenon' (1942), in *Alchemical Studies*, trans. R. F. C. Hull, Princeton University Press, Princeton, 1983, p. 120.
16. ibid., p. 121.
17. Pagel, p. 92.
18. Jacobi, p. 16.
19. ibid.
20. ibid., p. 17.
21. Pagel, p. 104, n. 271.
22. E. J. Holmyard, *Alchemy*, Dover, New York, 1990, p. 223.
23. B. Vickers, 'On the function of analogy in the occult', in I. Merkel and A. G. Debus (eds.), *Hermeticism in the Renaissance*, Associated Universities Presses, Cranbury, New Jersey, 1988, p. 283.
24. O. Temkin, 'The elusiveness of Paracelsus', *Bulletin of the History of Medicine*, **26** (1952), p. 211.
25. Goodrick-Clarke, p. 78.
26. A. M. Stoddart, *The Life of Paracelsus*, John Murray, London, 1911, p. 196.
27. Goodrick-Clarke, p. 86.
28. Stoddart, p. 115.
29. Goodrick-Clarke, p. 87.
30. Stoddart, pp. 199–200.
31. Goodrick-Clarke, pp. 82–3.
32. Stoddart, pp. 120–1.
33. ibid., p. 121.
34. Jacobi, pp. 92, 84.
35. ibid., p. 45.
36. Goodrick-Clarke, p. 81.
37. Stoddart, pp. 213–14.
38. P. Pomet, *A Compleat History of Druggs*, in C. J. Thompson, *Alchemy: Source of Chemistry and Medicine*, Sentry Press, New York, 1974, p. 169.

39. ibid., p. 170.
40. Goodrick-Clarke, p. 84.
41. ibid.
42. Paracelsus, *De natura rerum* (1537), I, xi, 344, in Goodrick-Clarke, p. 184.
43. Jung, p. 186.
44. Stoddart, p. 216.
45. Pagel, pp. 144–5.
46. J. B. van Helmont, 'De lithiasi', in *Ortus medicinae* (1648), Cap. III, in Pagel, p. 198.
47. Paracelsus, *Opus paramirum* (1530–31), I, ix, 125, in Goodrick-Clarke, p. 90.
48. Goodrick-Clarke, p. 90.
49. ibid., pp. 90–1.
50. ibid., p. 92.
51. Pagel, p. 161.
52. Pachter, p. 212.
53. ibid., p. 212.
54. Paracelsus, *De caduco matricis*, I, 8:345, in H. E. Keller, 'Seeing "Microcosma"', in G. S. Williams and C. D. Gunnoe, Jr (eds.), *Paracelsian Moments*, Truman State University Press, Kirksville, Missouri, 2002, p. 95.
55. Jacobi, p. 35.
56. Ephesians 5:22–23.
57. J. R. Hale, *Renaissance Europe*, University of California Press, Berkeley, 1977, p. 127.
58. Paracelsus, *Paramirum* I, ix, 200, in Goodrick-Clarke, p. 94.
59. H. C. Agrippa, *Female Pre-eminence, or the Dignity and Excellency of that Sex Above the Male* (1529), trans. H. Cane. London, 1670, in M. Baigent and R. Leigh, *The Elixir and the Stone*, Viking, Harmondsworth, 1997, p. 149.
60. Paracelsus, *Opus Paramirum*, I, ix, 192–3, in Goodrick-Clarke, p. 93.
61. ibid., pp. 196–8, in Goodrick-Clarke, p. 94.
62. Jacobi, p. 27.
63. ibid., p. 27.
64. ibid., p. 30.
65. Paracelsus, *Das Buch von der Gebärung der empfindlichen Dinge in der Vernunft*, 3.1 (*c.*1520), in Goodrick-Clarke, p. 58.
66. ibid., pp. 58–9.
67. Jacobi, p. 36–7.
68. Pagel, p. 88.
69. Pachter, p. 216.
70. ibid., p. 217.
71. Stoddart, p. 221.
72. Paracelsus, *Opus paramirum*, I, ix, 219–20, in Goodrick-Clarke, p. 95.

Chapter 15: Star and Ascendant

Epigraph: W. Gilbert, *De Magnete*, London (1600), trans. P. Fleury Mottelay, New York, 1893, in M. Boas Hall (ed.), *Nature and Nature's Laws*, Macmillan, London, 1970, p. 49.

1. E. Vansteenberghe, *Le cardinal Nicolas de Cues*, Paris, 1920, p. 247. See L. Thorndike, *History of Magic and Experimental Science*, vol. IV, Columbia University Press, New York, 1934, p. 387.
2. N. Culpepper, *Pharmacopoeia Londinensis: or the London Dispensatory*, sigs. A3^v–A4 (1654), in K. Thomas, *Religion and the Decline of Magic*, Penguin, Harmondsworth, 1991, p. 394.
3. J. Swift, *The Starr-Prophet anatomiz'd and dissected*, p. 29 (1675). Quoted in Thomas, p. 398.
4. *Poor Robin* (1664), sig. A6. Quoted in Thomas, p. 398.
5. G. Chaucer, *The Canterbury Tales* (*c.*1372–87), trans. N. Coghill, Penguin, Harmondsworth, 1951, p. 30.
6. J. Gadbury, *The Doctrine of Nativities*, ii, London, 1658, p. 235.

7. J. Chamber, *A Treatise against Iudicial Astrologie*, London, p. 102 (1601), in Thomas, p. 426.
8. *The Notebooks of Leonardo da Vinci*, ed. and trans. E. MacCurdy, Reynal and Hitchcock, New York, 1938, in H. E. Boynton (ed.), *The Beginnings of Modern Science*, Walter J. Black, Roslyn, NY, 1948, p. 343.
9. Paracelsus, *Volumen medicinae paramirum*, I, i, 184 (*c*.1520), in N. Goodrick-Clarke, *Paracelsus: Essential Readings*, North Atlantic, Berkeley, 1999, p. 48.
10. ibid.
11. ibid.
12. ibid., p. 47.
13. ibid., pp. 48–9.
14. ibid., p. 121.
15. ibid.
16. ibid.
17. ibid.
18. ibid., p. 134.
19. Paracelsus, *Astronomia magna* I, xii, 23 (1537–8), in Goodrick-Clarke, p. 114.
20. Paracelsus, *Volumen medicinae paramirum*, I, i, 180 (*c*.1520), in Goodrick-Clarke, p. 46.
21. Stoddart, p. 186.
22. Paracelsus, *Opus paramirum*, I, ix, 114–16 (1530–31), in Goodrick-Clarke, pp. 88–9.
23. A. G. Debus, *Man and Nature in the Renaissance*, Cambridge University Press, Cambridge, 1978, p. 98.
24. ibid., p. 98.
25. Nicolaus Copernicus, *On the Revolution of Heavenly Spheres*, in S. Hawking (ed.), *On the Shoulders of Giants*, Running Press, Philadelphia, 2002, p. 25.
26. ibid., p. 32.
27. ibid., p. 25.

Chapter 16: Demons of the Mind

Epigraph: W. Shakespeare, *A Midsummer Night's Dream*, act V, scene I (*c*.1594).

1. B. de Telepnef, *Paracelsus: A Genius amidst a Troubled World*, Banton Press, Largs, 1991, p. 66.
2. *Paracelsus. Sämtliche Werke: Theologische und religionsphilosophische Schriften*, Abt. II, ed. W. Matthiessen, Munich, 1923, p. 98, in H. M. Pachter, *Magic Into Science*, Henry Schuman, New York, 1951, p. 254.
3. J. Strebel (ed.), *Paracelsus' Sämtliche Werke in zeitgemässer kurzer Auswahl*, vol. I, St Gallen, 1944, p. 73, in Pachter, p. 253.
4. A. M. Stoddart, *The Life of Paracelsus*, John Murray, London, 1911, p. 223.
5. Pachter, p. 257.
6. J. R. Hale, *Renaissance Europe*, University of California Press, Berkeley, 1977, p. 37.
7. J. Hale, *The Civilization of Europe in the Renaissance*, Fontana , London, 1994, p. 146.
8. Telepnef, p. 48.
9. Paracelsus in Matthiessen (ed.), p. 168.
10. Strebel (ed.), vol. II, p. 128.
11. F. Strunz, *Paracelsus, Idee und Problem seiner Weltanschauung*, Leipzig, 1937, p. 173.
12. J. Jacobi (ed.) *Paracelsus: Selected Writings*, Princeton University Press, Princeton, 1988, p. 115.
13. Paracelsus, *Liber prologi in vitam beatam*, II, i, 76, 83 (1533), in N. Goodrick-Clarke, *Paracelsus: Essential Readings*, North Atlantic, Berkeley, 1999, pp. 149–51.
14. ibid., II, i, 71–2, in Goodrick-Clarke, p. 148.
15. J. Sprenger [attributed] and H. Kramer, *Malleus maleficarum* (1486), ed. P. Hughes, trans. M. Summers, Folio Society, London, 1968, p. 100.
16. S. B. Meeche (ed.), *The Book of Margery Kempe*, Early English Text Society, Oxford, 1940, p. 105.
17. Georgius Agricola, *De re metallica* (1556), trans. H. C. Hoover and L. H. Hoover, Dover, New York, 1950, p. 217.

18. Georgius Agricola, *De animantibus subterraneis*, Basle (1549), in Hoover and Hoover, p. 217.
19. L. Thorndike, *A History of Magic and Experimental Science*, vol. V, Columbia University Press, New York, 1934, p. 88.
20. Pachter, p. 73.
21. H. E. Sigerist, *Paracelsus: Four Treatises*, Johns Hopkins Press, Baltimore, 1941, p. 229.
22. ibid., p. 230.
23. ibid., p. 229.
24. ibid., pp. 244–5.
25. ibid., p. 238.
26. ibid., p. 239.
27. Sprenger and Kramer, p. 80.
28. Pachter, p. 74.
29. C. G. Jung, 'Paracelsus as a spiritual phenomenon' (1942), in *Alchemical Studies*, trans. R. F. C. Hull, Princeton University Press, Princeton, 1983, p. 161.
30. Pachter, p. 232.
31. ibid., p. 232.
32. G. Mora (ed.), *Witches, Devils, and Doctors in the Renaissance*, Medieval Renaissance Texts and Studies Vol. 73, State University of New York at Binghamton, 1991, p. 113.
33. ibid., p. 347.
34. Sigerist, p. 16.
35. R. Burton, *Anatomy of Melancholy*, i, p. 69 (1621), in K. Thomas, *Religion and the Decline of Magic*, Penguin, Harmondsworth, 1991, p. 15.
36. Sigerist, p. 142.
37. ibid., p. 144.
38. ibid., p. 147.
39. ibid., p. 167.
40. ibid., pp. 187–8.
41. M. Luther, *Werke: Tischreden*, no. 455, Weimar, 1883, in J. Weyer, *On Witchcraft*, eds. B. G. Kohl and H. C. E. Midelfort and trans. J. Shea, Pegasus Press, Asheville, NC, 1998, p. xxiii.
42. I. Galdston, 'The psychiatry of Paracelsus', *Bulletin of the History of Medicine*, 24 (1950), p. 205.
43. Sigerist, p. 153.
44. ibid., p. 177.
45. ibid., p. 156.
46. ibid., p. 157.
47. ibid., p. 19.
48. ibid., p. 182.
49. ibid., pp. 157–8.
50. ibid., p. 181.
51. Paracelsus, *De generatione stultorum* (*The Begetting of Fools*) (*c.*1530), trans. P. F. Cranefield and W. Federn, *Bulletin of the History of Medicine* **41** (1967), pp. 63 and 66.
52. K. H. Weimann, *Theophrast von Hohenheim, gen. Paracelsus: Die Kärntner Schriften*, Amt der Kärntner Landesregierung, Klagenfurt, 1955, p. 287.
53. Paracelsus, *De generatione stultorum*, p. 64.
54. ibid., p. 69.
55. ibid., p. 70.
56. ibid., p. 70.
57. Sigerist, pp. 56–7.
58. Stoddart, pp. 226–7.

Chapter 17: The Little Man

Epigraph: M. Warner, *Fantastic Metamorphoses, Other Worlds*, Oxford University Press, Oxford, 2002, p. 77.

1. H. M. Pachter, *Magic Into Science*, Henry Schuman, New York, 1951, p. 263–4.

2. W. Pagel, *Paracelsus: An Introduction to Philosophical Medicine in the Era of the Renaissance*, 2nd edn, Karger, Basle, 1982, p. 26.
3. Pachter, p. 276.
4. ibid., p. 275.
5. B. de Telepnef, *Paracelsus: A Genius amidst a Troubled World*, Banton Press, Largs, 1991, p. 71.
6. Paracelsus, *Labyrinthus medicorum errantium*, I, xi, 190 (1538), in N. Goodrick-Clarke, *Paracelsus: Essential Readings*, North Atlantic, Berkeley, 1999, p. 104.
7. ibid.
8. C. D. Gunnoe, Jr, *Thomas Erastus in Heidelberg: A Renaissance Physician during the Second Reformation, 1558–1580*, Ph.D. Dissertation, University of Virginia, 1998, p. 268.
9. H. E. Sigerist, *Paracelsus: Four Treatises*, Johns Hopkins Press, Baltimore, 1941, p. 12.
10. ibid., p. 20.
11. ibid., p. 33.
12. ibid., p. 34.
13. ibid., pp. 34–5.
14. ibid., p. 37.
15. Telepnef. p. 72.
16. Paracelsus, *Astronomia magna*, preface (1537–8), in Goodrick-Clarke, p. 109.
17. ibid., pp. 115–16.
18. ibid., p. 111.
19. ibid.
20. ibid., p. 144.
21. ibid., p. 114.
22. ibid., pp. 115 and 114.
23. Pachter, pp. 277–8.
24. Paracelsus, *De natura rerum*, I, xi, 312–13 (1537), in Goodrick-Clarke, pp. 174–5.
25. ibid., p. 175.
26. ibid.
27. W. Newman, 'The homunculus and his forebears' in A. Grafton and N. Siraisi (eds.), *Natural Particulars: Nature and the Disciplines in Renaissance Europe*, MIT Press, Cambridge, Mass., 1999, p. 322.
28. J. W. von Goethe, *Faust, Part 2* (*c.*1806–32), trans. P. Wayne, Penguin, Harmondsworth, 1959, p. 101.
29. Newman, p. 333.
30. ibid., p. 334.
31. Telepnef, p. 78.
32. Pachter, p. 289.

Chapter 18: The White Horse

Epigraph: Erasmus, *In Praise of Folly* (1511), trans. B. Radice, Penguin, Harmondsworth, 1993, p. 134.

1. A. M. Stoddart, *The Life of Paracelsus*, John Murray, London, 1911, p. 289.
2. ibid., p. 290.
3. F. Hartmann, *The Life of Paracelsus*, Kegan Paul, Trench, Trubner and Co., London, 1887, p. 10.
4. R. E. Schlueter, 'Fact and fiction in the names and titles of Paracelsus', *Annals of Medical History* **7** (1935), p. 274.
5. W. Pagel, *Paracelsus: An Introduction to Philosophical Medicine in the Era of the Renaissance*, 2nd edn, Karger, Basle, 1982, p. 347.

Chapter 19: Work With Fire

Epigraph: L. Thorndike, *History of Magic and Experimental Science*, vol. V, Columbia University Press, New York, 1941, pp. 629–30.

1. A. von Bodenstein, foreword dedicated to P. G. Schencke zu Schweinsburgk in

Paracelsus, *The Diseases That Deprive Man of His Reason* (1567), in H. E. Sigerist, *Paracelsus: Four Treatises*, Johns Hopkins Press, Baltimore, 1941, p. 137.
2. Bodenstein, in W. Pagel, *Paracelsus: An Introduction to Philosophical Medicine in the Era of the Renaissance*, 2nd edn, Karger, Basle, 1982, p. 126.
3. H. Trevor-Roper, 'The Paracelsian Movement', *Renaissance Essays*, Secker & Warburg, London, 1985, p. 153.
4. Thorndike, vol. V, p. 604.
5. Severinus, *Idea Medicinae*, Basel cap VII, p. 73 (1571), in Trevor-Roper, p. 161.
6. J. Spedding, R. L. Ellis and D. D. Heath (eds.), *The Works of Francis Bacon*, vol. 3, Longman and Co., London, 1857, p. 533.
7. Thorndike, p. 653.
8. ibid.
9. T. Erastus, *Disputationes de medicina nova Philippi Paracelsi*, Basle, p. 118, in D. P. Walker, *Spiritual and Demonic Magic from Ficino to Campanella*, Warburg Institute, London, 1958, p. 163.
10. Pagel, p. 324.
11. T. Erastus, letter to H. Bullinger, 29 October 1570, in C. D. Gunnoe, Jr, *Thomas Erastus in Heidelberg: A Renaissance Physician during the Second Reformation, 1558–1580*, Ph.D. dissertation, University of Virginia, 1998, p. 251.
12. Erastus, *Disputationes*, II, 2, in Gunnoe, p. 271.
13. Thorndike, p. 664.
14. A. G. Debus, *Man and Nature in the Renaissance*, Cambridge University Press, Cambridge, 1978.
15. Thorndike, p. 644.
16. Pagel, p. 49.
17. C. Webster, *From Paracelsus to Newton*, Cambridge University Press, Cambridge, 1982, p. 5.
18. C. G. Jung, 'Paracelsus as a spiritual phenomenon' (1942), in *Alchemical Studies*, trans. R. F. C. Hull, Princeton University Press, Princeton, 1983, p. 129.
19. J. Weyer, *De praestigiis daemonum* (1563). Published in abridged form as *On Witchcraft*, eds. B. G. Kohl and H. C. E. Midelfort and trans. J. Shea, Pegasus Press, Asheville, NC, 1998, p. 76.
20. ibid., p. 77.
21. F. Herring, *A Modest Defence of the Caveat given to the Wearers of Impoisoned Amulets, as Preservatives from the Plague*, London, pp. 32–33 (1604), in A. G. Debus, *The English Paracelsians*, Franklin Watts, New York, 1966, p. 78.
22. J. Hart, *The Anatomie of Urines* (1625), in Pagel, p. 197.
23. Trevor-Roper, p. 175.
24. J. Read, *From Alchemy to Chemistry*, Dover, New York, 1995, p. 101.
25. ibid., p. 105.
26. ibid., p. 107.
27. ibid., p. 108.
28. M. G. Kim, *Affinity, That Elusive Dream*, MIT Press, Cambridge, Mass., 2003, p. 217.
29. ibid., p. 282.

Chapter 20: Philosopher's Gold

Epigraph: F. Bacon, *Novum Organum*, book I (1620), in *Advancement of Learning and Novum Organum*, Willey Book Co, New York, 1944, p. 315.
1. S. Ward, *Vindiciae Academiarum*, p. 240 (1654), in H. Trevor-Roper, *Renaissance Essays*, Secker and Warburg, London, 1985, p. 191.
2. A. G. Debus, *Man and Nature in the Renaissance*, Cambridge University Press, Cambridge, 1978, p. 123.
3. ibid., p. 124.
4. W. Rawley, introduction to *Essay VII, the Summe of My Lord Bacon's New Atlantis*, in *Essays on Several Important Subjects in Philosophy and Religion*, John Glanvill, London (1676), in G. W. Steeves, *Francis Bacon: A Sketch of His Life, Works and Literary Friends*, Methuen, London, 1910, p. 79.

5. Debus, *Man and Nature*, p. 103.
6. J. Carey (ed.), *The Faber Book of Utopias*, Faber and Faber, London, 1999, pp. 63–4.
7. ibid., p. 64.
8. C. Webster, *From Paracelsus to Newton*, Cambridge University Press, Cambridge, 1982, p. 10.
9. *Fama fraternitatis*, trans. T. Vaughan, 1615. Reprinted in *Rosicrucian Digest*, **82(1)** (2004), p. 14.
10. Debus, *Man and Nature*, p. 120.
11. L. Thorndike, *A History of Magic and Experimental Science*. vol. VI, Columbia University Press, New York, 1941, p. 243.
12. H. Trevor-Roper, p. 184.
13. Thorndike, p. 243.
14. E. von Meyer, *Geschichte der Chemie* 4th edn, Veit, Leipzig, 1914, pp. 70-71. Quoted in Thorndike, p. 242.
15. D. Sennert, *De chymicorum cum Aristotelicis et Ganeicis consensu ac dissensu liber cui accesit appendix de constitutione chimiae*, Wittenberg (1619). The version here is *Chymistrie Made Easie and Useful. Or, The Agreement and Disagreement of the Chymists and Galenists* trans. N. Culpeper and A. Cole, Peter Cole, London, 1662, p. 15.
16. D. Sennert, *De chymicorum*, 3rd edn, Paris, 1633, p. 39. See W. Pagel, *Paracelsus: An Introduction to Philosophical Medicine in the Era of the Renaissance*, 2nd edn, Karger, Basle, 1982, p. 335.
17. W. L. Hine in B. Vickers (ed.), *Occult and Scientific Mentalities in the Renaissance*, Cambridge University Press, Cambridge, 1984.
18. Debus, *Man and Nature*, p. 23.
19. A. G. Debus, *The English Paracelsians*, Franklin Watts Inc., New York, 1966, p. 88.
20. Debus, *Man and Nature*, p.125.
21. ibid., p.127.
22. C. J. S. Thompson, *The Lure and Romance of Alchemy*, George G. Harrap and Co., London, 1932, p. 212. Reprinted as *Alchemy and Alchemists*, Dover, New York, 2002.
23. ibid.
24. J. B. van Helmont, 'De lithiasi', in *Oriatrike; or, Physick Refined*, trans. J. Chandler, II, 13, London, p. 838 (1662), in Pagel, p. 164.
25. Van Helmont, ibid., in B. Vickers, 'On the function of analogy in the occult', in I. Merkel and A. G. Debus (eds.), *Hermeticism and the Renaissance*, Associated University Presses, Cranbury, NJ, 1988, p. 285.
26. Van Helmont, pp. 235–7.
27. J. Read, *From Alchemy to Chemistry*, Dover, New York, 1995, p. 111.
28. ibid.
29. G. W. Leibniz, *Sämtliche Schriften*, ser. 1, vol. 2, letter to Duke Johann Friedrich, June, p. 176, 1679.
30. C. Philipp, letter to Leibniz, 27 March/6 April (1680), in Leibniz, ser. 2, vol. 3, pp. 368–9.
31. C. Philipp, letter to Leibniz, 10/20 April (1680), in Leibniz, p. 386.
32. J. R. Partington, *A History of Chemistry*, vol. 2, Macmillan, London, 1961, p. 369.
33. Paracelsus, *De natura rerum* bk. IV, I, xi, 331 (1537), in N. Goodrick-Clarke, *Paracelsus: Essential Readings*, North Atlantic, Berkeley, 1999, p. 179.
34. Paracelsus, *Volumen medicinae paramirum* I, I, 234 (*c.*1520), in Goodrick-Clarke, p. 57.

Bibliography

Abu'l-Qasim Muhammad ibn Ahmad al-'Iraqi (Abulcasis), *Book of Knowledge Acquired Concerning the Cultivation of Gold*, ed. and trans. E. J. Holmyard, Paul Geuthner, Paris, 1923

Agricola, G., *De re metallica* (1556), trans. H. C. Hoover and L. H. Hoover, Dover, New York, 1950

Agrippa, H. C., *Of the Vanitie and Uncertaintie of Artes and Sciences* (1530), ed. C. M. Dunn, California State University, Northridge, Calif., 1974

—, *Three Books of Occult Philosophy*, trans. J. French, Moule, London, 1651.

—, *Three Books of Occult Philosophy*, ed. D. Tyson, Llewellyn Publications, St Paul, Minnesota, 1994

Atkinson, D. T., *Magic, Myth and Medicine*, World Publishing Co., Cleveland and New York, 1956

Baigent, M., and R. Leigh, *The Elixir and the Stone*, Viking, Harmondsworth, 1997

Bainton, R. H., *Here I Stand: A Life of Martin Luther*. Penguin, London, 2002

Baylor, M. G. (ed. and trans.), *The Radical Reformation*, Cambridge University Press, Cambridge, 1991

Bayon, H. P. 'Paracelsus: personality, doctrines and his alleged influence in the reform of medicine', *Proceedings of the Royal Society of Medicine*, **35** (1942), p. 69

Benzenhöfer, U. (ed.) *Paracelsus*, Wissenschaftliche Buchgesellschaft, Darmstadt, 1993

Boas, M. *The Scientific Renaissance*, Collins, London, 1962

Boas Hall, M. (ed.), *Nature and Nature's Laws*, Macmillan, London, 1970

Boorstin, D. J., *The Discoverers*, Vintage, New York, 1985.

Borges, J. L., *Collected Fictions*, trans. A. Hurley, Penguin, Harmondsworth, 1999

Boynton, H. (ed.), *The Beginnings of Modern Science*, Walter J. Black, Roslyn, NY, 1948

Brann, N. L., *Trithemius and Magical Theology*, State University of New York Press, Albany, 1999

Braun, L., *Paracelse*, SV International/Schweizer, Zurich, 1998

Browning, R. P., 'Paracelsus', (1835), in *The Poetical Works of Robert Browning*, eds. I. Jack and M. Smith, vol. I, Clarendon Press, Oxford, 1983

Bullough, V. L., *Universities, Medicine and Science in the Medieval West*, Ashgate Publishing, Aldershot, 2004

Burckhardt, T., *Alchemy*, trans. W. Stoddart, Penguin, Baltimore, 1971

Burr, C. W., 'Galen', *Annals of Medical History*, **3**, p. 209

Butler, E. M., *The Myth of the Magus*, Cambridge University Press, Cambridge, 1948

Caelius Aurelianus (fifth century), *On Acute Diseases and On Chronic Diseases*, ed. and trans. I. E. Drabkin, University of Chicago Press, Illinois, 1950

Carey, J. (ed.), *The Faber Book of Utopias*, Faber and Faber, London, 1999

Cavendish, R., *A History of Magic*, Weidenfeld and Nicolson, London, 1997

Cennini Cennino, *Il libro dell'Arte* (*c*.1390), trans. D. V. Thompson, Jr as *The Craftsman's Handbook*, Dover, New York, 1954

Cooper, J. C., *Chinese Alchemy*, Sterling, New York, 1990
Cornfield, T., *Luther, Paracelsus, and the Spirit*, thesis, University of Chicago, Illinois, 1996
Craddock, P. (ed.), *2000 Years of Zinc and Brass*, British Museum Occasional Paper No. 50, British Museum Press, London, 1998
Crombie, A. C., *Robert Grosseteste and the Origins of Experimental Science 1100–1700*, Clarendon Press, Oxford, 1953
Cunningham, A. and O. P. Grell, *The Four Horsemen of the Apocalypse*, Cambridge University Press, Cambridge, 2000
Davies, N., *Europe: A History*, Pimlico, London, 1997
Dawson, P. M., 'The heritage of Paracelsus', *Annals of Medical History*, **10** (1928), p. 258
Debus, A. G., *The English Paracelsians*, Franklin Watts Inc., New York, 1966
—, *The Chemical Philosophy*, Science History Publications, New York, 1977
—, *Man and Nature in the Renaissance*, Cambridge University Press, Cambridge, 1978
—, 'History with a purpose: the fate of Paracelsus', *Pharmacy in History* **26** (1984), p. 83
Debus, A. G. (ed.), *Science, Medicine and Society in the Renaissance: Essays to Honour Walter Pagel*, Science History Publications, New York, 1972
Deetjen, C., 'Witchcraft and medicine', *Institute for the History of Medicine Bulletin*, **11**, p. 164
De Ruette, M., 'From *conterfei* and *speauter* to zinc: the development of the understanding of the nature of zinc and brass in post-medieval Europe', in D. R. Hook and D. R. M. Gaimster (eds.), *Trade and Discovery: The Scientific Study of Artefacts from Post-Medieval Europe and Beyond*, British Museum Occasional Paper No. 109, British Museum Press, London, 1995
Dickens, A. G., *Reformation and Society in Sixteenth-Century Europe*, Thames and Hudson, London, 1966
Dopsch, H., K. Goldammer and P. F. Kramml (eds.), *Paracelsus: Keines Andern Knecht*, Verlag Anton Pustet, Salzburg, 1993
Eamon, W., *Science and the Secrets of Nature: Books of Secrets in Medieval and Early Modern Culture*, Princeton University Press, Princeton, 1994
Englander, D., D. Norman, R. O'Day and W. R. Owens (eds.), *Culture and Belief in Europe 1450–1600: An Anthology of Sources*, Blackwell, Oxford, 1990
Fama fraternitatis, trans. T. Vaughan, 1615. Reprinted in *Rosicrucian Digest*, **82(1)** (2004), p. 243
Fellmeth, U., and A. Kotheder (eds,), *Paracelsus Theophrastus von Hohenheim: Naturforscher, Arzt, Theologe*. Wissenschaftliche Verlagsgesellschaft, Stuttgart, 1993
Galdston, I., 'The psychiatry of Paracelsus', Bulletin of the History of Medicine, **xxiv** (1950), p. 205.
Gantenbein, U. L., 'The magic and theology of Paracelsus', in *Dictionary of Gnosis and Western Esotericism*, eds. W. J. Hanegraaff, A. Faivre, R. van den Broek and J.-P. Brach Brill, Leiden (in press)
Gibson, T., 'Theodore Turquet de Mayerne', *Annals of Medical History*, **5** (1933), p. 315
Gilruth, J. D., 'Chiron and his pupil Ascelpius', *Annals of Medical History*, **1**, (1939), p. 158.
Gimpel, J., *The Medieval Machine*, Pimlico, London, 1992
Godwin, W., *Lives of the Necromancers*, Chatto and Windus, London, 1876.
Goldammer, K., 'Paracelsus, Osiander and the theological Paracelsism in the middle of the sixteenth century', in *Science, Medicine and Society in the Renaissance*, ed. A. G. Debus, vol. I, Science History Publications, New York, p. 105
Goodrick-Clarke, N., *Paracelsus: Essential Readings*, North Atlantic, Berkeley, Calif., 1999
Grafton, A., and N. Siraisi (eds.) *Natural Particulars: Nature and the Disciplines in Renaissance Europe*, MIT Press, Cambridge, Mass., 1999
Grant, E., *The Foundations of Modern Science in the Middle Ages*, Cambridge University Press, Cambridge, 1996
Green, V. H. H., *Renaissance and Reformation*, Edward Arnold, London, 1952
Grell, O. P. and A. Cunningham (eds.), *Medicine and the Reformation*, Routledge, London, 1993
Guicciardini, F., *History of Italy*, ed. J. R. Hale, Washington Square Press, New York, 1964
Gunnoe, Jr, C. D., *Thomas Erastus in Heidelberg: A Renaissance Physician During the Second Reformation, 1558–1580*. Ph.D. dissertation, University of Virginia, 1998
Hale, J. R., *Renaissance Europe*, University of California Press, Berkeley, 1977
—, *War and Society in Renaissance Europe 1450–1620*, Fontana, London, 1985

—, *The Civilization of Europe in the Renaissance*, Fontana, London, 1994
Hargrave, J., *The Life and Soul of Paracelsus*, Victor Gollancz, London, 1951
Harley, D., 'Rychard Bostok of Tandridge, Surrey (*c*.1530–1605), M.P., Paracelsian propagandist and friend of John Dee', *Ambix* **47** (2000), p. 29.
Harrison, J., *The Library of Isaac Newton*. Cambridge University Press, Cambridge, 1978
Hartmann, F., *Paracelsus*. Kegan Paul, Trench, Trubner and Co., London, 1887
Hays, D., *Europe in the Fourteenth and Fifteenth Centuries*, 2nd edn, Longman, London, 1989
Haynes, R. D., *From Faust to Strangelove: Representations of the Scientist in Western Literature*, Johns Hopkins University Press, Baltimore, 1994
Henderson, B., *Paracelsus: A Poem in 40 Parts*, The Porcupine's Quill, Inc., Erin, Ontario, 1977
Hendrickson, G. L., 'The "Syphilis" of Girolamo Fracastoro', *Institute for the History of Medicine Bulletin*, **II** (1934), p. 515
Henry, J., *Knowledge is Power*, Icon Books, Cambridge, 2002
Holmyard, E. J., *Alchemy*, Dover, New York, 1990
Huizinga, J., *Erasmus of Rotterdam*, Phaidon, London, 1952
Hutchison, K., 'What happened to occult qualities in the Scientific Revolution?' *Isis* **73** (1982), p. 233
Jacobi, J., *Paracelsus: Selected Writings*, Princeton University Press, Princeton, 1979
Jardine, L., *Wordly Goods*, Macmillan, Basingstoke, 1996
Johns, A., *The Nature of the Book*, University of Chicago Press, Chicago, 1998
Jonson, B., *The Alchemist* (1612), in *Three Comedies*, ed. M. Jamieson, Penguin, Harmondsworth, 1966
Jung, C. G., 'Paracelsus as a spiritual phenomenon' (1942), in *Alchemical Studies*, trans. R. F. C. Hull, Princeton University Press, Princeton, 1983
Kim, M. G., *Affinity, That Elusive Dream*, MIT Press, Cambridge, Mass., 2003
Kinsman, R. S. (ed.), *The Darker Vision of the Renaissance*, University of California Press, Berkeley, 1974
Kudlien, F., 'Some interpretive remarks on the antisemitism of Paracelsus', in *Science, Medicine and Society in the Renaissance*, ed. A. G. Debus, vol. I, Science History Publications, New York, 1972, p. 121
Leicester, H. M., *The Historical Background of Chemistry*, Dover, New York, 1971
Loudon, I. (ed.), *Western Medicine: An Illustrated History*, Oxford University Press, Oxford, 1997
Lynch, M., *Mining in World History*. Reaktion Books, London, 2002
Mackenney, R., *Sixteenth-Century Europe: Expansion and Conflict*, Macmillan, London, 1993
Maland, D., *Europe in the Sixteenth Century*, 2nd edn, Macmillan, Basingstoke, 1982
Manchester, W., *A World Lit Only by Fire*, Macmillan, Basingstoke, 1993
McCall, A., *The Medieval Underworld*, Book Club Associates, London, 1979
Merkel, I., and A. G. Debus (eds.), *Hermeticism and the Renaissance*, Folger Books, Washington, 1988
Mez-Mangold, L., *A History of Drugs*, Hoffmann-LaRoche, Basle, 1972
McIntosh, C., *The Rosicrucians*, Samuel Weiser Inc., York Beach, Maine, 1997
Mora, G. (ed.), *Witches, Devils, and Doctors in the Renaissance*, Medieval Renaissance Texts and Studies Vol. 73, State University of New York, Binghamton, 1991
Multhauf, R., 'Medical chemistry and "the Paracelsians"', *Bulletin of the History of Medicine*, **XXVIII** (1954), p. 101
—, 'John of Rupescissa and the Origin of Medical Chemistry', *Isis*, **XLV** (1954), p. 364
—, 'The significance of distillation in Renaissance medical chemistry', *Bulletin of the History of Medicine*, **XXX** (1956), p. 329
—, *The Origins of Chemistry*, Gordon and Breach, Langhorne, Penn., 1993
Müller-Jahnke, W. D., 'The attitude of Agrippa von Nettesheim (1486–1535) towards alchemy', *Ambix*, **22** (1975), p. 134
Mulryne, J. R., and M. Shewring (eds.), *War, Literature and the Arts in Sixteenth-Century Europe*, Palgrave Macmillan, London, 1989
Nauert, Jr, C. G., *Agrippa and the Crisis of Renaissance Thought*, University of Illinois Press, Urbana, 1965
Neusner, J., E. S. Frerichs and P. V. M. Flesher (eds), *Religion, Science, and Magic: In Concert and in Conflict*. Oxford University Press, Oxford, 1989

Newman, W., 'The homunculus and his forebears', in A. Grafton and N. Siraisi (eds.), *Natural Particulars: Nature and the Disciplines in Renaissance Europe*, MIT Press, Cambridge, Mass., 1999, p. 321
Norpoth, L., 'Paracelsus – a mannerist?' in *Science, Medicine and Society in the Renaissance*, ed. A. G. Debus, vol. I, Science History Publications, New York, 1972, p. 127
Olson, R., *Science Deified and Science Defied*, vol. 2. University of California Press, Berkeley, 1990
Pachter, H. M., *Paracelsus: Magic Into Science*, Henry Schuman, New York, 1951
Pagel, W., 'Paracelsus and the Neoplatonic and Gnostic tradition', *Ambix*, **8** (1960), p. 125
—, 'The prime matter of Paracelsus', *Ambix*, **9** (1961), p. 117
—, *Paracelsus: An Introduction to Philosophical Medicine in the Era of the Renaissance*, 2nd edn, Karger, Basle, 1982
—, 'From Paracelsus to van Helmont', in *Studies in Renaissance Medicine and Science*, ed. M. Winder, Variorum Reprints, London, 1986
Pagel, W., and P. Rattansi, 'Vesalius and Paracelsus', *Medical History*, **8** (1964), p. 309
Palmer, P. M., and R. P. More, *Sources of the Faust Tradition from Simon Magnus to Lessing*, Oxford University Press, New York, 1936
Paracelsus, *Sämtliche Werke*, ed. K. Sudhoff, Barth, Munich, 1922–33
—, *De generatione stultorum* (The Begetting of Fools) (*c.*1530), trans. P. F. Cranefield and W. Federn, *Bulletin of the History of Medicine* **41** (1967), 56 and 161
Park, K., 'Medicine in the Renaissance', in I. London (ed.), *Western Medicine: An Illustrated History*, Oxford University Press, Oxford, 1997
Partington, J. R., 'Trithemius and alchemy', *Ambix*, **2** (1938), p. 53.
—, *A History of Chemistry*, vols I and II, Macmillan, London, 1961-70
Patai, R., *The Jewish Alchemists*, Princeton University Press, Princeton, 1994
Platter, F., *Beloved Son Felix: The Journal of Felix Platter, a Medical Student in Montpellier in the Sixteenth Century, 1552–7*, trans. S. Jennett, Muller, London, 1961
Principe, L. M., *The Aspiring Adept: Robert Boyle and his Alchemical Quest*. Princeton University Press, Princeton, 1998
Quétel, C., *History of Syphilis*, Polity Press, Cambridge, 1990
Randell, K., *Luther and the German Reformation 1517–55*, 2nd edn, Hodder and Stoughton, London, 2000
Rattansi, P. M., 'Newton's alchemical studies', in *Science, Medicine and Society in the Renaissance*, ed. A. G. Debus, Science History Publications, New York, 1972, p. 167
—, 'Art and science: the Paracelsian vision', in J. W. Shirley and F. D. Hoeniger (eds.), *Science and the Arts in the Renaissance*, Associated University Presses, Cranbury, NJ, 1985
Rawcliffe, C., *Medicine and Society in Later Medieval England*, Sandpiper, London, 1999
Read, J., *From Alchemy to Chemistry*, Dover, New York, 1995
Riesman, D., 'Life in a medieval university', *Annals of Medical History*, **8** (1936), p. 395
Rocke, A. J., 'Agricola, Paracelsus, and chymia', *Ambix*, **32** (1985), p. 38
Rosen, G., 'Some recent European publications dealing with Paracelsus', *Journal of the History of Medicine*, **II** (1947), p. 537
Russell, B., *A History of Western Philosophy*, Unwin, London, 1984
Sarton, G., *Six Wings: Men of Science in the Renaissance*, Bodley Head, London, 1958
Schlueter, R. E., 'Fact and fiction in the names and titles of Paracelsus', *Annals of Medical History*, **7** (1935), p. 274
Schneider, W., 'Chemistry and iatrochemistry', in *Science, Medicine and Society in the Renaissance*, ed. A. G. Debus, vol. I, Science History Publications, New York, 1972, p. 141
Schumaker, W., *Natural Magic and Modern Science: Four Treatises 1590–1657*, Medieval and Renaissance Texts and Studies Vol. 63, State University of New York, Binghamton, 1989
Sherlock, T. P., 'The chemical work of Paracelsus', *Ambix*, **3** (1948), p. 33
Sigerist, H. E., 'Karl Sudhoff, the man and the historian', *Institute for the History of Medicine Bulletin*, **II** (1934), p. 3
—, *Paracelsus: Four Treatises*, Johns Hopkins Press, Baltimore, 1941
—, 'Laudanum in the works of Paracelsus', *Bulletin for the History of Medicine* **9** (1941), p. 530
—, 'Paracelsus in the light of four hundred years', in *Henry E. Sigerist: On the History of Medicine*, ed. F. Marti-Ibanez, M. D. Publications, New York, 1960, p.162

Singer, C., *A Short History of Medicine*, Oxford University Press, Oxford, 1928
Smith, P. H., *The Business of Alchemy: Science and Culture in the Holy Roman Empire*, Princeton University Press, Princeton, 1994
Sprenger, J. [attributed], and H. Kramer, *Malleus maleficarum* (1486), ed. P. Hughes and trans. M. Summers, Folio Society, London, 1968
Stahl, W. H., *Roman Science*. University of Wisconsin Press, Madison, 1962
Stillman, J. M., *Paracelsus*, Open Court Publ. Co., Chicago, 1920
Steeves, G. W., *Francis Bacon: A Sketch of His Life, Works and Literary Friends*, Methuen, London, 1910
Stoddart, A. M., *The Life of Paracelsus*, John Murray, London, 1911
Strathern, P., *Mendeleyev's Dream*, Penguin, London, 2001
Sudhoff, K., 'Faust in the Rhineland', in *Essays in the History of Medicine*, eds. K. Sudhoff and F. H. Garrison, Medical Life Press, New York, 1926, p. 391
Szydlo, Z., *Water Which Does Not Wet Hands*, Polish Academy of Sciences, Warsaw, 1994
Taylor, F. S., *The Alchemists*, Paladin, St Albans, 1976
Telepnef, B. de, *Paracelsus: A Genius Amidst a Troubled World*, Banton Press, Largs, 1991
Temkin, O., 'Karl Sudhoff, the rediscoverer of Paracelsus', *Institute for the History of Medicine Bulletin* **II** (1934), p. 16
—, 'The elusiveness of Paracelsus', *Bulletin for the History of Medicine* **26** (1952), pp. 201.
Theophilus, *De diversis artibus* (*c.*1122), trans. J. G. Hawthorne and C. S. Smith as *On Divers Arts*, Dover, New York, 1979
Thomas, K., *Religion and the Decline of Magic*, Penguin, London, 1991
Thompson, C. J. S., *The Lure and Romance of Alchemy*, George G. Harrap and Co., London, 1932. Reprinted as *Alchemy and Alchemists*, Dover, New York, 2002
—, *Alchemy: Source of Chemistry and Medicine* (1897), reprinted by Sentry Press, New York, 1974
Thorndike, L., *A History of Magic and Experimental Science*, vols I–VI. Columbia University Press, New York, 1923–41
Tracy, J. D., *Holland under Habsburg Rule, 1506–1566*, University of California Press, Berkeley, 1990
Trevor-Roper, H., *Renaissance Essays*. Secker and Warburg, London, 1985
Vansteenberghe, E., *Le Cardinal Nicolas de Cues*, Pris, 1920
Vickers, B. (ed.), *Occult and Scientific Mentalities in the Renaissance*, Cambridge University Press, New York, 1984
—, 'On the function of analogy in the occult', in Merkel and Debus (eds.) *Hermeticism and the Renaissance*, p. 265
Waite, A. E., *Alchemists through the Ages*, Banton Press, Largs, 1990
—, *Hermetic and Alchemical Writings of Paracelsus the Great*, Alchemical Press, Edmonds, Wash., 1992
Walker, D. P., *Spiritual and Demonic Magic from Ficino to Campanella*, Warburg Institute, London, 1958
Walsh, J., 'Galen's writings and influences inspiring them', *Annals of the History of Medicine*, **6** (1934), p. 1
Webster C., *From Paracelsus to Newton*, Cambridge University Press, Cambridge, 1982
—, 'Paracelsus confronts the saints: miracles, healing and the secularisation of magic', *Social History of Medicine*, **8** (1995), p. 403
—, 'The nineteenth-century afterlife of Paracelsus', in *Studies in the History of Alternative Medicine*, ed. R. Cooter, Macmillan, Basingstoke, 1988, p. 78
—, 'Paracelsus on natural and popular magic', *Atti del Convegno Internazionale su Paracelso, Rome 1993*, Edizioni Paracelso, Rome, 1994, p. 89
Weeks, A., *Paracelsus: Speculative Theory and the Crisis of the Early Reformation*, State University of New York Press, Albany, 1997
Weyer, J., *De praestigiis daemonum* (1563), published in abridged form as *On Witchcraft*, eds. B. G. Kohl and H. C. E. Midelfort, trans. J. Shea, Pegasus Press, Asheville, NC, 1998
Williams, G. A., 'Michael Servetus, physician and heretic', *Annals of Medical History* **10** (1928), p. 287
Williams, G. H., *The Radical Reformation*, 3rd edn, Sixteenth-Century Journal Publishers, Kirksville, Mo., 2002

Williams, G. S., and C. D. Gunnoe, Jr (eds.) *Paracelsian Moments*. Truman State University Press, Kirksville, Mo., 2002
Yates, F. A., *The Rosicrucian Enlightenment*, Routledge and Kegan Paul, London, 1972
—, *Ideas and Ideals in the North European Renaissance*, Routledge and Kegan Paul, London, 1984

Index

Copenhagen
York
Lübeck
London
Bruges
Oxford
Hamburg
Rostock
Calais
Geldern
Antwerp
Louvain
Paris (1518–19)
Valladolid
Montpellier (1517)
Innsbruck
León
Milan (1513)
TYROL
Santiago
Burgos
Genoa
Toulouse
Marseilles
Venice
Jaca
Salamanca
Saragossa
Lisbon
Bologna
Barcelona
Cordoba
Rome
Seville
Naples
Granada
Cartagena (1517–18)
Salerno
Almeira
Algiers
Oran
Travels of Paracelsus
0
200
400 miles
0
200
400
600 kilometres